The Book of the Light of the Great Magistery of John of Rupescissa

Convinced that the antichrist was "at our gates," the Franciscan friar, alchemist, and prophet John of Rupescissa (ca. 1310–1366) wrote *The Book of the Light of the Great Magistery* to help his future brethren restore a devastated Christendom after the antichrist's inevitable downfall. Written in 1354, John's text provides a detailed process for making the philosophers' stone, the substance able to transmute base metals into gold and silver. With this alchemical gold and silver, the true and faithful Franciscans, the "poor men of the Gospel," could rebuild the Church and help usher in a millennium of peace and prosperity.

This volume presents the first critical edition, modern English translation, and focused study of John's *Book of the Light*. Based on over one hundred surviving manuscripts, many recently discovered, this critical edition restores substantial original text that was omitted from the standard editions first printed in the sixteenth century and corrects many longstanding textual errors. In the accompanying study, Lawrence M. Principe explores John's sources and ideas, especially his striking theories of matter and material change, and follows the transmission, reception, vernacularization, and multiple modifications of John's text as copyists and reader-practitioners engaged with it over the next three centuries. Modern laboratory reworkings, fully illustrated and explained, go hand in hand with the textual analyses, providing both a vivid picture of the friar's practices and observational skills and a deeper understanding of his text.

John of Rupescissa stands revealed as an innovative theorist and observant practitioner; he was not an "armchair" alchemist whose knowledge came only from textual sources. The clear evidence of his alchemical practice presented here provokes, in turn, a close reevaluation of the conditions of his sixteen years of confinement at a papal prison in Avignon, where he learned and practiced much of his alchemy before writing his *Book of the Light*. Historians of science and technology, medievalists, book historians, manuscript scholars, and many others will find new and important information in this third volume of the *Ambix* series *Sources of Alchemy and Chemistry*.

Lawrence M. Principe is the Drew Professor of the Humanities in the Department of the History of Science and Technology and the Department of Chemistry at Johns Hopkins University.

SOURCES OF ALCHEMY AND CHEMISTRY
SIR ROBERT MOND STUDIES IN THE HISTORY OF EARLY CHEMISTRY

AMBIX VOLUME 73 SUPPLEMENT 1 2026

The Book of the Light of the Great Magistery of John of Rupescissa

(*Liber lucis magisterii magni*)

Lawrence M. Principe

Routledge
Taylor & Francis Group
NEW YORK AND LONDON

Designed cover image: Montpellier, Bibliothèque Universitaire Historique de Médecine, MS H 493, fol. 236r

First edition published 2026
by Routledge
605 Third Avenue, New York, NY 10158

and by Routledge
4 Park Square, Milton Park, Abingdon, Oxon, OX14 4RN

Routledge is an imprint of Taylor & Francis Group, LLC

ISBN: 978-1-041-12117-6 (hbk)
ISBN: 978-1-041-12777-2 (pbk)
ISBN: 978-1-003-66654-7 (ebk)

DOI: 10.1201/9781003666547

Typeset in Sabon
by codeMantra

CONTENTS

ACKNOWLEDGEMENTS

Like most scholarly projects, this one has taken longer to complete than I had planned, delayed by a global pandemic, travel restrictions, needless departmental upheavals, and all the usual and varied distractions from the scholarly life so lamentably inescapable in the twenty-first century. Nevertheless, a wonderful network of fellow scholars has helped this project all along its way to completion. The project was encouraged and assisted by multiple "Sources Workshops" that have been central to the development of this series of publications, the *Sources of Alchemy and Chemistry: Sir Robert Mond Studies in the History of Early Chemistry*, of which this is the third volume. Jennifer Rampling, co-editor for this series, organizer of several Sources Workshops, and active contributor to all of them, has been crucial for the production of this volume, as she has been for the previous installments. Jenny also provided me with photos of roughly a dozen of the manuscripts needed for this project. Sébastien Moureau, who organized a Sources Workshop in Louvain-la-Neuve, made available to me early versions of his remarkable software ChrysoCollate for manuscript collation and preparing critical editions. Now in version 1.3.1, ChrysoCollate revolutionizes the making of critical editions; this volume would have taken at least twice as long to prepare without it. Sébastien was also a constant source of advice about making critical editions, and he put his impressive technical talents to work in helping prepare layouts for the critical Latin text. The circle of alchemy scholars around him at Louvain-la-Neuve also provided helpful advice: Marion Dapsens, Guillaume Delmeulle, and Doriane Moenaert. At Sources Workshops held in several European venues, I received much help along the way from fellow participants, particularly Matteo Martelli, Gabriele Ferrario, Vincenzo Carlotta, Sebastian Richter, and Marco Bellini. Very special thanks and gratitude go in particular to Robert and Olivia Temple, whose generosity and support as outstanding patrons of textual scholarship and publishing have sponsored this series and encouraged its authors from its inception.

Many other scholars have helped at various junctures by sharing their expertise and by their willingness to let me fill their ears with news and hypotheses about John of Rupescissa. William Newman read and offered helpful advice on several parts of this volume, as did Jennifer Rampling. For collegial help in many ways, I thank Meagan Allen, Charles Burnett, Antoine Calvet, Elizabeth Casteen, Nuno Castel-Branco, William Jordan, Didier Kahn, Daryn Lehoux, Robert E. Lerner, Francesco Luzzini, Gerasimos Merianos, Gianna Pomata, Lucia Raggetti, Salam Rassi, and Mike Zuber. I'm also grateful to James Voelkel and Ron Brashear at the Science History Institute, who purchased and put at my disposal two of the manuscripts used in this study. I benefited also from the helpful feedback I received when speaking about this project from participants at the Università di Bologna, Uppsala University, All Souls College Oxford, the Domus Comeliana in Pisa, the Vrije Universiteit Brussel, the Virginia Fox Stern Center for the History of the

Book at Johns Hopkins, the Warburg Institute, and the Philology Seminar at the Max Planck Institut für Naturwissenschaftsgeschichte Berlin. Time and opportunity for organizing the material for this book were provided by wonderful fellowships at the Herzog August Bibliothek Wolfenbüttel, thanks to Ute Friesch, and at the Università di Bologna, thanks to Matteo Martelli and his ERC AlchemEast project. I thank also my departmental colleagues Alison McManus, Lijing Jiang, Joris Mercelis, Nathaniel Comfort, and Lan Li for their unwavering support during some very difficult times.

To be sure, a project that relies on examining over one hundred medieval and early modern manuscripts preserved at dozens of repositories in fifteen countries could not have been carried out without the generous help and kindness of archivists and librarians too numerous to mention individually here. I am also grateful to those at Taylor & Francis who cooperated in producing a volume with a great number of "special requests," especially Randi Slack, Ed Curtis, Paolo Pitari, and Justin Robinson. Under the aegis of the Society for the History of Alchemy and Chemistry, Frank James, Anna Simmons, and Jennifer Rampling smoothed the way for this publication in multiple and invaluable ways.

Last but certainly not least, my husband Kip Kuntz read through multiple versions of the book manuscript and offered corrections, improvements, and many exploratory conversations. He provided statistical analyses of the manuscript witnesses, resolved sporadic panics over computer issues, painstakingly compared the claims of my stemmata with those in my manuscript descriptions, gladly combined manuscript hunting expeditions with vacation plans, and provided a wonderful place looking out over the sea during the 2024 "Charge-Exchange in the X-Ray Universe" meeting in Volos, Greece where I wrote up most of Sections 2 and 3.

Finally, during my academic formation and career I have had the privilege of learning from a remarkable array of distinguished scholars. I would like to acknowledge the debt and gratitude I owe to two mentors during and following my graduate work. When I was a PhD student in the chemistry department at Indiana University from 1983 until 1988, I used to sneak out of the laboratory to attend classes in the history and philosophy of science department, where I completed the equivalent of a master's degree. One of several outstanding professors in that department from whom I learned so much was Edward Grant. I took his courses on ancient and medieval science and on science and religion, each of which established foundations for courses by the same name that I myself would later teach. At that time I hoped to work on medieval alchemy, and I remember his telling me that if I wanted to "earn my stripes" as a medievalist I had to prepare a critical edition of a previously unedited medieval text. It took forty years of wandering through the early modern period, but here at last is that critical edition of a medieval alchemical text. I hope Professor Grant (whom I could never bring myself to call "Ed," even after "Prof." was prefixed to my own name) would be pleased.

After I left Indiana and settled in Baltimore, from 1989 until 1996 I was a PhD student in the history of science department at Johns Hopkins, where I had the great good fortune to have Robert Kargon as my advisor. Once my degree was completed, Bob was instrumental in arranging a tenure-track position for me, split between the chemistry and history of science departments, in which position I have remained ever since. Bob thereafter continued to give me sound advice as a senior colleague, a valued relationship that continues to this day. Bob Kargon has remained a central figure in our department as a scholar, instructor, and innovator for an extraordinary sixty years. Time and time again, I witnessed (and often benefited from) his amazing ability to cut through confusion and needless complications to get right to the heart of a historical problem and then deftly and succinctly define the important questions to be asked, a talent I have endeavored to imitate but have never equaled. As a small token of my continuing gratitude to my teachers, advisors, colleagues, and friends Ed Grant and Bob Kargon, I would like to dedicate this book to them.

ABBREVIATIONS

AFH	*Archivum franciscanum historicum*
BCC	*Biblioteca chemica curiosa*, ed. Jean Jacques Manget, 2 vols. (Geneva, 1702; reprint ed., Sala Bolognese: Arnoldo Forni, 1976).
Benzenhöfer	Udo Benzenhöfer, *Johannes de Rupescissa* Liber de consideratione quintae essentiae omnium rerum *deutsch* (Stuttgart: Franz Steiner Verlag, 1989).
BNF	Bibliothèque nationale de France
Broszinski, *Handschriften*	Hartmut Broszinski, *Die Handschriften der Universitätsbibliothek Kassel Landesbibliothek und Murhardsche Bibliothek der Stadt Kassel: Manuscripta chemica in quarto* (Wiesbaden: Harrassowitz, 2011)
Calvet, *Oeuvres alchimiques*	Antoine Calvet, *Les oeuvres alchimiques attribuées à Arnaud de Villeneuve: Grand oeuvre, médecine et prophétie au Moyen-Âge*, *Textes et Travaux de Chrysopoeia* 11 (SÉHA: Paris, 2011).
Corbett, *Catalogue*	James Corbett, *Catalogue des manuscrits alchimiques latins 1: Manuscrits des bibliothèques publiques de Paris* (Brussels: Union Académique Internationale, 1939); *Catalogue des manuscrits alchimiques latins 2: Manuscrits des bibliothèques publiques des départements français* (Brussels: Union Académique Internationale, 1951).
De confectione (1561)	John of Rupescissa, *De confectione veri lapidis philosophorum*, in *Verae alchemiae artisque metallicae*, ed. Guglielmo Gratarolo, 2 vols. (Basel, 1561), 2:226–31. [The "Basel edition" of *Liber lucis*]
De famulatu	John of Rupescissa, *De famulatu philosophie evangelio Domini Jesu Christi et pauperibus evangelicis viris* [Frequently cited elsewhere under the title of its first book: *De consideracione quinte essencie*]

Halleux, "Rupescissa"	Robert Halleux, "Ouvrages alchimiques de Jean de Rupescissa," in *Histoire litteraire de la France* 41 (1981):241–77.
LO	John of Rupescissa, *Liber ostensor quod adesse festinant tempora, Édition critique*, sous la direction d'André Vauchez par Clémence Thévenaz Modestin et Christine Morerod-Fattebert (Rome: École Française de Rome, 2005).
LSE	Robert E. Lerner and Christine Morerod-Fattebert, *Johannes de Rupescissa*, Liber secretorum eventuum: *Édition critique, traduction et introduction historique* (Fribourg: Éditions Universitaires, 1994).
Moorat, *Catalogue*	S. A. J. Moorat, *Catalogue of Western Manuscripts on Medicine and Science in the Wellcome Historical Medical Library*, 2 vols. (London: Wellcome Institute for the History of Medicine, 1962–73).
MS	Manuscript
Newman, *Summa*	William R. Newman, *The Summa perfectionis of Pseudo-Geber: A Critical Edition, Translation, and Study* (Brill: Leiden, 1990).
Thorndike, *HMES*	Lynn Thorndike, *A History of Magic and Experimental Science*, 8 vols. (New York: Columbia University Press, 1934–58).
VM (Kaup)	Matthias Kaup, *John of Rupescissa's* Vade mecum in tribulacione *(1356): A Late Medieval Eschatological Manual for the Forthcoming Thirteen Years of Horror and Hardship* (New York: Routledge, 2017).
VM (Tealdi)	John of Rupescissa, *Vade mecum in tribulatione: Edizione critica a cura di Elena Tealdi; introduzione storica a cura di Robert E. Lerner e Gian Luca Potestà* (Milan: Vita e Pensiero, 2015).

LIST OF ILLUSTRATIONS

Part 1

Introduction

1 JOHN OF RUPESCISSA AND HIS *LIBER LUCIS MAGISTERII MAGNI*

The fourteenth century was a period of significant development for alchemy in Latin Europe, and accordingly, it is home to some of the most prominent names in the entire history of the subject, such as Ramon Lull and Arnald of Villanova—which is not to say that either of those two authors actually wrote any of the alchemical texts that appeared under their names. An equally prominent name of the fourteenth century is that of John of Rupescissa, known in his native Occitan as Jean de Roquetaillade, a Franciscan friar, prophet, alchemist, and prisoner. John is unusual, and particularly significant, among fourteenth-century alchemical authors for several reasons. First, he actually *did* write the two alchemical treatises that bear his name, which allows for a more straightforward contextualization of these works in time and place than can be done for the more plentiful pseudonymous texts of the period.[1] This richer contextualization has been greatly enhanced by a wealth of studies over the past thirty years that have done so much to correct and elaborate John's biographical details, many of which had previously been quite befuddled, to the point that much of the earlier literature did not agree even on his country of origin.[2] The present critical edition, English translation, and study of the shorter of his two alchemical treatises, the *Liber lucis magisterii magni*, adds further to our understanding of John, his thought, his practical experience, and the nature of his long imprisonment. John's alchemical authorship, however, represents only a small fraction of his total literary output—most of his writings involve prophecies about the imminent arrival of the antichrist.[3] John's alchemy is nevertheless integral with his larger prophetic and exegetical project. He wrote his alchemical works in preparation for the apocalyptic events he was certain would soon befall Christendom, and they propose the means for surviving and recovering from the antichrist's horrific persecutions. Second, John of Rupescissa's influence on the history of chymistry is both substantial and well established. Although only two of John of Rupescissa's eleven surviving works (of the roughly forty he is known to have written) involve alchemy, these two

[1] John's authorship of the two alchemical works that bear his name has often been questioned in the past; their authenticity is demonstrated in Section 1.5.

[2] John has previously been called Provençal, Catalan, Spanish, and even English; his nationality and many other previously befuddled biographical claims were collected, presented, and replaced by the pioneering study by Jeanne Bignami-Odier, *Études sur Jean de Roquetaillade* (Paris: Vrin, 1952), 26–38. Past claims—and some reasons behind them—for John's origins are treated also in Graziana Ciola, "Giovanni di Rupescissa: autobiografia, profezia e leggenda," in *The Medieval Legends of Philosophers and Scholars*, Micrologus 21 (Florence: SISMEL, 2013), 517–78, on 546–54.

[3] Actually, *antichrists*, since John predicted several antichrists who would appear at different times.

DOI: 10.1201/9781003666547-2

became in due course his most widely circulated texts. They were copied, translated, excerpted, imitated, and plagiarized. They exerted enormous influence on subsequent developments in the history of science and the history of medicine for more than two centuries after they were written. These works document John's remarkable ideas about matter, material change, and the role of experiment and reveal several unexpected features of medieval alchemical practice and natural philosophy.

Although John's position in the history of science and the significance of his two alchemical works is widely acknowledged, no critical edition of either has yet been produced, despite the attention they have otherwise received from numerous scholars.[4] The lack of a critical edition is far more troubling than most readers would suspect; in fact, it is more troubling than even I believed when I began this project. Until now, historians have relied almost exclusively upon the versions of John's two alchemical texts that were printed in the sixteenth century, or, less frequently, upon a single manuscript witness. But much can happen to a text if it circulates in manuscript, copied over and over by multiple hands, for two hundred years before being fixed in print. Sixteenth-century editors and printers were often not especially discerning when choosing the manuscripts they published. They also frequently and often silently altered the text in multiple ways, sometimes to "update" or streamline it, or to make it more appealing to buyers. John's *Liber lucis* illustrates the extreme extent to which such modifications, both intentional and unintentional, can reach. In comparison with the critical edition presented here, the first publication of this text, the Basel edition of 1561, omits *nearly fifty percent* of John's original text and appends or inserts substantial new material with no relationship to John. The second publication of John's book, printed in Cologne in 1579, provides an alternate version of the text, which, although much of its gist corresponds roughly with that of the 1561 edition, does not share *a single complete sentence in common* with the earlier edition and is even further from John's original than the 1561 edition. Clearly then, a critical edition of John's *Liber lucis* based on the most reliable surviving manuscript materials is crucial for correctly representing the content of the text and for understanding John's extraordinary thought and innovations and how his alchemy actually relates to his parallel roles as a prophet, exegete, and Franciscan friar.[5]

[4] Thorndike, *HMES*, 3:347–69, 722–40 (who, almost uniquely, paid attention to and began to analyze divergent readings); Robert P. Multhauf, "John of Rupescissa and the Origin of Medical Chemistry," *Isis* 45 (1954): 359–67; Michela Pereira, "I Francescani e l'alchimia," *Convivium Assisiense* 10 (2008): 117–57, on 154–57; Leah DeVun, *Prophecy, Alchemy, and the End of Time: John of Rupescissa in the Late Middle Ages* (New York: Columbia University Press, 2009); Chiara Crisciani, "Giovanni di Rupescissa: sapere, alchimia e profezia," 239–79, in *I Francescani e le scienze: Atti del XXXIX Convegno internazionale* (Spoleto: Centro Italiano di Studi Sull'Alto Medioevo, 2012); and Zachary Matus, *Franciscans and the Elixir of Life: Religion and Science in the Later Middle Ages* (Philadelphia: University of Pennsylvania Press, 2017), 92–98.

[5] The same is to be said of John's other alchemical text, which was also first published in Basel in 1561, under the title of *On the Consideration of the Quintessence* (*De consideratione quintae essentiae*). Many parts of that sole edition are nearly as corrupt as the printed editions of the *Liber lucis*. I am currently completing its critical edition.

Producing this critical edition yielded not only a reliable text but also gave answers, sometimes quite unexpected, to several questions of considerable interest. The unexpected recovery of over one hundred surviving manuscripts of the *Liber lucis*, dating from the start of the fifteenth century to the end of the eighteenth, demonstrates that the work circulated much more widely and for a longer time than previously thought. Comparison of these manuscripts with one another reveals important information about how, where, and when the text was disseminated and how and by whom it was used and modified, thus providing an example of how natural philosophical knowledge could circulate—and be modified—in the Late Middle Ages and into the early modern period. As a fundamentally *practical* text intended to guide laboratory operations, successive generations of reader-practitioners reworked and rewrote the *Liber lucis* to reflect their own interests and experience, and these reworkings are variously preserved in the numerous manuscripts. These adaptations include multiple translations into seven vernacular languages, the reorientation of John's ideas to fit later theoretical frameworks, the insertion of new laboratory experiences, the removal of less practical content, and even the coopting of parts of John's text to create new, original works, some of which attempted to pass themselves off as John's own compositions. In all these ways, the labor of creating critical editions not only provides a reliable text that is essential for improving our historical understanding but also contributes to studies of book history, of the dissemination of knowledge, and of reading practices. In short, while the critical edition presented here endeavors to come as close as possible to the *Urtext*—what John himself wrote—the evidence of pathways of diffusion and the modifications made by later reader-practitioners that are preserved in the subsequent manuscript tradition are of no less importance or interest. This parallel information is cited both throughout the following analysis of the text and especially in the descriptions of each manuscript provided in Section 6.

In preparing this edition and study, I have combined traditional philological techniques with the more innovative methodology of experimental reconstruction; that is, I brought the critical text of the *Liber lucis* into my chemical laboratory in order to rework the practices it describes. The results, presented in both text and photographs in Section 2, not only bore lively witness to John's own practical and observational skill but at several junctures also guided my choice of readings for the critical edition and how best to translate them. The laboratory results also forced a serious reconsideration of the status of John's long confinement in Avignon, a topic treated in Section 2.9. Thus, the two seemingly disparate historical methodologies—one traditional and one more innovative—worked together in surprising ways both to pose and to answer questions that enrich our understanding of John and his work.[6]

6 As an early proponent of reworking historical processes in the chemical laboratory, I am equally gratified and surprised that this technique, initially criticized through the 1980s and 1990s, has become widely accepted in the history of science during the past twenty years. See Hjalmar Fors, Lawrence M. Principe, and H. Otto

Critical editions of several of John's surviving prophetic works have been produced during the last twenty-five years, during what Robert E. Lerner has called a "Rupescissa Renaissance."[7] This fresh attention to John's ideas and authorship has been augmented by the discovery in 2008 of a previously unknown and substantial work of his, the *Sexdequiloquium*, whose sole surviving copy was found, virtually miraculously, at the bottom of an armoire in a country house near Nancy.[8] The more discoveries that modern scholarship makes about John of Rupescissa, the more we learn about his life and work, the more remarkable he appears, amply justifying the moniker bestowed upon him of "John the Astonishing."[9] Given the attention now being paid to the Occitan friar, critical editions of his two alchemical works should join the critical editions of his prophetic works already produced or now in production. The restored text of the *Liber lucis* presented and analyzed here most certainly makes John and his ideas all the more astonishing, particularly to historians of science and especially to historians of chemistry, and provides a richer and more accurate understanding of his practical knowledge and of the relationships between his theological and natural philosophical pursuits.

1.1 John of Rupescissa's Biography

John of Rupescissa's life can hardly be called a pleasant one.[10] He was born in Marcolès, a few miles southwest of Aurillac, in Auvergne, central France, around 1310. Little is known of his family origins, despite extensive research; he may have come from a knightly family. In 1326/7, he went to the University of Toulouse, where he studied for five years. In 1332, he joined the Franciscan Order and continued his studies in the Franciscan *studium* thereafter for another five years. By 1340, he was living at the Franciscan friary at Aurillac, where the direction of his life changed definitively. He later recalled that on 8 September 1340, the Nativity

Sibum, "From the Library to the Laboratory and Back Again: Experiment as a Tool for Historians of Science," *Ambix* 63 (2016): 85–97, and for an early example of the technique applied to alchemy specifically, Lawrence M. Principe, "Chemical Translation and the Role of Impurities in Alchemy: Examples from Basil Valentine's *Triumph-Wagen*," *Ambix* 34 (1987): 21–30.

7 Robert E. Lerner, "Life, Works, and Fortune of the Prophet Immured," 25–39, in Giovanni di Rupescissa, *Vade mecum in tribulatione*, ed. Elena Tealdi (Milan: Vita e Pensiero, 2015), on 25. The works currently available in modern critical editions include *LSE*: *Liber secretorum eventuum* (1349); *RP*: *Reverendissime pater* (1350, see Elizabeth Casteen, "John of Rupescissa's Letter *Reverendissime pater* in the Aftermath of the Black Death," *Franciscana* 6 [2004]: 139–84); *LO*: *Liber ostensor quod adesse festinant tempora* (1356); and *VM*: *Vade mecum in tribulatione* (1356); see the list of abbreviations for full bibliographical details. Others, such as the *Sexdequiloquium* (1352–53) are currently in preparation; see the following note.

8 Sylvain Piron, "Un traité inconnu de Jean de Roquetaillade," *Revue d'histoire des textes* 4 (2009): 299–307; the special edition of *Oliviana* 3 (2009), *Le manuscrit trouvé dans l'armoire*, contains multiple articles devoted to this document; see https://journals.openedition.org/oliviana/322. Additionally, a previously known text, the *Verba fratris Johannis de Rupescissa*, thought to have been a later extract from the *Vade mecum in tribulatione*, has been shown to be an independent composition by John himself, see Robert E. Lerner, "Yet Another Work by John of Rupescissa," *Oliviana* 5 (2016); https://journals.openedition.org/oliviana/825

9 Robert E. Lerner, "John the Astonishing," *Oliviana* 3 (2009); http://oliviana.revues.org/335

10 John's biographical details were reliably studied and presented first in Bignami-Odier, *Études*, and then enriched by further discoveries and analysis in Robert E. Lerner, "Historical Introduction," 13–85, esp. 15–29 in *LSE*, and "Life, Works, and Fortune," 25–39 in *VM* (Tealdi), as well as in Sylvie Barnay's and Hélène Millet's contributions to *LO*, 1–12.

of the Virgin, while chanting the *Te deum laudamus* at Matins, he received a sudden insight into the meaning of Holy Scripture that revealed details about the advent of the antichrist, specifically that he would come to power in mid-1366. John's interest in the antichrist and prophecy, however, well predated this providential insight. In the first year of his Franciscan novitiate, he had a dream that the antichrist had been born in a far-off city called Zayton. Believing it to be an imaginary place, he gave little credence to his dream until two or three years later, when he read a letter from a fellow Franciscan who bore the title bishop of Zayton. Thus John learned that he had dreamed of a real place (now Quanzhou on the Chinese coast of the Taiwan Strait), and now, taking his dream more seriously as a vision, started telling his brethren about it. Soon thereafter, in 1337, "a certain young friar," identifiable as John, spoke up at the general chapter of the Franciscans in Cahors to announce the birth of the antichrist.[11] Apocalyptic ideas and prophecies about the antichrist were extremely widespread throughout the south of France in the early fourteenth century and were especially popular among many within the Franciscan Order; it would have been impossible for John not to have come into contact with them.

John began speaking and/or writing about his interpretative insight of 1340 and the arrival of the antichrist, and it is presumably this activity that brought him into conflict with Guillaume Farinier (d. 1361), the Franciscan Provincial of the Aquitaine from 1343 until 1348, and thereafter Minister General of the Order.[12] A related source of concern on Farinier's part was surely John's status as a Franciscan "rigorist" that linked him with the Spiritual branch of the Franciscans and their lay followers who embraced a strict interpretation of St. Francis' teaching and Rule.[13] The Spirituals preached radical poverty, criticized ecclesiastical wealth and what they saw as laxity (or even betrayal) in regard to St. Francis' ideals by their more settled and non-Spiritual brethren, and were particularly caught up in a fervor for apocalyptic prophecies. Indeed, radical poverty and apocalyptic prophecy were virtually inseparable for the Spirituals. The Spirituals caused serious tensions within the Franciscan Order. Their views provoked extensive theological debate and multiple papal pronouncements, and on multiple occasions they were suppressed, sometimes violently. Thus John's affinities with the Spirituals in terms of poverty, prophecy, and the criticism of Church authorities could not have been welcome to a Franciscan Provincial interested in maintaining order in a province that had already witnessed so much disruption from strife within the Order.[14] Whatever the exact cause (and John claimed not to know what it was, or

[11] Lerner, "Historical Introduction," 22–25; *LO*, 372–73.

[12] On Farinier, see Hugues Dedieu, "Les ministres provinciaux d'Aquitaine, des origines à la division de l'ordre," *AFH* 76 (1983): 129–214 and 646–70; on 202–7.

[13] See David Burr, *The Spiritual Franciscans* (University Park, PA: Pennsylvania State University Press, 2001) and *Franciscains d'Oc: Les Spirituels ca. 1280–1324, Les cahiers de Fanjeux* 10 (Toulouse: Privat, 1975).

[14] The causes behind Farinier's seeming vendetta against John are explored in Lerner, "Historical Introduction," 23–29.

simply that Farinier's accusations were false), the Provincial had John arrested on 2 December 1344 and imprisoned at the Franciscan friary at Figeac. He remained confined there for a little over a year in what he describes as horrible conditions. Thereafter, he was transferred to a series of other imprisonments in the friaries at Martel, Brive, and Limoges, and finally—after some unspecified agreement was made with Farinier—he was sent to the friary at Saint-Junien where he once again "walked freely among the friars." But after five months, the pact was broken (it remains unclear in what way or on whose side), and John was once again imprisoned and then shuffled from one prison to another in quick succession—Cahors, Montcuq, Penne, and Saint-Antonin—remaining in each place for as little as two days before being transferred to the next.[15]

Finally, in summer 1346, John was brought to Toulouse to be examined by the Dominican inquisitor for Languedoc, and later cardinal, Jean de Moulins (d. 1353). De Moulins found John orthodox and set him at liberty, but this new freedom lasted a mere twenty days. Even though the inquisitor reportedly ordered, under pain of excommunication, that John not be confined again, Farinier, obviously displeased by the Dominican's verdict, ordered John imprisoned once more at the Franciscan friary in Toulouse. He remained there in "the most horrific and most forsaken prison" for a year until he was relocated to the "deadly prison" (*mortiferus carcer*) in the friary at Rieux, a few miles southwest of Toulouse, where on 1 August 1347, he was locked "in the belly of a brick wall under the dormitory staircase."[16]

John was still imprisoned there when the Black Death reached the friary in 1348. The plague killed off most of the friars, leaving John often forgotten and without food or water. Suffering from a respiratory illness that made breathing difficult for much of his first year at Rieux, John then fell ill, perhaps of the plague, but at last recovered. Finally, in July 1349, as he later recalled, he received an angelic message that he would shortly be set free, and indeed on 1 August 1349 (the feast of St. Peter in Chains, as he carefully notes) two years to the day after his imprisonment at Rieux and three after his reimprisonment at Toulouse, John was released from under the staircase. This liberation, made on the order of Raoul de Cornac (d. 1361), who had replaced Farinier as Provincial of the Aquitaine in 1348, was however only for the purpose of transferring John to yet another prison, this time in Castres, about seventy-five miles to the east. Nevertheless, John somehow convinced the friar who was escorting him there to take him yet further east to the papal court in Avignon, where he could lay his case before the pope. John arrived in Avignon on 17 August 1349 and appeared before Pope Clement VI and the papal curia at a consistory that took place on 2 October.[17] (Presumably, he

[15] *LO*, 518: "inter fratres libere ambulam," and 519.

[16] *LO*, 520: "in ventre muri latericii sub scala dormitorii." The fullest account of John's imprisonments appears in *LO*, 517–25.

[17] *LO*, 521–22.

spent the intervening six weeks in the Franciscan friary in Avignon, but under what sort of confinement is not known.)

Remarkably, a notarial record of John's appearance before the papal court actually survives, both in the original Latin version, of which the first half is regrettably lost, and in German and Czech translations that contain condensed versions of the material now lost in Latin.[18] According to this eyewitness account, a Franciscan superior appeared at the consistory along with John and accused him of various crimes and heresies. After hearing the accusations and collecting several of John's writings that he had carried with him to Avignon, the Pope asked John to respond. John then preached boldly before the curia. He defended his actions and ideas, upbraided the cardinals for their wealth and the curia for its corruption, and called Clement VI the "herald of the antichrist." According to John's later account of the event, Clement, at least at one point of the proceedings, defended John's views about poverty against a Dominican procurator who was also present.[19] But before John had finished and a conclusion was reached, the consistory was forced to break up unexpectedly when an irate priest from Montpellier entered and hurled a papal document he had received back at the Pope, the heavy seal of which struck Clement across the face. John and the priest with such good aim were arrested, and John was placed in the *carcer soldani*, or "Sultan's prison," so called because its warden bore the title of *soldanus*.[20]

Three days later, on 5 October—the day after the feast of St. Francis of Assisi—Clement VI gathered a group of scholars to consider John's case. They concluded that John was a learned and honest cleric: "nothing bad was found in his writings, or from himself, but instead he was found to be an outstanding master of astronomy, theology, and philosophy."[21] The Pope then assembled a more formal committee to examine the matter further. This committee was headed by the Cistercian cardinal Guillaume Court (d. 1361) who set John to writing out an account of his ideas about the coming antichrist, which he completed on 11 November 1349 as the *Liber secretorum eventuum* (*The Book of the Secrets of*

[18] The Latin original is Wolfenbüttel, Herzog August Bibliothek, MS Guelf 876 Helmstedt, fols. 1r–v; the German translations are Munich, Bayerische Staatsbibliothek, Cgm 409, fols. 306r–8v and Strasbourg, Bibliothèque nationale et universitaire, MS 2119, fols. 292r–97r. These accounts are reproduced in Bignami-Odier, *Études*, 221–32, but note that the French *traduction abregée* of the German text provided on 229–31 is quite faulty—e.g., *Ketzerey* (heresy) is rendered as *sorcellerie* (sorcery)—and omits significant information. A revised edition and English translation of the German text, however, is available in *John of Rupescissa, Vade mecum in tribulatione Translated into Medieval Vernaculars*, eds. Robert E. Lerner and Pavlina Rychterová (Milan: Vita e Pensiero, 2019), 314–23. The origin, coherence, and content of the accounts are closely studied in Robert E. Lerner, "Prophet and Pope: John of Rupescissa Confronts Clement VI," *Franciscana* 23 (2021): 97–122.

[19] *LO*, 588; this claim is shown to be unlikely on logical grounds by Lerner, "Prophet and Pope," 115–17.

[20] John's entry into the *carcer soldani* on 2 October 1349 is recorded in Vatican Archives, *Collectoriae* 50, fol. 91r; see *LO*, 3 n.6 and Lerner, "Historical Introduction," 30 n.41. The *carcer soldani* was one of several places of confinement used by the papacy; on the various papal prisons, see Karl Heinrich Schäfer, *Die Ausgaben der apostolischen Kammer unter Johann XXII* (Paderborn: Schöningh Verlag, 1911), 411–14. A closer treatment of the *carcer soldani* and John's time there appears in Section 2.9, below.

[21] MS Guelf 876 Helmstedt, fol. 1v: "nichil mali tam in libris quam ab ipso est repertum sed egregius magister astronomie theolo[g]ye et phylosophie est inventus." Also in Bignami-Odier, *Études*, 225.

Things to Come).[22] The book did much more than interpret Scripture in order to predict the arrival of the antichrist. It is filled with precise details—names, dates, and places—of the wars and the political and ecclesiastical upheavals that will precede, accompany, and follow upon the antichrist's rise to power in 1366. John names this antichrist as King Louis of Sicily and further describes his destruction by divine intervention in 1370 and the age of peace to follow, built by a holy pope, a righteous emperor, and the true Franciscans. But corruption would begin to set in again around 2115, according to John's insights, leading to the arrival of a final antichrist in 2370, whose annihilation would be followed by the end of time and the Last Judgment. The papal commission was slow to reach a verdict, perhaps delayed by what John calls "four appeals by the adversary party" (presumably from Franciscan officials keen to have the embarrassing friar back in their own custody), but finally in October 1350 John was declared free of heresy. Judging from John's later comments, he seems to have been judged *fantasticus*—that is, more or less, "wildly imaginative." He was not however free to leave; he was kept confined (in one form or another) in Avignon or its environs for the rest of his life.[23]

John wrote that he considered it a divine grace that he "had found mercy in the sight of Clement VI, and that he did not turn me over to the prisons of the Friars Minor as they wanted, but ordered me to remain in his Sultan's prison, under the protection of the apostolic See."[24] The exact nature of John's imprisonment at Avignon remains a subject of scholarly discussion, a topic to which I must return in Section 2.9, assisted by new information offered by the critical edition and analysis of the *Liber lucis*. According to John's own accounts, however, despite what he calls the mercy and protection of the papacy, he endured terrible conditions for many years in the papal prison. He recounts, for example, that toward the end of 1351, the lunatic Simon Legat, an apostate English Benedictine priest, was confined in an adjacent space.[25] For the following years, Legat shouted curses at John or hissed at him in English "heretic, tic, tic, heretic" whenever the friar tried to pray, study, or sleep. At other times he threw his feces at John in a "new type of stoning," and on one occasion in mid-1355 he nearly succeeded in stabbing John fatally with a sharpened bone.[26] In August 1354, John was again brought before Guillaume Court—who earlier that year had excommunicated two Spiritual Franciscans who were subsequently executed—to be questioned anew in front of an assembly of scholars and clerics. John shrewdly answered Court's questions that were designed to entrap him regarding the executed friars, whereupon

[22] John himself states that the book was written while he was "in the Sultan's prison" (*LSE*, 218).

[23] Vatican Archives, *Collectoriae* 50, fol. 91v records that at this time John was "relaxatus," but the meaning of this term is ambiguous; for further treatment, see Section 2.9.

[24] *LO*, 522.

[25] Legat's arrival in the *carcer soldani* on 14 November 1351 is recorded in Vatican Archives, *Collectoriae* 50, fol. 100r.

[26] *LO*, 523–25.

(according to John's report) Court threw copies of the friar's books on the ground and laughed derisively at him and his ideas.[27]

Details of John's life become scanty after 1356 when he completed the lengthy *Liber ostensor* (from which many of his biographical details have been drawn) and his most widely disseminated prophetic work, the *Vade mecum in tribulacione*, as well as his last known text, the brief *Verba fratris*, compiled late in that year. The chronicler Jean le Bel (c. 1290–1370) reports that around 1360 John was being held at the castle in Bagnols-sur-Cèze, some twenty miles upriver from Avignon.[28] Some time thereafter, he was glimpsed in the papal complex by the Carthusian writer and theologian Heinrich Egher de Kalkar (1328–1408) during his visit to Avignon. From June through December 1365, John was ill and being cared for in the Franciscan friary at Avignon, at which time the papal accounts indicate that his expenses were being paid by Urban V on a monthly basis.[29] Since the last payment was made on 31 December 1365, it is probable that John died there in early 1366—ironically, just shortly before the date he had predicted the antichrist would come to power.

1.2 John's Alchemical Authorship: Medicine and Chrysopoeia

John of Rupescissa wrote both of his two alchemical treatises during his first five years in Avignon. The first of these twi treatises, which popularly goes under the title *On the Consideration of the Quintessence* (*De consideratione quintae essentiae*), but is more correctly titled *On the Service of Philosophy to the Gospel of the Lord Jesus Christ and to the Poor Men of the Gospel* (*De famulatu philosophie evangelio domini Jhesu Christi et pauperibus evangelicis viris*), is by far the longer of the two and has received much more scholarly attention.[30] Indeed, this work has long featured as an important reckoning point—if occasionally oversimplified or decontextualized—in narratives of the history of science and medicine in regard to its significant impact on the development of chemical pharmacy.[31] *De famulatu* is divided into two books. The first deals with the nature and production of the "fifth essence" or "quintessence" from a range of substances using a variety of laboratory processes. John pays attention first of all to the quintessence of wine, identifiable as distilled, purified, and somewhat elaborated alcohol (ethanol).

[27] *LO*, 446–50 and 530–31; Lerner, "Historical Introduction," 31–32. On the trial and execution of the two Spirituals, see Alexander Patschovsky, *Ein kurialer Ketzerprozeß: Die Verurteilung der Franziskanerspiritualen Giovanni di Castiglione und Francesco d'Arquata* (Wiesbaden: Harrassowitz Verlag, 2018).

[28] Jean le Bel, *Chronique*, ed. Jules Viard and Eugène Déprez, 2 vols. (Paris: Renouard, 1905), 2:273–74.

[29] Karl Heinrich Schäfer, *Die Ausgaben der apostolischen Kammer unter den Päpsten Urban V und Gregor XI* (Paderborn: Schöningh Verlag, 1937), 330–34, on 330 (14 June 1365): "pro expensis fr. Johannis de Ruppescissa et elemosina facta sibi...quia est infirmus"; and *LO*, 4.

[30] *De consideratione* is the title of only the first book of the treatise; the second is titled *De remediis*, and the ensemble, *De famulatu*. This point will be discussed further in my forthcoming critical edition of that work. I will continue to refer to this text here with the title *De famulatu*, which, besides being more correct, provides a stronger sense of John's inherently religious motivations in writing it.

[31] For example, Multhauf, "Rupescissa and Medical Chemistry."

John then uses this alcoholic distillate to prepare various medicines and describes the production of other quintessences from substances such as blood, herbs, metals, and minerals. The second book describes the preparation of twenty remedies for particular illnesses. *De famulatu* is clearly linked to John's prophetic concerns about the arrival of the antichrist; faithful Christians, and in particular the true Franciscans, will require better and more powerful medicines in order to survive the time of tribulation. The book's full and correct title underscores the religious and eschatological driving force that lies behind its composition: the service that philosophy—in this case meaning *alchemy* specifically—offers to the mission of the Gospel and to the crucial future role of the "poor men of the Gospel" (*pauperes viri evangelici*), meaning John's fellow rigorist Franciscans for whom he wrote it.

The text of *De famulatu* gives internal evidence for its date of composition. In the last section of the first book, John notes that he "has been thrown into chains by spiteful detractors—against God and justice, as the Day of Judgement will show—for now seven whole years."[32] Assuming that John's reckoning begins with his imprisonment in Figeac on 2 December 1344, "seven whole years" brings the composition of *De famulatu* somewhere between December 1351 and December 1352, during his third or fourth year in Avignon.[33] It would be out of place to analyze the contents of that work in detail here, and it would be injudicious to do so before its critical edition is complete given the textual corruption present in its only printed version, which appeared in Basel in 1561 at the hands of Guglielmo Gratarolo, who would also publish the *Liber lucis* just two months later.[34] Nevertheless, some consideration of its content is essential at present because that book paved the way for John's second alchemical composition, the much shorter treatise commonly known by the two disparate titles under which it was published in the sixteenth century: *On the Preparation of the True Philosophers' Stone* (*De confectione veri lapidis philosophorum*) and *The Book of Light* (*Liber lucis*). While *De famulatu* describes the use of alchemy to prepare improved pharmaceuticals, the *Liber lucis* describes the use of alchemy to produce the philosophers' stone, the material agent of metallic transmutation capable of converting base metals into silver or gold. No less than *De famulatu*, the *Liber lucis* is integrally tied to

[32] *De famulatu*, "Excusacio pauperis": "conjectus contra deum et justiciam, sicut dies judicii ostendet, ab emulis in vinculis integris jam septem annis." Since the printed edition is unreliable, I quote here and throughout this study from a preliminary stage of the critical edition I am preparing, at this point predominantly a *lectio plurium codicum* based upon the twenty most reliable fifteenth-century witnesses selected from across multiple families of descent. Any quotations used here may be amended in the final edition. Further, since the numbering of the sections of *De famulatu* is widely variable, I will here use as citations the opening words of each section, which are relatively stable.

[33] Two manuscript witnesses of *De famulatu* give the date of composition as 1350, but this dating is less reliable than the internal evidence of the text; see, for example, Padua, Biblioteca universitaria, MS 1380, fol. 27r and Munich, Bayerische Staatsbibliothek, Clm 617, fol. 2r.

[34] *Joannis de Rupescissa...de consideratione quintae essentiae rerum omnium* (Basel, 1561); there were several subsequent reprintings of this edition. John's work was actually first printed in a French translation: *La vertu et propriété de la quinte essence de toutes choses, faite en latin par Joannes de Rupescissa et mise en françois par Antoine du Moulin* (Lyon, 1549).

John's prophetic concerns about the antichrist and the role that his fellow rigorist Franciscans must play in rebuilding the Church and Christendom after the antichrist's destruction. The two books together not only cover the two major goals of alchemy, the production of precious metals and the preparation of medicines but also address the twin needs of future Christians and the Church grappling with the onslaught and aftermath of the antichrist, namely wealth and health.

The contents of the *Liber lucis* and *De famulatu* are closely linked. Most significantly, *De famulatu* provides a brief outline of the same laboratory process that reappears in greater detail as the first four "operations" of the *Liber lucis* and it mentions the construction and use of the same "dry stove" (*stupha sicca*) that is more minutely described in the final section of the *Liber lucis* (a section absent from both printed versions).[35] What has been largely overlooked or misinterpreted, however, is John's claim in the final section of the first book of *De famulatu* that he knows the secret of metallic transmutation and his not-too-subtle hint, presumably to the ecclesiastical authorities in Avignon, that he should be given special permission to write a full exposition of it.

John's turn to transmutation in an otherwise predominantly pharmacological work fits neatly into the book's organization. The first book of *De famulatu* progresses in order through various materials and how to prepare them alchemically—beginning with wine, then passing to blood, meat, and eggs; thence to fruits, roots, and herbs; and then to metals and minerals. After having read the sections about metals and minerals and John's description of how to separate gold and silver using mineral acids (another section absent from the printed Latin edition), a contemporaneous reader would reasonably have expected to encounter the *making* of precious metals next. Obviously aware of this expectation, John provides his "excuse of the poor man why he does not reveal the sublime arcana of the marvelous philosophy" in order to explain why he brings the first book to "an end here where the operations or considerations of alchemy could or should be added."[36]

John begins his "excuse" by confirming that the "beyond-marvelous medicines" revealed to him during his imprisonment "not only cure our bodies almost miraculously from all evil diseases, but also transmute the imperfect metals into gold and silver in the blink of an eye." Nevertheless, he will not reveal the latter knowledge "out of reverence for the statutes of our Order and obedience to the prelates of the holy church of God" since "treating alchemical operations is not allowed to

[35] This overlap was first noted by Halleux, "Rupescissa," 267; see below, Section 1.5; these parallel sections are presented for reference in Appendix 1.

[36] *De famulatu*, "Excusacio pauperis quia archana sublimia mirabilis philosophie ulterius non revelat"; "ibi... finem constituo ubi operaciones vel consideraciones alchimie possent vel deberent misceri." Note from this line (and others) that it is not true that John "never mentioned alchemy except in a pejorative manner," cf. DeVun, *Prophecy*, 63–64.

our religion."[37] John almost certainly refers here to two specific things. The first are the prohibitions in the Franciscan *Constitutiones* (regulations of the Order) that forbid the friars to practice or teach alchemy or to write, copy, or even own books about it. These strictures appeared for the first time in the *Constitutiones* of either 1307 or 1310 and were repeated for the rest of the century, including in 1354, the very year in which John wrote the *Liber lucis* and when the chapter was overseen by his chief persecutor Farinier, who had by that time become Franciscan Minister General. By 1331, the punishment for violating the prohibition included imprisonment and automatic excommunication. Given the abundance of surviving alchemical texts authored by fourteenth-century Franciscans, the friars regularly ignored these regulations; the continuing profusion of Franciscan alchemists may explain the Order's need to repeat the prohibitions.[38]

John clearly alludes also to John XXII's decretal *Spondent quas non exhibent* ("They Promise What They Do Not Deliver") issued in 1317. This decree was aimed specifically against alchemists who turn to cheating practices when they fail to achieve their goals, deceptively selling false gold and silver and counterfeiting coins.[39] John echoes the wording of this papal decretal when a little later in the "Excusacio" he mentions that frustrated chrysopoeians, when they fail to attain the knowledge of transmutation, can become "counterfeiters of coins" and "turn themselves towards vain adulterations and generally come to a bad end."[40]

John therefore warns his audience not to apply themselves too rashly to the pursuit of transmutational alchemy. On the one hand, alchemical endeavors may sometimes be demonically encouraged, which is the "road to perdition."

[37] *De famulatu*, "medicinas supra mirabiles quasi miraculose corpora nostra sanarent ab omnibus malis morbis sed ipsa metalla imperfecta in aurum et argentum in ictu oculi transmutarent"; "propter reverenciam statutorum ordinis nostri et obedienciam prelatorum ecclesie sancte dei"; "quia de alchimicis operacionibus religioni nostre pertractare non licet."

[38] See Lawrence M. Principe, "Friars Minor and *Opus Magnum*: Why Did So Many Medieval Franciscans Practice Alchemy?," forthcoming in *Religious Experience and the Foundations of Modern Science*, ed. Nuno Castel-Branco (Cambridge: Cambridge University Press, 2026); *Constitutiones generales Ordinis Fratrum Minorum*, ed. Cesare Cenci and Georges Mailleux, 2 vols. (Grottaferrata: Quaracchi, 2007–10), 2:11–12 (Padua, 1310). Some secondary literature claims that this prohibition appeared in 1279; however, its mention in a now-lost manuscript of the 1279 *Constitutiones* was a later marginal addition intended to "update" the older version; see Cenci, "Le Constituzioni Padovane del 1310," *AFH* 76 (1983): 505–88, esp. 515. For the 1354 *Constitutiones* overseen by Farinier at Assisi, see Michael Bihl, "Statua generalia ordinis edita in capitulo generali an. 1354 Assisii celebrato," *AFH* 35 (1942): 35–112 and 177–253, on 187–88, and St. Bonaventure University, NY, St. Bonaventure University MS 20, fols. 31r–57v, on fol. 46. The large number of Franciscan alchemists was noted already by Thorndike, *HMES*, 3:347–48. For important recent treatments, see Antoine Calvet, *L'alchimie au Moyen Âge: XIIe-XVe siècles* (Paris: Vrin, 2018), 125–35; Pereira, "Francescani e l'alchimia"; Chiara Crisciani, "Note sull'alchimia 'francescana' nel sec. XIII" in *Atti del XXV Congresso Nazionale di Filosofia* (Rome, 1980), 214–20; Crisciani, "Alchimia e potere: presenze francescane (secoli XIII-XIV)," in *I Francescani e la politica*, *Franciscana* 13 (Palermo: Officina di Studi Medievali, 2007), 223–35.

[39] See Lawrence M. Principe, *The Secrets of Alchemy* (Chicago: University of Chicago Press, 2013), 61–62, and for the full text of the decretal, Robert Halleux, *Les textes alchimiques* (Turnhout: Brepols, 1979), 124–26.

[40] *De famulatu*, "Excusacio": "efficiuntur falsificatores monete," and "quando non possunt ad scienciam perfectam pervenire, ad sophisticaciones inutiles se convertunt et communiter malo fine clauduntur." The phrase *sophisticaciones inutiles* possibly refers to the cheating practice of "multiplication," that is, extending the weight of precious metals through debasement with other metals; such debased metals could then be used to produce counterfeit coins.

This connection may depend upon the use of alchemy by several thirteenth-century scholastic theologians (St. Thomas and St. Bonaventure among them) as an example of human artifice in discussions of the power of demons to induce changes in material objects.[41] John further points out that no author has written about transmutation "except falsely and under metaphors that cannot be understood by human ingenuity." Therefore, no one, he asserts, attains "the sublime things of the art of alchemy" unless his mind has been prepared by deep contemplation and holy life "so that he may know not only all the inner things of the nature [of metals] but also how to transmute the transmutable nature."[42] The slightly obscure way in which John phrases this qualification strongly suggests that he was familiar with, and is here responding to, the alchemically infamous Latin text known as the *Sciant artifices* ("Let the artificers know"), excerpted and translated from Ibn-Sīnā's *Kitāb al-Shifā'*. There the Persian philosopher states that alchemists can indeed make lesser metals *look like* gold or silver by changing their external appearances but cannot produce true gold or silver because that would require changing their "specific differences," and those differences are simply unknown: "if the differences are not known, how can one know whether to remove them or not, or know how they could be removed?"[43] Thus for Ibn-Sīnā, transmutation cannot be accomplished because of human ignorance of both the internal nature of the metals and the practical means of changing those natures. In apparent response, John emphasizes the need for the alchemist to know both the "interior things" of the nature of metals and how to transmute that nature. In contrast to Ibn-Sīnā, who implies that such knowledge simply cannot be humanly obtained, John writes that "it is conceded to very few." John then clearly claims himself to be one of those few, having gained this knowledge "through sublime theories and heavenly illuminations." Nevertheless, he continues, he "will remain silent about those most stupendous works of philosophy." Crucially, however, he concedes that his silence may not be permanent, promising that "hereafter, if God wills, I will reveal [this knowledge] most fully as I will be able, in separate books

41 William R. Newman, *Promethean Ambitions: Alchemy and the Quest to Perfect Nature* (Chicago: University of Chicago Press, 2004), 49–53. The Franciscan *Constitutiones* of 1310 list alchemical operations alongside necromancy and sorcery—although this juxtaposition (interestingly enough) disappears from the later *Constitutiones* starting in 1316, leaving alchemy as the only explicitly forbidden practice. This part of John's text is quoted, and misinterpreted as an argument against transmutation, by Athanasius Kircher in his treatment of the philosophers' stone in his *Mundus subterraneus* (Amsterdam, 1678), part 2, 294; reprinted in *BCC*, 1:54–112, on 75–76.

42 *De famulatu*, "Excusacio": "nullus philosophorum scripsit veritatem in libris nisi ficte et sub parabolis que non valent ab humano ingenio comprehendi," and "nullus valet ad sublimia pervenire artis alchimice nisi fuerit per altissimam contemplacionem et sanctissimam vitam deificata mens ejus ita ut noverit non solum omnia interiora nature sed eciam transmutabilem transmutare naturam."

43 Elisa Rubino and Samuela Pagani, "Il *De mineralibus* di Avicenna tradotto da Alfredo di Shareshill," *Bulletin de philosophie médiévale* 58 (2016): 23–87, on 43 (contains a critical edition of the text). For more on this important text, see Jean-Marc Mandosio and Carla DiMartino, "La *Météorologie* d'Avicenne (*Kitāb al-Shifā'* V) et sa diffusion dans le monde latin," *Miscellanea mediaevalia* 33: *Wissen über Grenzen: Arabisches Wissen und lateinisches Mittelalter* (2006): 406–24; Newman, *Summa*, 2–5; and Athanasios Rinotas, "The *Sciant artifices* in the Work of Albert the Great: Towards Two Kinds of Transmutations?," *Early Science and Medicine* 27 (2022): 57–82. For a brief introduction to the text, see Principe, *Secrets*, 58–59.

to you men of the Gospel for whom I write."[44] Whether John received the green light from his earthly superiors or from his heavenly Father remains unknown, but roughly two years later, he fulfilled the qualified promise he made in *De famulatu* when he wrote the *Liber lucis*.

1.3 Date of Composition of the *Liber lucis*

Unlike *De famulatu*, the *Liber lucis* contains no clear internal means for dating its composition. To determine its date, Robert Halleux cautiously pointed to the years given by the explicits in three manuscripts known to him: 1330, 1354, and 1350.[45] While these dates disagree with one another, he concluded that they at least "intersect" with the dating of *De famulatu* to 1351–52, although he left open the possibility that these dates are later copyists' additions rather than authentic material traceable to John. Since Halleux's publication, the date 1354 has often been cited as a "probable" year of composition. Now, with the benefit of many more manuscript witnesses and their organization into a coherent stemma, this date can be considered definite and made more precise. There are now nine known manuscript witnesses bearing dated explicits:

family π:[46]
- **P²**: 14 October 1354
- **Ko²**: 14 September 1354
- **G**: 4 October 1350

family γ:
- **B**: 14 October 1380
- **Z**: 17 October 1370
- **S** and **Hb²d**: 14 October 1370

family ε²:
- **X²**: 4 October 1350

undetermined family (possibly π):
- **Br**: 14 October 1380

Three of the witnesses that bear these dates—**P²**, **Ko²**, and **B**—are also three of the earliest and most reliable. Moreover, the dated explicit occurs in members of at least three different manuscript families, meaning that it must have been present in these three families' common ancestor, and since these families diverged very early in the transmission of the text, that ancestor was close to (or might have

[44] *De famulatu*, "Excusacio": "Reliqua autem opera philosophie summe stupenda que cum his que hic subticeo... per sublimes theoricas et illuminaciones celestes didici, inferius per libros singulos, si deus voluerit vobis viris evangelicis quibus scribo plenissime revelabo ut potero."

[45] Halleux, "Rupescissa," 267, citing **B**, **Ko²**, and **X²**, respectively; he erred in reading the date given in **B** as 1330—it is in fact 1380.

[46] See the stemma, Section 4, for the location of these families and witnesses in the lines of descent.

been) John's autograph.[47] Hence, these dates cannot be readily dismissed as later copyists' intrusions. Judging from John's other surviving works, he often included an explicit that gives the day of completion, and so it is reasonable to believe that he wrote a similar one for the *Liber lucis*.

Although the years and days of the month diverge significantly, October is common to all but one, thus giving that month a high degree of reliability and underscoring the common origin of these multiple explicits. The day and year are more discrepant, but their variations are readily explicable as simple copyist's errors. Six witnesses give the day as 14, two as 4, and one as 17. The erosion of 14 into 4 is easily understandable as the omission of a single cipher, whether an Arabic 1 or a Roman x. The corruption of 14 into 17 is even easier to explain; it results from the well-known "minim problem," in this case reading *xiiii* as *xuii*.[48] As for the year, the specific variations suggest that an early ancestor, perhaps John's autograph, gave the year in Arabic rather than Roman numerals, since to copyists not well familiar with the early forms of these numerals, 5 looks much like a later 7, and 4 like a later 0. Thus, 1354 could easily be misinterpreted as 1350 or 1370. We can therefore conclude, with a high degree of confidence, that John's *Liber lucis magisterii magni* was completed on 14 October 1354.

Given John's typically swift rate of composition and the very short length of the book—just over 3500 words—the writing could have been completed within a few weeks.[49] This means that it could have been started just shortly after John was interrogated by Court a second time in August 1354. While this timing might be no more than mere coincidence, it might also be possible that John found this an opportune moment to advertise more explicitly his potentially lucrative knowledge of the secrets of gold-making when his life seemed once again in danger. While the prologue to the *Liber lucis* states that the book is intended for the *viri evangelici* to help them rebuild a Church despoiled by the future tribulations of the antichrist, John may have intended his book to perform a double duty by first helping one particular *vir evangelicus* during his own present tribulations in Avignon. There is no shortage of later prisoners who endeavored to help their situation (or at least prevent it from worsening) by advertising their knowledge of transmutational alchemy; if this motivation was also present in John's authorship, the *Liber lucis* would be the earliest known example.[50]

47 Given the presence of the dated explicit in both **Ko²** and **P²**, it must have been present in their first common source, π; given its presence in **Z** as well as in **B**, **S**, and **Hb²d**, it must have been present in γ. If present in both π and γ, it must have been present in their common fourteenth-century ancestor. I ignore **X²** here because of its late date and contamination, and **Br** because I have not been able to localize it convincingly in the stemma.

48 A minim is a single vertical stroke of the pen—a single minim gives the letter *i*, two minims produce *n*, *u*, and *ii*, which are often difficult, even impossible, to distinguish in Gothic scripts; three minims produce *m* and also the combinations *ni*, *in*, *ui*, and *iu* as well as the numeral *iii*. The dot over the letter i is a medieval innovation intended to address this problem of legibility, although the dot was not always used; Germanic writers often added a squiggle over *u* to distinguish it from *n*.

49 On John's rapid rate of composition, see Lerner, "Life, Work, and Fortune," in *VM*, 35.

50 Jennifer M. Rampling, *The Experimental Fire: Inventing English Alchemy, 1300–1700* (Chicago: University of Chicago Press, 2020), 301ff. and Lawrence M. Principe, *The Transmutations of Chymistry: Wilhelm Homberg and the Académie Royale des Sciences* (Chicago: University of Chicago Press, 2020), 299–325.

One possible piece of supporting evidence, although admittedly highly conjectural, for this hypothesis appears in a hitherto-unrecognized German translation of the *Liber lucis* published in 1610 as part of a miscellany of short alchemical treatises. The ragged state of the German text suggests both that its Latin exemplar was difficult to read and/or heavily damaged and that the translator's Latin was poor. Nevertheless, it is the text's attribution that could be potentially significant, for it is titled *Pope Innocent's Treatise on the Philosophers' Stone*.[51] It is extremely difficult to explain why a copy of the *Liber lucis* would be fathered upon a pope named Innocent—no Pope Innocent is otherwise known as an alchemical author or to have had an interest in alchemy. But in 1354, when John wrote the *Liber lucis*, the reigning pope was in fact Innocent VI; thus, it is conceivable that the title and authorial attribution of this German text preserves a garbled indication from its Latin ancestor that the treatise was originally *written for* or *presented to* Innocent VI, a pontiff whose reign was, incidentally, especially impecunious.

1.4 Title of the *Liber lucis*

The correct title for John's second alchemical composition requires some discussion. The two sixteenth-century printed editions offer two completely different titles, and the now-known manuscripts provide twenty-one more.[52] While the titles of medieval books are often rather fluid, precedence should be given to the title (if any) assigned by the author, often in explicits, rather than to the wide variety of headings provided by later copyists or editors. In this case, six of the dated explicits cited above, plus another three undated ones, altogether representing at least one witness from each of the reliable manuscript families, provide the same title uniformly: *Liber lucis magisterii magni*, that is, *The Book of the Light of the Great Magistery*.[53] John himself seems to refer to this title for his work when he modifies a Biblical quotation to assert the need to protect "this book of true light" (ll. 229–30) from potential abusers. Based on the manuscript evidence then, this title should be considered the correct one, although of course *Liber lucis* remains the useful short form.

In terms of the printed Latin editions, the 1579 Cologne version provides the title *Liber lucis*, adopted from the manuscript from which its text was taken, while the 1561 Basel version bears the longer title *De confectione veri lapidis philosophorum* (*On the Preparation of the True Philosophers' Stone*).[54] This latter

[51] *Tractat Bapsts Innocentii, Vom Stein der Philosophorum*, 227–32 in Joachim Tancke, *Promptuarium alchemiae* (Leipzig, 1610). The state of the German text is quite interesting and is further treated below in Section 3.4.

[52] Halleux, "Rupescissa," 262–63, lists fifteen, to which can be added several more: *Lumen amenitatis magni*, in the explicit to **Z**; *De arte benedicta* (**Be**); *Opus lune* (**Po**); *Via universalis* ($\mathbf{W^7}$); *Practica arboris philosophie* (**S**); and *De sulphure philosophorum* ($\mathbf{Y^2}$).

[53] Family π: $\mathbf{P^2}$ and $\mathbf{Ko^2}$; family γ: **B**, $\mathbf{Hb^2{}_d}$, **V**, and **Z**; family γ^2: $\mathbf{Ko^1}$; family γ^3: $\mathbf{Ka^2}$.

[54] For the origin of the text for the Cologne edition, see Lawrence M. Principe, "Bibliotheca Philosophica Hermetica MS 338: The 'Missing Link' in the Transmission of John of Rupescissa's *Liber lucis*," *Quaerendo* 53 (2023): 187–97, and below, Section 3.3.

title originates in a lengthy incipit added to John's text after it circulated into Germanic lands:

> Here begins the book of the magistery *of the preparation of the true philosophers' stone*, written clearly and without any concealment by friar John of Rupescissa of the Order of Friars Minor, a man surely illuminated by God, who composed this little book for the purpose of alleviating the want of the pope and clergy in the time of the coming tribulation of the Church, being forewarned of it by divine revelation.[55]

This incipit, which appears with only minor variations in most members of the large Germanic manuscript family designated γ, must have been added sometime before 1423, the date of the earliest known manuscript of this family. The Basel edition preserves this added incipit, from which the editor derived his title, and since this incipit does not appear in any manuscripts outside of family γ, its presence indicates that the Basel edition was printed from a manuscript in that family.

1.5 Authenticity of the *Liber lucis*

John's authorship of the *Liber lucis* has sometimes been called into question. Indeed, the manuscripts examined for this study provide an array of supposed authors besides John of Rupescissa. Many are attributed to Ramon Lull or to Arnald of Villanova, the most frequently cited alchemical "authors" of the fourteenth century. English copyists sometimes attributed it to one of their own countrymen, either Roger Bacon or George Ripley. A great many were simply left anonymous. Doubts about the authenticity of *De famulatu* have been less frequently raised because that much longer work contains more biographical details and more references to topics known to be of concern to John.[56] But doubts about John's authorship of the *Liber lucis* were raised already in the seventeenth century when some writers considered that a Franciscan with a fervent commitment to poverty could not have written a treatise on something as lucrative as gold-making. In contrast, we now know that a great many early Franciscans were devoted to chrysopoeia. Moreover, as John states in his prologue and elsewhere in the text, his motive for promoting gold-making was not for personal enrichment but rather so that faithful Franciscans would have the means to rebuild a Church despoiled by the antichrist. This motive remained hidden from many later readers, however, because the explanatory prologue about the use of alchemy to repair the damage caused by the antichrist was omitted from the Basel edition. Likewise, both printed versions

55 "Incipit liber magisterii *confeccionis veri lapidis philosophorum* clare et absque quacumque palliacione editus per fratrem Johannem de Rupescissa ordinis minorum virum utique a deo illustratum quem libellum composuit ad sublevandas inopias pape et cleri tempore future tribulacionis ecclesie eciam ad hoc divina premonitus revelacione." Italics added here and above. The text given here for the added incipit is a composite of the most reliable versions, each of which bears one or more obvious but minor copyist's errors.

56 There now can be no possible doubt of *De famulatu*'s authenticity. It gives what was the sole reference to an otherwise unknown work of John's called the *Directorium simplicium electorum*, but now the recently discovered *Sexdequiloquium*, whose authorship is unquestionable, references the same lost title.

deleted almost all of a lengthy section of Operation Six (ll. 220–51), which I term the "Credas" from its opening word, expressing John's hopes for the use of his work by future *viri evangelici* only to "help the afflicted Church and the holy ones in the time of the tribulation or of general want" (ll. 250–51). Once restored to the text, these sections render arguments against John's authorship less tenable, for they link the *Liber lucis* clearly to John's well-recognized prophetic works and concerns about the antichrist.

The overlaps in technical content between *De famulatu* and the *Liber lucis* have already been mentioned; Halleux considered them a "decisive argument" in favor of the authenticity of the latter.[57] Yet not all scholars have been convinced, and thus occasional references to the author of the *Liber lucis* as "pseudo-John of Rupescissa" still appear. Such skepticism is not entirely unwarranted, given the prevalence of pseudonymous fourteenth-century alchemical authorship and the obvious ability of a forger to crib material from an authentic work. Indeed, some texts spuriously attributed to John use terms like *quintessentia* and *viri evangelici* in order to look like the Occitan friar's own compositions, although such works also introduce ideas (often anachronistic ones) that are clearly alien to John's authentic writings, which the *Liber lucis* does not.[58]

The chemical analysis and laboratory reproduction of John's process, described below, provide new and powerful evidence of John's authorship. The two parallel procedures for preparing the "quintessence of mercury and vitriol" found in *De famulatu* and the *Liber lucis* are not in fact identical; they vary in terms of the ingredients to be used. Significantly, *neither* recipe as written can possibly give the results described, and only when the two recipes are *combined* can the desired product be produced. Thus, John here deploys the classical alchemical technique of the "dispersion of knowledge," whereby the details of an important process are separated and put in different places in order to safeguard the knowledge.[59] In this case, John divided the full ingredient list between his two books, a move that not only recalls the popular alchemical maxim "one book opens another" but also indicates the common authorship of the two works.

The critical edition also affords new linguistic evidence of John's authorship. In a general sense, the wealth of Biblical quotations and allusions typical of John's writing, largely obscured or omitted in the highly edited published versions, leaps out from the pages of the critical edition. It likewise displays vocabulary and particular turns of phrase that John uses commonly in his prophetic works, such as "demonstration to sense/to the eyes" (*demonstratio ad sensum/ad oculum*), "experience teaches" (*experientia docet*), and "at the gates" (*in januis*) to describe the antichrist's imminent arrival. The reasons for brevity stated at the start of the *Liber lucis* reappear in similar language at the start of the *Vade mecum* and in

[57] Halleux, "Rupescissa," 267.
[58] See Appendix 2 for spurious works forged to look like authentic ones.
[59] For further explanation of this literary technique and John's use of it, see Section 2.2.2.

several (even the longest) of John's other writings.[60] More specifically and more strikingly, there are at least three very well-defined linguistic connections between John's prophetic texts and the *Liber lucis*. First, the prologue to the *Liber lucis* states that although "the little ship of Peter shall be tossed about by powerful waves, it shall nevertheless be liberated at the end of the days of tribulation," thus quoting nearly verbatim a line from the so-called Tripoli prophecy that John quotes again in the *Liber ostensor*.[61] While the Tripoli prophecy was well known to John, it is extremely unlikely to have been known and quoted by an alchemical forger or, in fact, by virtually any other alchemical author. Second, the *Liber lucis* quotes Ephesians 5:27 (at ll. 227–28) about the Church having "neither spot nor wrinkle" but expands the Pauline verse with the additional words "of mortal crime, nor...of heresy or of error." The same verse appears expanded in the same way three times in the *Liber ostensor*.[62] Finally, the *Vade mecum* contains the puzzling line that "the head shall become red and a drop of warm dew of the innocent paupers shall be sought." While the exact meaning of this line in the context of John's prophecy of the curia's flight from Avignon remains unexplained and difficult to divine, it nevertheless alludes to the distillations described in the *Liber lucis*. There, John uses the term *stilla roris* ("drop of dew") as an alternate name for the distillate from liquefied mercury quintessence and cites the reddening of the head (of an alembic) as a practical reckoning point for gauging the heat needed for the distillation of his aquafort.[63] Taken all together, these various pieces of evidence remove any lingering doubts about John of Rupescissa's authorship of the *Liber lucis magisterii magni*.

[60] *Liber lucis*, ll. 16–17, and *VM* (Tealdi, 218; Kaup, 136–38); on John's promises of "brevity" and his style more generally, see Christine Morerod-Fattebert, "Un style, un homme? L'exemple de Jean de Roquetaillade," *Oliviana* 5 (2016), https://journals.openedition.org/oliviana/846

[61] *Liber lucis*, ll. 5–6, and *LO*, 429: "navicula Petri jactabitur in validis fluctibus, sed dominabitur in fine dierum." Noted first by Robert E. Lerner, *The Powers of Prophecy: The Cedar of Lebanon Vision from the Mongol Onslaught to the Dawn of the Enlightenment* (Berkeley: University of California Press, 1983), 137–38; see also Lerner, "Prophet and Pope," 120–22.

[62] *Liber lucis*, ll. 227–28, and *LO*, 229, 331, and 443. This seems to be a favorite verse for John in *LO*; he uses it in its unelaborated form also at 193, 205, 281, 341, 438–41, 444, and 446.

[63] *VM* (Tealdi, 228 and n. 39; Kaup, 166): "Caput autem rubificabitur, et stilla roris tepidi innocentium pauperum requiretur" and *Liber lucis*, ll. 89–90 and 126.

2 THE CONTENT AND CONTEXT OF THE *LIBER LUCIS MAGISTERII MAGNI*

John of Rupescissa's *Liber lucis magisterii magni* is a highly practical text, offering its readers instructions on how to prepare the philosophers' stone. Yet it is much more than a recipe, although some later readers and copyists tried to pare it down to that format. For the text also records not only John's ideas about alchemical secrecy, the right use of alchemical knowledge, and the role alchemy can play in the apocalyptic events that John believed were on the horizon, but also the friar's approaches to experimentation and his emphasis on testing and methods of demonstration. These latter features suggest that he held conceptualizations of matter and material transformations that contrast with widely diffused natural philosophical principles and practices of the High Middle Ages. The critical text also points toward some of John's sources, a significant evolution in his thinking about chrysopoeia, and connections of his alchemy with prophecy and Biblical exegesis. This section closely examines John's route to making the philosophers' stone, contextualizes it within developments of fourteenth-century alchemy, and analyzes it further with the help of modern laboratory reworkings. Such experimental reworkings both assist in understanding John's text more accurately and provide compelling evidence that John was not a "literary" or "armchair" alchemist who read, gathered, and modified existing textual sources in order to create the *Liber lucis*. On the contrary, John was an experienced practitioner who performed the operations he describes and a novel thinker who struck out in new directions in the quest to achieve metallic transmutation.

2.1 The Structure of the *Liber lucis* and John's Prologue

John divides his *Liber lucis* into a prologue and eight "operations." The prologue is either omitted or highly abridged in the printed editions of the text—as it is in many of the manuscripts, especially the later ones—thus leaving many readers unaware of the religious motivation behind John's chrysopoeia, namely, as preparation for the coming of the antichrist announced in John's other writings. The prologue first reminds readers of the antichrist's imminent arrival, his persecution of Christians and despoiling the Church of all its temporal possessions. To address the future needs of the Church and the faithful, John promises to lay aside the invidious "custom of the philosophers" (l. 11) and instead reveal the secret of making the philosophers' stone "with only a straightforward narration of words" (ll. 14–15) rather than with riddles, obscurities, and metaphors. John had already remarked in his earlier *De famulatu* on the virtual impossibility of understanding

DOI: 10.1201/9781003666547-3

typical chrysopoetic texts since "no author wrote the truth in his books except falsely and in metaphors that cannot be understood by human ingenuity."[1] John, however, promises to be candid and incorporates a line from St. Luke's Gospel where Christ tells the disciples that "to you it is given to know the mystery of the kingdom of God; but to the rest in parables." John redirects this line to describe his own revelation of the mystery of transmutational alchemy to his own chosen "disciples," namely the true Franciscans, the *viri evangelici*, of the post-antichrist future (ll. 8–9). John keeps his promise to be clear and straightforward…most of the time. Later in the *Liber lucis* he returns to the issue of alchemical secrecy, his chosen Franciscan audience, and why he is not afraid to be candid with such powerful and potentially dangerous knowledge (see Section 2.8).

The first six of the eight operations that comprise the bulk of the text describe how to prepare the transmuting agent in both its white (argyropoetic, or silver-producing) and red (chrysopoetic, or gold-producing) forms. The seventh describes how to use these prepared materials to transmute base metals into silver or gold, and the eighth gives instructions on how to construct a special furnace designed especially for preparing the philosophers' stone.

2.2 Operation One: Starting Materials and Methods for Making the Philosophers' Stone

The first operation is the longest and most detailed of all the operations, and so requires careful comment. It contains important information about John's sources and the ways in which he transformed the alchemical principles that he gleaned from the existing Latin tradition. More strikingly, John insists on proving his claims using measurements, signs (*signa*), and experimental demonstrations (*demonstrationes*), and these features provide intriguing indications of his theories of matter and material transformations that guided his practices.

Making the philosophers' stone requires the discovery of two secrets: the identity of the proper starting material(s) and the correct procedure for elaborating such material(s) into the completed stone. In the very first line, John hints at the first secret by repeating an alchemical commonplace that dates back to Arabic sources, namely, that the starting material is "a single thing, of low price, and able to be found everywhere." But John immediately clarifies that this substance is nothing other than "common quicksilver as it is extracted from the Earth" (ll. 18–19). He ridicules foolish alchemists who seek the starting material in excrement because they read too literally the alchemical authorities who write that the material is found in *locis vilibus* (low/despicable places). Correct interpretation of the written word is crucial across all of John's work, whether the authoritative text is alchemical about making the stone or prophetic about the coming of the antichrist.

[1] *De famulatu*, "Excusacio pauperis."

John's opening declaration that common mercury is the "one thing" needed for making the stone suggests that he is a proponent of the "mercury alone" school of chrysopoeia, a relatively new position of fourteenth-century Latin alchemy, which held that the stone was to be prepared from mercury alone.[2] This viewpoint makes its debut in Geber's *Summa perfectionis* but is more fully developed by pseudo-Arnald of Villanova, particularly in his *Rosarius philosophorum*, during the first part of the fourteenth century, and thereafter continued by many others. Pseudo-Arnald is, in fact, the authority that John quotes by name the most frequently in the *Liber lucis*. But John's method diverges in important ways from all earlier expressions of the "mercury alone" school, indicating a significant level of both theoretical and practical originality.

John declares that the practitioner must transform common mercury in two ways: "by removing something, namely the earthiness which it has, and by adding something, namely, the sulphur of the philosophers" (ll. 23–25). This mention of mercury's earthiness and a later reference to its excessive humidity (l. 43) probably derive from Geber's *Summa*, which cites earthiness and wateriness as the two impurities found in common mercury.[3] Likewise, John probably drew his description of mercury as a "thick water" (*aqua viscosa*) from the same source.[4] The need to add a "sulphur of the philosophers" initially recalls pseudo-Arnald's *Rosarius philosophorum* which stipulates the need to add gold, frequently called "sulphur" by pseudo-Arnald, to the purified mercury. Indeed, when John states that this sulphur is the "tincture of redness," readers familiar with fourteenth-century alchemical theory would readily conclude that John is referring to gold under the cover name of "sulphur of the philosophers," a widely used Deckname for the precious metal. Although Geber uses the term "tincture of redness" (*tinctura rubedinis*) in reference to several different materials capable of imparting a red color, pseudo-Arnald and other later authors view it as a particular substance found *only* in gold, and a necessary ingredient for the philosophers' stone.[5]

For pseudo-Arnald, as for many other Latin authors (such as the early fourteenth-century alchemist John Dastin who was himself in Avignon in the generation before John), the philosophers' stone cannot bring about transmutation until it has been "fermented" with a precious metal.[6] Mercury, treated in one way or

2 On the origins of the "mercury alone" theory, see Newman, *Summa*, 204–8, and for its elaboration by pseudo-Arnald of Villanova, Petrus Bonus, and others, see Thorndike, *HMES*, 3:57–58, 160, and 624.

3 The notion that common mercury needs to be purified of excess humidity has a long tradition before Geber, dating back to the Arabic tradition; on this point, see William R. Newman, "Mercury and Sulphur among the High Medieval Alchemists: From Rāzī and Avicenna to Albertus Magnus and Pseudo-Roger Bacon," *Ambix* 61 (2014): 327–44.

4 Newman, *Summa*, 462–65, 721–23; 333, 669.

5 Newman, *Summa*, 339 (in gold), 335, 345, 348 (in other substances); pseudo-Arnald of Villanova, *Rosarius philosophorum*, in Calvet, *Oeuvres alchimiques*, 284–87, on 284: "Aurum namque...est tinctura rubedinis."

6 John Dastin, *Epistola ad Johannem XXII*, in C. H. Josten, "The Text of John Dastin's Letter to Pope John XXII," *Ambix* 4 (1949): 34–51, on 38: "quoniam [aurum] est fermentum elixir, sine quo medicina philosophorum nequaquam peragitur, quemadmodum pasta absque fermento non fermentatur" ("because it [gold] is the ferment of the elixir, without which the medicine of the philosophers' will not act in the least, just like dough

another, is sealed in a flask and gradually heated at increasing temperatures until it first turns black, then white, and finally red.[7] The white stage is an immature form of the philosophers' stone that can potentially produce silver, but for it to do so successfully, it must contain the "tincture of whiteness" (*tinctura albedinis*) found uniquely in silver. The final red stage, the fully mature state of the stone, is potentially able to produce gold, but again, it can do so only if combined with the tincture of redness (*tinctura rubedinis*) found naturally in gold and only in gold. In short, silver—or more precisely, the tincture found within it—is needed to produce more silver, and gold is needed to produce more gold. The idea behind fermentation is that the small amount of these tinctures found within the two precious metals can be multiplied in quantity or potency within the white or red stone, thus rendering these stones able to produce a much greater quantity of new silver or gold than was used up to provide the initial tincture. The term fermentation is analogical. It is borrowed from the common household practice in which a small quantity of leaven or of vinegar can be multiplied many times over in quantity through fermentation by being mixed with unleavened bread dough or fresh wine, respectively.[8]

Despite his evident familiarity with Geber, pseudo-Arnald, and other alchemical authors, John now makes a surprising and unique turn. He tells readers exactly where to find the crucial tincture of redness, and surprisingly, it is *not* in gold: "I shall name it for you with its own proper name: it is for certain the spirit of Roman vitriol" (ll. 26–27)—that is, the "spirit" of what we today call ferrous sulphate.[9] The fact that John employs no gold at all in his process and stipulates no fermentation is unusual. This absence of an aurous ferment in fact provoked skepticism in several of his later readers. One fifteenth-century manuscript contains a marginal gloss questioning the viability of John's process precisely because there is no fermentation with precious metals and cites authorities such as pseudo-Arnald

without leaven is not leavened"). To Dastin is also attributed a *Rosarius philosophorum* (incipit: "Desiderabile desiderium") sometimes conflated with the one by the pseudo-Arnald; see Calvet, *Oeuvres alchimiques*, 128–30 and 147–52, and Thorndike, *HMES* 5:85–102.

7 Authors differ as to whether the precious metal should be combined with the mercury before or after its digestion into the white or red stage.

8 In both these cases, the "fermentation" that converts the entire mass of dough into leaven or large quantity of wine into vinegar is the result of microbial action. The end result, however, is summed up neatly by pseudo-Avicenna in *De anima*; see Sébastien Moureau, *Le* De anima *alchimique du pseudo-Avicenna*, 2 vols. (Florence: SISMEL, 2016), 2:913: "Fermentum reducit massam in qua ponitur ad suam naturam" ("A ferment reduces the mass in which it is put into its own nature").

9 When I provide modern chemical "equivalents" for John's terms, it is to help readers who have some chemical expertise to follow the operations more clearly. I never imply that the modern equivalents are identical to what John used; his ingredients, prepared or collected from naturally occurring sources, would generally have been more impure and of variable composition. In the present case, for example, John's Roman vitriol—produced from the slow air oxidation of pyrites—probably contained a percentage of other metal sulphates besides that of iron, possibly of copper or zinc, as well as other materials, but was nevertheless *predominantly* iron sulphate. I have shown elsewhere the critical role that impurities play in alchemical operations; see Lawrence M. Principe, "Chemical Translation and the Role of Impurities in Alchemy: Examples from Basil Valentine's *Triumph-Wagen*," *Ambix* 34 (1987): 21–30, and "Chymical Exotica in the Seventeenth Century, or, How to Make the Bologna Stone," *Ambix* 63 (2016): 118–44.

as evidence.[10] The unknown sixteenth-century compiler of the German translation of the *Liber lucis* embedded in *Thesaurus mundi* felt obliged to address this point explicitly. He writes that John's process will indeed work, but it requires much more time than if he had added gold or silver. To prove that John's method still works even without a ferment, the compiler supplements his translation with accounts of successful transmutations performed in the fifteenth century using the method of the *Liber lucis*.[11] The chymist and physician of Schmalkalden, Ortloph Marold (1526–1585), who showed great interest in John's process, left a marginal tag in the notes he took from the *Liber lucis* that reads "some believe, by the testimony of everyone, that a ferment must be added."[12] Even the printed Basel edition highlights the lack of a ferment and succinctly advises its readers to "apply a ferment, as you know to do."[13] John leaves the reasons behind his innovation in omitting gold unexplained, but one possibility is that he felt obliged to find a less expensive source for the tincture of redness than gold, given that he predicted the future poverty of the Church and assumed the strict poverty of the true Franciscans, the "poor men of the Gospel," to whom he explicitly addresses his treatise and whom he expects to prepare and manage the stone for the benefit of Christendom.

John then makes a second innovation, or diversion from the "mercury alone" concept, besides locating the necessary tincture of redness in vitriol rather than in gold. A key tenet of the "mercury alone" theory is that the liquid metal contains within itself its own sulphur. This internal sulphur is sufficient for coagulating the mercury, under the right conditions, into a solid: either a solid metal or the philosophers' stone. According to the mercury-sulphur theory central to alchemical thought, sulphur is the coagulant of mercury, like rennet is to milk, and is exemplified by the way in which common sulphur heated with liquid mercury quickly turns it into solid cinnabar (mercuric sulphide, HgS, in modern parlance). Sulphur is the principle of dryness and solidity, while mercury is the principle of moistness and liquidity.[14] But John combines the functions of the tincture of redness and the coagulating sulphur into one, arguing that the needed sulphur is not present in common quicksilver; it must be added. He will later clarify his position by saying that quicksilver does have some of its own tincture of redness—as Geber himself asserted—but it is too earthy and in too small a quantity for making the

[10] **Mp**[1], fol. 289v: "Sed judico immo, ista medicina hoc modo non valet, quia est sine fermento" ("But I indeed judge that this medicine has no effect in this way because it is without a ferment").

[11] *Thesaurus mundi*, 294–320, in *Heptas alchymica, das ist, Auszerlesene philosophische Tractätlein von dem magisterio lapidis*, ed. Theophilus Neander (Halle, 1621), on 317–20. On this German text, see below, Section 3.4.

[12] Kassel, Universitätsbibliothek Kassel-Landesbibliothek und Murhardsche Bibliothek der Stadt Kassel, 4° Ms. chem. 29, fols. 76r–77r, on 77r: "Quidam putant addendum fermentum omnium testimonio..." On Marold, see Hartmut Broszinski, "Spuren klösterlicher Alchemie in Kasseler Handschriften des 15. und 16. Jahrhunderts," in *Nova de veteribus: Mittel- und neulateinische Studien für Paul Gerhard Schmidt*, ed. Andreas Bihrer and Elisabeth Stein (Munich: Saur, 2004), 824–47, on 834n56.

[13] *De confectione* (1561), 230.

[14] For more on this point and about the mercury-sulphur theory, see Principe, *Secrets of Alchemy*, 35–37 and 122.

philosophers' stone, thus showing that he knew the standard literature but chose to diverge from it (ll. 143–45). John therefore uses the terms "invisible sulphur," "sulphur of the philosophers," and "tincture of redness" interchangeably to refer to the "spirit" within Roman vitriol that must be incorporated with common quicksilver in the first operation: "the invisible sulphur of the philosophers which is the tincture of redness, and is the spirit of Roman vitriol" (ll. 46–47). While he is clear that this red tincture will become manifest in the completed red philosophers' stone, he is equally clear that this same substance is what "congeals the mercury" (l. 49) in the first operation.

2.2.1 Organic Versus Metallic Starting Materials

John of Rupescissa thus begins his process by prescribing both a metal (mercury) and a mineral (vitriol) as the starting materials. This choice of exclusively metallic and mineral starting materials marks a significant change from John's earlier thinking about alchemy. John's only known reference to alchemy outside of the *Liber lucis* and *De famulatu* cites instead a wide array of possible starting materials, with a clear emphasis on organic, rather than metallic or mineral, substances. In his lengthy prophetic work, the *Commentary on the Oracle of Cyril*, written predominantly between 1347 and 1349 (while imprisoned under the staircase at Rieux), John provides the following definition of alchemy:

> The science of alchemy...teaches how to make gold and silver artificially from nearly any of the substances of this world, for example from blood, from plants or from eggs, or from fruits, or from any animal, or from stones and from all the metals.[15]

John's citation of this wide diversity of starting materials for making gold signals that at this date, prior to his arrival in Avignon, he drew his knowledge of alchemy either from Arabic sources or from the texts of the earliest Latin alchemy that depended upon such Arabic materials. Many texts of the large and highly regarded corpus of writings attributed to Jābir ibn-Hayyān, which date from the late ninth and tenth centuries, for example, indicate that the stone can be made from mineral, vegetable, or animal materials, with a preference for organic starting materials, especially those coming from animals. These texts, such as the *Liber de septuaginta* (*Book of the Seventy*), propose the preparation of various "elixirs" by combining the four elements (fire, air, earth, and water) in precise ratios according to the "principle of balance."[16] Since the four elements are present in all substances,

[15] BNF, MS Lat. 2599, fol. 156v: "sciencia alquimie que docet artificialiter facere aurum et argentum de quibuscumque corporibus fere orbis hujus puta de sanguine de plantis vel de ovis vel de fructibus vel de quibuscumque animalibus aut de lapidibus et de universis metallis." On the content of John's *Commentary on the Oracle of Cyril*, see Bignami-Odier, *Études*, 53–112; on its dating, see Lerner, "Life, Work, and Fortune," in *VM*, 35. See also Marc Boilloux, *Étude d'un commentaire prophetique du XIVe siecle: Jean de Roquetaillade et l'Oracle de Cyrille* (Thesis presented at l'École des Chartes, Paris, 1993).

[16] On the principle of balance, see Principe, *Secrets*, 37–42.

any starting material will work as a source of the elements, such that after their isolation they can be combined in the correct ratios to produce transmutational elixirs. Organic materials probably seemed preferable to these Islamic alchemists because they are easier to decompose practically into their presumed elements than are more thermally resistant metallic and mineral materials. Pseudo-Avicenna's *De anima*, written originally in Arabic in Iberia during the late eleventh or twelfth century and translated into Latin in the 1220s or 1230s, is based upon this Jabirian alchemy and thus continues the use of a wide array of starting materials with special attention to organic materials: human blood, hair, and chicken eggs.[17] John's fellow Franciscan, Roger Bacon, writing in the second half of the thirteenth century, adopted the use of such materials from Arabic sources, particularly from *De anima*, and accordingly cites many of the same organic starting materials.[18]

John was clearly of the same opinion at the end of the 1340s when he wrote the Cyril commentary. But by the time he wrote *De famulatu* in 1351–52 (where he describes the first steps of the *Liber lucis* process without explicitly indicating its chrysopoetic purpose), and obviously before 1354 when he completed the *Liber lucis*, he had completely changed his mind. He had replaced the Arabic notions of multiple "elixirs" producible from a wide variety of mostly organic materials with the more "up-to-date" Latin alchemy of the early fourteenth century—a development pioneered by the late thirteenth-century Geber—that advocated the exclusive use of mineral and metallic ingredients. Significantly, this timing implies that John learned this newer approach to alchemy and developed his mature theory and practice of alchemy during quite a short timeframe *after* his arrival in Avignon in August 1349. This intellectual trajectory is of the greatest importance both for assessing the roots of John's alchemical activity and ideas and for answering the question to what extent John was actually learning and practicing alchemy while under confinement at Avignon, a topic to which I return in Section 2.9.

2.2.2 John's Secrecy: Dispersion of Knowledge and *Decknamen*

After revealing the proper starting materials for the stone as mercury and Roman vitriol, John proceeds to a detailed description of the laboratory operation that will remove the mercury's "superfluities" and incorporate the vitriol's "invisible sulphur" or "tincture of redness" with the purified mercury. Here John displays both practical experience and the ability to manage alchemical secrets. Although John has been true to his promise to write clearly and without secrecy thus far, when he comes to "reveal" the second major secret needed for making the philosophers' stone—that is, *how* to treat the starting materials—he silently, and

17 On the *De anima*, see Moureau, *De anima* and "*Elixir atque fermentum*: New Investigations about the Link between Pseudo-Avicenna's Alchemical *De anima* and Roger Bacon: Alchemical and Medical Doctrines," *Traditio* 68 (2013): 277–325.

18 William R. Newman, "An Overview of Roger Bacon's Alchemy," in *Roger Bacon and the Sciences: Commemorative Essays*, ed. Jeremiah Hackett (Leiden: Brill, 1996), 317–36, and Moureau, "*Elixir atque fermentum*."

certainly intentionally, omits crucial information. The first step involves grinding saltpeter (potassium nitrate) and vitriol, drying them gently at low heat, and then mixing mercury with them. This mixture is then strongly heated for twelve hours to produce a snow-white sublimate. In this process, John explains, the mercury's earthiness is left behind as dregs at the bottom of the sublimatory vessel and its excess humidity "by means of which [the mercury] runs quick [i.e., as a liquid]" (l. 43) is dried out such that, together with the coagulating properties of the invisible sulphur, the mercury congeals into a dry white solid.

Despite John's detailed instructions, when tried in the laboratory his process does not—in fact, *cannot*—work. In the past, such failure might not have been unexpected, since according to an earlier and now-discredited historiography, alchemical processes were widely thought to be little more than the literary juggling and repurposing of earlier texts, the exercise of imagination, or a metaphor for non-material topics or transformations, rather than the result of actual practical experience with material substances.[19] While there is without a doubt a certain proportion of both imagination (or what I prefer to call "alchemical optimism") and literary topoi present both here and in other alchemical texts, we now know that a substantial body of laboratory operations and experience stand behind a great many, perhaps most, alchemical texts, including even some of the most allegorical, secretive, and seemingly outlandish.[20] In the present case, readers familiar with inorganic chemistry can recognize a white sublimate produced from mercury as one or both of the mercury chlorides—most likely mercuric chloride ($HgCl_2$, formerly known as corrosive sublimate) and less likely mercurous chloride (HgCl, calomel). In either case, a source of chloride is required to produce such a sublimate, and none is present in the mixture that John prescribes.[21]

[19] It remains true that *some* alchemical texts are the result of literary juggling, but certainly not all nor even a majority, as recent scholarship has abundantly shown. For a contrary view, see, for example, the writings of Brian Vickers and their cogent analysis in William R. Newman, "Brian Vickers on Alchemy and the Occult: A Response," *Perspectives on Science* 17 (2009): 482–506. For a broader treatment of several earlier interpretations of alchemy, see Lawrence M. Principe and William R. Newman, "Some Problems with the Historiography of Alchemy," in *Secrets of Nature: Astrology and Alchemy in Early Modern Europe*, ed. Newman and Anthony Grafton (Cambridge, MA: MIT Press, 2001), 385–431. In terms specifically of John, DeVun, *Prophecy*, 62–63, states that John "never practiced alchemy" and that his writings were based on "alchemical authorities, rather than....his own experience." This claim was not unreasonable at the time, given usual assumptions about John's imprisonment, but the present study clearly indicates John's practical experience and originality and consequently forces a careful reexamination (presented in Section 2.9) of the conditions of his confinement in Avignon.

[20] On "alchemical optimism," see Principe, *Secrets*, 158; I will return to this concept in Section 2.5.2 as an explanation for the background to some of the later sections of the *Liber lucis*. For examples of laboratory experiences expressed in obscure allegories, see Principe, *Secrets*, 143–58, and William R. Newman, *Gehennical Fire: The Lives of George Starkey, and American Alchemist in the Scientific Revolution* (Cambridge, MA: Harvard University Press, 1994), 115–33.

[21] Previously (Principe, *Secrets*, 65), I hypothesized that John's saltpeter was poorly purified and thus contained sufficient chloride to produce the sublimate. I subsequently had to reject this hypothesis because in *De famulatu* ("Sciencia separacionis auri ab argento," a section omitted from the printed edition) John distills an aquafort from saltpeter that dissolves silver but not gold, correctly citing a property of the nitric acid produced by this method. Had his saltpeter been contaminated with enough chloride to produce the white sublimate from mercury, however, the process in *De famulatu* would have provided an aqua regia rather than an aquafort, which would have dissolved gold not silver.

The solution, however, appears in the parallel process John described roughly two years earlier in *De famulatu*. There the two salts he prescribes for the same operation are vitriol and common salt (sodium chloride) rather than vitriol and saltpeter.[22] But the *De famulatu* process does not work any more than does the *Liber lucis* process, in this case because of the lack of an oxidizing agent, namely the saltpeter. Only when the *two processes are combined*, that is, by using *all three* salts mentioned across the two separate books—vitriol, saltpeter, and common salt all together—can one obtain the result that John describes. This is a beautifully clear example of the "dispersion of knowledge," a well-documented technique for maintaining secrecy in alchemical texts by dividing up and scattering the full details of a process in two or more places.[23] Some later readers recognized the omission in the *Liber lucis*. A marginal note in an early manuscript reads "some put in one part of salt."[24] In the 1561 Basel edition, an "annotatio" attached to the end of the text and written (presumably) by the editor Guglielmo Gratoriolo seems intended to give a hint to the reader about the missing ingredient: it cites at length the virtues of the "salt of food" (*sal cibi*), which is the crucial ingredient silently missing from the *Liber lucis*. This annotation concludes, with a wink and a nod, that "the whole secret is in salt."[25]

Much alchemical literature is famous for its frequent—indeed virtually ubiquitous—use of Decknamen, that is, cover-names for chemical substances. The practice of using Decknamen dates back to some of the earliest texts of alchemy from the Graeco-Egyptian period and was widespread through much of the Arabic literature. John, however, never uses Decknamen and signals this choice by telling readers explicitly that he names vitriol "for you with its own proper name." Such candor however seems to have disquieted some later copyists who decided to encode their copies of the *Liber lucis* with traditional (and occasionally unique) Decknamen. All the descendants of π^2, for example, replace "Roman vitriol" with "green lion" (*leo viridis*).[26] The fifteenth-century German translation $\mathbf{Bg_d}$ uses *Anatron* for saltpeter and *weisses Leben* for vitriol, while the two descendants of λ substitute "the sounding stone" (*lapis sonans*) for saltpeter, "May dew" (*ros maii*) for mercury, and both "white serpent" (*serpens album*) and "green lion" (*leo viridis*) for vitriol.[27] Most peculiar, however, is the use by **G** and its descendant **T**

[22] *De famulatu*, "Gloriosus deus"; see Appendix 1.

[23] This method, called *tabdid al-`ilm*, is acknowledged in the Jabirian corpus, as noted by Paul Kraus, *Jābir ibn Hayyān: Contribution à l'histoire des idées scientifiques dans l'Islam, Volume I: Le Corpus des écrits jābiriens, Mémoires de l'Institut d'Égypte* 44 (1943), on xxvii–xxxiii. The fourteenth-century friar Efferrarius (or Ferrarius) describes the technique clearly: "they did not hand over the whole in any one place, but instead they divided it up and delivered it in pieces scattered here and there through their books" (*nec in aliquo loco totum tradiderunt: sed per eorum libros sparsim & per partes diviserunt*); see *De lapide philosophorum*, in *Verae alchemiae*, 2 vols. (Basel, 1561), 2:232–37, on 232. See also Principe, *Secrets*, 44–45.

[24] $\mathbf{P^1}$, fol. 29v: "aliqui ponunt 1 partis salis."

[25] *De confectione* (1561), 231.

[26] I.e., **M**, **Rc**, $\mathbf{W^1}$, $\mathbf{Fn^3{}_i}$, and from ε^2, $\mathbf{Cu^2}$ and $\mathbf{X^4}$.

[27] $\mathbf{Hb^2{}_d}$ and **S**; "sounding stone" or "the stone that makes noise" may refer to the sounds of deflagration made by saltpeter when thrown on a burning coal, or to its explosive power in gunpowder. The use of the adjective

of *lubricus*, or "earthworm," as a Deckname for mercury.[28] It seems strange that later copyists would feel obliged to introduce a previously absent level of concealment into a text that had been circulating widely without Decknamen for nearly a century or more. It may possibly have been an attempt to accommodate contemporaneous expectations for alchemical literature, which tended to be more secretive in general in the fifteenth and sixteenth centuries than in the mid-fourteenth.

2.2.3 Mercury as the Mortified Grain of Wheat

Once the common salt mentioned in *De famulatu* is added to the mixture stipulated in the *Liber lucis*, John's process works very well. Actually performing the process in the laboratory, however, revealed aspects of John's practical experience that could not have been gleaned solely from reading the text. His method of adding mercury to the mixture of dried salts deserves comment. When first attempting the process, I added the mercury all at once to the salts and attempted to incorporate the two by stirring and grinding. The attempt failed miserably. The liquidity and weight of the mercury caused it always to separate from the salts and run back together into large puddles. John however advises adding the mercury through a linen cloth held over the salts, and then by merely stirring the mixture, "the quicksilver will be drunk up in the aforesaid material so that it is not visible in it" (ll. 33–34). In practice this method worked perfectly, just as John describes, because the cloth breaks the mercury into droplets so small they remain suspended within the dry mixture of salts. This reworking simultaneously gave deeper understanding of John's connection of this process to an allegorical description of making the stone couched in the words of the Gospel and borrowed from a "metaphorical treatise" by pseudo-Arnald of Villanova:[29]

> "Unless the grain of wheat falling into the earth has died, it remains a single grain; but if it has died, it brings forth much fruit." Understanding the grain of wheat which is to die in the earth as the quicksilver which is to die in the earth of saltpeter and Roman vitriol, and there having been mortified and sublimed with fire, the quicksilver brings forth much fruit because it brings forth the great stone that all the philosophers seek.[30]

The visual experience of adding mercury to the salts using John's method immediately recalls this image strikingly. As is visible in Figure 2.1, the tiny but heavy droplets of mercury emerging through the weave of the linen and falling onto the dry mixture produce tiny pockets in the surface of the powder, such that the whole resembles soft earth upon which seeds have been broadcast. The salts themselves, in the process of being dried, turn into a light brown or tan powder, similar in

"white" for Roman vitriol in two unrelated witnesses is rather odd because the substance is greenish in color, as is expressed more literally in the more common Deckname "green lion."

28 Perhaps because mercury slithers underground like an earthworm?

29 On this treatise, cited several times in the *Liber lucis*, see below, Section 2.3.2.

30 Jn 12:24–25 and *Liber lucis*, ll. 55–60.

FIGURE 2.1 The dried salt mixture after the mercury has been added, showing the pockets made in the surface by the falling mercury droplets. Author's laboratory.

appearance to fine soil. The mercury, called "*living* silver" in Latin (*argentum vivum*) because of the liquid metal's high mobility, becomes immobilized in this powdery mixture such that it can no longer run together into large globules and can thus be said to have "died" in the earth of the salts.[31] Indeed, as John writes, the added mercury easily and completely disappears within the salt mixture.

John's word choice further reinforces the connection with the Gospel, for he uses the word *seminare* to describe the mercury's passing through the pores of the cloth onto the salts, a word whose root meaning is "to sow seeds," even though John uses it here in the sense of *disperse* or *disseminate* (the latter English word sharing the same origin in *semen*, or seed). Thus, the Biblical metaphor John borrows from pseudo-Arnald also has a *literal* basis in the visual appearance of the materials used, and John further enhances this connection by his verb choice. While there is no question that the Biblical text bore special significance to a Franciscan like John, expressing as it does Christ's Resurrection, there exists in parallel an immediate visual experience of the materials as literally *seen* in the laboratory.

John's closely detailed instructions suggest his direct experience with the process, but further evidence appears in his following description of the sublimation

[31] Cf. pseudo-Aristotle, *De perfecto magisterio*, *BCC* 1:645 where the author uses the imperative *mortifica* ("mortify") to stipulate the incorporation of mercury into a different powdery mixture.

FIGURE 2.2 The mercury–salts mixture before heating in a sublimatory. Author's laboratory.

(Figure 2.2). He is clearly aware of the quantity of water present in the salts and the need to remove it before the sublimation. Iron vitriol is ordinarily a heptahydrate, meaning that roughly 45% of its weight is water, and this water needs to be removed before the mixture is heated or it will interfere with the sublimation or, if the vessel is closed, burst the flask. Most of this water (about 85%) can be removed at low temperature (about 65 degrees C) "by stirring [the salts] in an earthenware dish on a gentle fire" (l. 29), as John prescribes. John emphasizes that either the salts have to be very carefully dried or the sublimatory must have a small aperture at the top for venting the vapor; he even suggests holding a cold metal blade over the aperture to monitor when the water vapor stops emerging (any vapor will condense on the metal and cloud its surface). Furthermore, before carrying out the process, I doubted that it would require over twelve hours of a "strong fire" to complete the sublimation, given that mercuric chloride sublimes quite easily at a relatively low temperature. However, when carrying out the process, the full twelve hours of strong heating was in fact required, because the salt mixture quickly hardens into a solid mass upon heating, and extricating the volatile mercuric chloride from that mass requires a surprisingly long period of time. The final result, with the sublimed mercury "white as snow," is shown in Figure 2.3.

FIGURE 2.3 The mercury-salts mixture before (left) and after heating (right). Author's laboratory.

2.2.4 Conservation of Weight and Proof of Composition by Analysis

From the point of view of the history of chemistry, John's most significant statements at this point occur when he endeavors to *prove* that the mercury has combined with the "invisible sulphur" of the vitriol to produce the white sublimate. John's persistent interest in proving his claims and theoretical assumptions is noteworthy; it signals the intellectual curiosity and commitment to understanding that he brings to alchemical experimentation. Simple direct observation cannot demonstrate that the "invisible sulphur" or "tincture of redness" has, in fact, combined with mercury to produce the white sublimate. Indeed, John considers it "the greatest miracle" that "although that invisible spirit has the nature of the tincture of redness, it nevertheless congeals the mercury during sublimation into the whiteness of snow" (ll. 47–50). Here John recognizes the fundamental chemical principle that when different substances combine, they do not yield something that is an average or mixture of their independent properties; despite being red, the tincture turns silvery mercury white as snow. Since direct application of the senses does not provide the evidence that John requires, he provides another means of proof. John cites a "sign" (*signum*) that reveals the unseen incorporation of the invisible sulphur into the mercury.[32]

> The sign that the spirit of vitriol is indeed incorporated into the sublimed mercury is that even though the sublimed mercury leaves behind many earthy dregs in the earth, nevertheless if you have put in one pound you will get the same amount back (ll. 60–63).

[32] For more on the use and language of "signs" in alchemical literature, see Jennifer M. Rampling, *The Experimental Fire: Inventing English Alchemy 1300–1700* (Chicago: University of Chicago Press, 2020), 296–301 and "Citrination and Its Discontents: Yellow as a Sign of Alchemical Change," *Ambix* 71 (2024): 73–97.

Clearly, John weighed his starting mercury and his white product and found their weights identical. Because some of the weight of the original one pound of mercury was left behind when it "deposited many earthy dregs," the sublimed mercury should weigh *less* than a pound. But since it still weighs a whole pound, *something* must have been added to it in order to compensate for the lost weight. That something, John concludes, is the "spirit of vitriol," which, although invisible, still has weight. He makes his implicit principle clear: "this [result] would be impossible unless the mercury, whiter than snow, has carried up with itself the purest spirit of the aforesaid vitriol, which is the invisible sulphur" (ll. 63–65). John could not see the invisible sulphur directly, so he followed it by means of weight determinations as it changed partners, moving invisibly out of the vitriol and into the mercury.[33]

It is easy for modern readers, accustomed to equating matter with mass, to overlook how significant a statement this is for John to express in 1354. It contradicts standard Aristotelian ideas in contemporaneous Scholastic natural philosophy, because for Aristotle, although *matter* is conserved, *weight* is not.[34] Prime matter has no inherent properties, such as weight. When water boils and turns into air, its weight vanishes as it transforms from a heavy element into a light one. The substantial form of water does include weight, along with fluidity, wetness, and so forth among its qualities, but once the form of water is abolished through evaporation or other substantial change, all of those formal qualities vanish as well. When two substances combine to form a new third substance, weight is no more conserved than color or any other sensible quality. Therefore, for anyone following this system, using weight to follow or monitor a substantial change is futile. But John's claims depend upon two interrelated principles: first, the conservation of weight during material transformations, and second, the ability to use this principle as a means of tracking otherwise imperceptible components. The way John uses these dual principles for his proof of the presence of the tincture of redness in the sublimate is fundamentally the same, although obviously much more roughly deployed, as the so-called balance-sheet method employed by Antoine Laurent Lavoisier (1743–1794) in the late eighteenth century to track the course of chemical reactions. Historians of science no longer believe that the use of "weight balance" (i.e., the principle that the weight of products must equal the weight of starting materials) as a probe of chemical transformations emerged only at the time of the so-called Chemical Revolution; its use by late sixteenth- and seventeenth-century

[33] In terms of modern chemistry, John should have obtained a *greater* weight of sublimate (by about a third) than of the mercury initially used, due to the incorporation of chloride. However, the practical yield is not 100%, and even with modern equipment, I also never obtained more sublimate than roughly the weight of the original mercury.

[34] Several medieval thinkers attempted to find a way to talk about the quantity of matter (e.g. Giles of Rome on *quantitas materiae*), often in the context of theological study of the transubstantiation of the Eucharist, but I have not located any who used weight as a straightforward means of expressing or defining the amount of matter. See Max Jammer, *Concepts of Mass* (Cambridge, MA: Harvard University Press, 1961), chap. 4.

chymists has now been well documented.[35] That said, there has not hitherto been any indication that weight balance already formed an operational principle for an alchemist in the middle of the *fourteenth* century, over two centuries earlier than previously recognized.[36]

Yet John apparently wanted even more evidence for the incorporation of the invisible sulphur than weight balance could provide; he evidently wanted to *see* it and to enable his readers to see it for themselves. So he provides an "ingenium" whereby the invisible sulphur or tincture of redness can be directly "demonstrated to sense" (l. 65) rather than inferred from weight comparisons. John's use of the word *ingenium* here deserves comment. In medieval and early modern Latin, *ingenium* ordinarily refers to an inborn power of the mind, an especially keen ability to understand and to discover new things through mental acuity.[37] I have already given an example of John's use of the word in this standard sense in his statement in *De famulatu* that allegorical alchemical texts cannot be understood by "human ingenuity" (*ab humano ingenio*). But now John uses *ingenium* in a very different way, not to denote a mental power but rather a physical process, what might be called a demonstrative experiment or technique, in this case one that is designed to disclose the hidden nature of the white sublimate. (He will use the word again in this way in the significant closing lines of the *Liber lucis*; see Section 2.8.) The use of *ingenium* to describe a physical process or "technique" is not unique to John; the word appears with a similar meaning in other medieval alchemical sources. Geber, for example, uses the word several times in the *Summa perfectionis* to describe practical methods for purifying or producing various substances, even while he also uses it in its more standard meaning of an acute quality of mind; it also appears with the same meaning in pseudo-Arnald.[38]

John's demonstrative *ingenium* involves first quenching red-hot iron in vinegar, then adding a quantity of the mercury sublimate, and leaving the mixture to stand overnight. In the morning, the liquid is to be filtered and evaporated, yielding the

[35] Principe, *Transmutations of Chymistry*, 102–3, 214–15, 263–64, et passim; William R. Newman and Lawrence M. Principe, *Alchemy Tried in the Fire: Starkey, Boyle, and the Fate of Helmontian Chymistry* (Chicago: University of Chicago Press, 2002), 54–56, 68–71, and 120–26; on Lavoisier, Frederic L. Holmes, *Antoine Lavoisier: The Next Crucial Year, or The Sources of His Quantitative Method in Chemistry* (Princeton: Princeton University Press, 1998).

[36] It was in fact reading a version of this line in the Basel edition some years ago that compelled me to produce a critical edition of the *Liber lucis*. The most likely explanation at the time was that the line was a later interpolation; however, the critical edition shows that it is authentically John's; see Principe, *Secrets*, 65–66 and notes.

[37] On the varied meanings of *ingenium*, see Jonathan Morton, "Making Sense of *Ingenium*: Translating Thought in Twelfth-Century Latin Texts on Cognition," in Katja Krause, Maria Auxent, Dror Weil, eds., *Premodern Experience of the Natural World in Translation* (New York: Routledge, 2022), 90–110, on 94: "*Ingenium* is some kind of power, activity, or function that leads to the discovery of new things"; Raphaële Garrod and Alexander Marr, eds., *Descartes and the* Ingenium*: The Embodied Soul in Cartesianism* (Leiden: Brill, 2021), esp. Igor Agostini, "*Ingenium* between Descartes and the Scholastics," 139–62. The need for *ingenium* (as clever insight) specifically in alchemy is treated in Jennifer M. Rampling, "Spirited Matter and Ingenious Nature: Accounting for Alchemical Change," 13–24 in *Ingenuity in the Making*, ed. Richard J. Oosterhoff, José Ramón Marcaida, Alexander Marr (Pittsburgh: University of Pittsburgh Press, 2021).

[38] Newman, *Summa*, 385, 420, 437, 584–85; compare 256 and 261–62 where the word is used instead for mental acuity as *ingenium naturale*; pseudo-Arnald of Villanova, in *BCC* 1:668, 697, and 701.

FIGURE 2.4 The separated tincture of redness (left) and the separated mercury (right). Author's laboratory.

isolated spirit of vitriol, that is, the previously hidden tincture of redness, now made manifest to sight. I followed John's instructions and obtained a bright red material that exactly, and quite beautifully, answers to his description of a *tinctura rubedinis* (Figure 2.4).[39] John concludes,

> And so you will have the mercury, rendered languid, separated on the one side, and the abovementioned invisible sulphur on the other, whereby it has been demonstrated to me [*michi est demonstratum*] that quicksilver sublimed with vitriol carries with itself the spirit of the vitriol which is called the invisible sulphur of the stone, the tincture of redness (ll. 73–77).[40]

It is difficult not to read in John's words an implicit opposition between the logical *demonstrationes* of the medieval university—what he called "the great din of inane words and clash of useless disputations" in *De famulatu*—and the "demonstration to sense" provided by his *ingenium* that successfully "demonstrated to me" the hidden presence of the red tincture within the snow-white sublimate.[41]

The other component of the snow-white sublimate, the mercury, remains at the bottom of the vessel from which the vinegar solution was filtered. Oddly, however, John calls this mercury *marcidus*, that is, languid, withered, or rotten. How this peculiar adjective could refer to mercury is not at all obvious. The word initially

39 Although John does not specify it, the vinegar he used must have been distilled so that it was colorless; some manuscripts (e.g. **Pd/L**^{5} and **Oth**^{2} [descendants of ι′] and **S**) add the adjective "distilled." I used red-hot iron nails dropped into either 8% acetic acid or distilled white wine vinegar. Note that the iron must be left in the vinegar after quenching. The heating and quenching cleans the iron surface, allowing a fresh metal surface to react easily with the mercuric chloride by a replacement reaction. The red material does not actually come out of the sublimate as John believed; it is produced (in part) from the iron.

40 Marold (see above, note 12), in his notes to this part of the *Liber lucis*, writes simply of John's claim about the presence of the hidden tincture that *id probat luculenter*: "he proves it brilliantly," Kassel, 4° MS chem 29, fol. 76v.

41 *De famulatu*, "Consideravi ergo": "strepitu multo inanium verborum et conflictu inutilium disputaciones."

suggested to me that the text was corrupt at this point; indeed, there is considerable variation at this point in the manuscripts, indicating that early copyists were also confused by this odd word choice. But repeating the process in the laboratory showed that John's choice of this little-used adjective is strikingly appropriate. Only by actually performing the experiment could one learn that the recovered mercury, rather than reappearing in its usual form as a highly mobile silvery liquid, emerges instead as a gray slime, which when dried looks like a dull gray powder (Figure 2.4). Its myriad tiny globules, which are visible as such only upon very close inspection, do not readily coalesce into large drops due to dusty residues on their surfaces, and therefore the recovered mercury no longer runs about in the dish quickly as mercury ordinarily does. Thus the separated mercury appears to be *precisely* "marcidus," that is, languid or withered.[42] John may have seen this result as coherent with the theory he stated earlier, namely, that the sublimation removes the excess humidity of the mercury "by means of which it runs quick" (l. 43). What is certainly clear is that neither he nor anyone else would ever have thought of using the adjective *marcidus* to describe this recovered mercury without having actually seen it. Nor could the striking difference in appearance between the initial and the recovered mercury have been predicted in advance. Thus, John's use of the term *marcidus* is strong evidence that John actually carried out this operation and saw the results first-hand.

Significantly, in terms of the broader horizon of the history of science, John's "demonstration to sense" represents a complete *cycle of synthesis and analysis* deployed for proving the composition of a substance. In order to prove that the white sublimate really results from the combination of the tincture of redness with common mercury, he subjects the sublimate to a process—his *ingenium*—that reseparates it into its original components. He thus isolates the tincture of redness as a material that is visibly bright red in color and recovers the original mercury as a gray powder. He even goes one step further by adding a negative proof: "you cannot extract the abovementioned spirit from quicksilver that is sublimed without vitriol" (ll. 77–78). In short, the same *ingenium* used upon mercury sublimed *without* vitriol does not provide a red material, thus proving that the red material comes from the vitriol.[43]

Once again, the possibility of such a procedure is contrary to much of Scholastic matter theory, at least according to the widespread Thomistic interpretation of Aristotelian thought. According to Aristotle's hylomorphism and theory of mixture, when two substances combine to form a third, their individual identities are entirely lost, and a new substantial form is imposed on their composite, according to

[42] Rendering *marcidus* as "rotten" does not seem as fitting as "languid," even though the dull, slimy appearance of the wet mercury might also have suggested rottenness. The important point is that the mercury looks very different after the vinegar and iron treatment than it did when mixed with the salts at the beginning.

[43] John is correct that if mercury is sublimed without vitriol, use of his *ingenium* does not produce a red material since the sublimate would be unchanged mercury not mercuric chloride.

the Aristotelian maxim "the generation of one thing is the destruction of another." In the thirteenth century, St. Thomas Aquinas and his followers argued that this new form, the *forma mixti* (form of the mixt), is imposed on the persistent prime matter while the previous substantial forms of the ingredients, in this case those of the red spirit of vitriol and the silvery liquid mercury, cease to exist because matter can hold only one substantial form at a time. This viewpoint is known as the "unicity of substantial form." There is, therefore, no possibility of separating out the original ingredients again because they no longer exist in the composite. Yet John is clearly operating with a different idea; for him, the original components *do* continue to exist within the compound formed from them, and for this reason, they *can* be separated out again from their combinations in their original forms. Indeed, what John is doing here in the middle of the fourteenth century is essentially a *reductio ad pristinum statum* (reduction to the pristine state), the demonstrative experiment employed by Daniel Sennert, Robert Boyle, and others in the seventeenth century to refute the Aristotelian doctrine of mixture and hylomorphism, in those cases in favor of their own particulate conceptions of matter.[44]

2.2.5 Plurality of Forms, Particles, and Franciscanism

How did John, working and writing in the mid-fourteenth century, come to believe in the possibility of reseparating components in their original form from a composite and trust the use of weight balance for monitoring chemical change? John furthermore uses these principles as demonstrative proofs in the *Liber lucis* without justifying or defending them to his readers. Did John have an audience that shared his views or that would not find them controversial? Three possible foundations of John's views can be proposed here in relatively short order, without attempting to decide among them (they are not mutually exclusive) or following them out exhaustively.

The first possibility is that John rejected the Thomistic unicity of forms and instead adhered to the "plurality of forms" doctrine. This viewpoint holds that multiple substantial forms, arranged hierarchically, can coexist within a given thing. Significantly, in the thirteenth and fourteenth centuries this perspective was connected particularly with Franciscan thinkers—from St. Bonaventure (1221–1274) through Richard of Middleton (1249–1308) to John Duns Scotus (1265/66–1308)—and so John of Rupescissa would undoubtedly have encountered it during his five years at the Franciscan *studium* in the 1330s.[45] Even Peter of John Olivi (1248–1298), one of John's key sources for his eschatology and for strict

44 William R. Newman, *Atoms and Alchemy: Chymistry and the Experimental Origins of the Scientific Revolution* (Chicago: University of Chicago Press, 2006), 98–104, and Newman and Principe, *Tried in the Fire*, 18–20.

45 Roberto Zavalloni, *Richard de Mediavilla et la controverse sur la pluralité des formes* (Louvain: Institut superieur de philosophie, 1951), 213–503; Richard Cross, *The Physics of Duns Scotus: The Scientific Context of a Theological Vision* (Oxford: Clarendon Press, 1998), 47–76; and Anneliese Maier, *An der Grenze von Scholastik und Naturwissenschaft* (Rome: Edizioni di Storia e Letteratura, 1952), 3–140.

views on poverty, wrote on the plurality of forms.[46] According to this perspective, the substantial forms—that is to say, the identity—of the tincture of redness (for example, its redness) and of the mercury (its metallicity and liquidity) persist after combination but are hidden beneath the dominant *forma mixti* of the snow-white sublimate, subordinated to that new form such that they do not appear externally. John seems to invoke this concept implicitly when he later writes that the tincture of redness is "hiding beneath the whiteness of the stone" (*latens sub lapidis albedine*, ll. 200–1). The form of the red tincture must likewise have existed "invisibly," as John says, under the visible external and dominant substantial form of Roman vitriol; otherwise, it could not have been transferred from the vitriol to the mercury during sublimation. John's *ingenium* using vinegar and iron then functions by stripping away the newly imposed and predominant *forma mixti* of the white sublimate, thereby allowing the hidden ingredients to reemerge in their original forms with their identities intact. In *De famulatu*, he describes in similar terms a "fifth essence hidden invisibly beneath the whiteness of the quicksilver."[47]

It is significant that the plurality of forms was a specifically *Franciscan* perspective, one rejected by the Dominican (and other) followers of St. Thomas. The debate between the Thomistic unicity and Franciscan plurality of forms represented a major theological controversy during the early fourteenth century. Franciscans championed the plurality of forms, particularly for discussions of the relationship between the human body and soul, and by extension, to other living systems, and in the case of Scotus, for assuring the identity of Christ's body under the forms of bread and wine in the Eucharist.[48] If the plurality of forms plays a role in John's conception of alchemical transformations, it would be a striking example of how a specific theological tenet directed the formulation of a natural philosophical principle—indeed, a natural philosophical principle that was applied in *practice*. The plurality of forms was not widely invoked for inanimate objects, such as chemical substances; some authors stated explicitly that a single substantial form suffices for inanimate things. Nevertheless, the plurality of forms clearly has significant application and consequences for material substances, their combinations and transformations.[49]

A second concept that may lie behind John's principles is the notion that matter is composed of tiny and persistent or semi-persistent particles. If John was thinking along such lines, then the tiniest parts, or *minimae partes*, of the tincture and

[46] Zavalloni, *Controverse*, 333–42. On John and Olivi, see Robert E. Lerner, "Historical Introduction," in *LSE*, 45–49.

[47] See Appendix 1: "invisibiliter quintam essenciam eorum absconsam infra albedinem argenti vivi predicti."

[48] Cross, *Physics of Scotus*, 55.

[49] It should be noted that Thomas's unicity of form was actually an innovation at the time and was condemned at Oxford in 1277; even his teacher St. Albert the Great held that the elements retained their identities in compound materials, see J. A. McWilliams, "St. Albert the Great and the Plurality of Forms," *The Modern Schoolman* 9 (1932): 43–44 and 59–61 and Berard Vogt, "The *forma corporeitatis* of Duns Scotus and Modern Science," *Franciscan Studies* 3 (1943): 47–62, on 53–54. See also Richard of Middleton's arguments in *De gradu formarum* on the presence of the elements in compounds, "Pluralitas formarum in quolibet mixto," given in Zavalloni, *Contraverse*, 119–32.

the tiniest parts of the mercury are simply mingled or juxtaposed beneath the level of sense perception without losing their identities. This concept essentially short-circuits potential problems with the persistence of substantial forms, because in this case, no substantial forms are created or destroyed, and therefore, one could readily conceive of both weight conservation and the possibility of reseparating the different particles into their original agglomerations. John does, in fact, at one point in reliable manuscripts of *De famulatu* refer to a material being "torn to pieces and converted into atoms" (*laniabitur et in athoma convertetur*) to describe how a substance, later recovered, is dissolved by a corrosive acid.[50] While the word *atom* in its medieval usage need not entail the ancient notion of *indivisible* particles, it does at least imply a concept of extremely tiny particles of matter.[51] Such a concept would, however, entail the rejection of a strict reading of Aristotle's doctrine of mixture expressed in *On Generation and Corruption*, where the Stagyrite argues that a mere juxtaposition of discrete particles (like grains of wheat and barley shaken together) does not count as a true combination because no alteration of the ingredients has taken place. While seeming homogenous to the ordinary mortal, the keen-sighted Lynceus would detect the heterogeneity, and for Aristotle, a true combination must produce homogeneity at every level.[52]

Significantly, like the plurality of forms doctrine, explanations of chemical transformation based on the coalitions and disaggregations of extremely tiny particles are linked particularly to Franciscan alchemical authors. The most thoroughly examined example appears in William Newman's studies of the late thirteenth-century lecturer at the Franciscan *studium* in Assisi, Paul of Taranto, author of the *Theorica et practica* and very probably of the enormously influential *Summa perfectionis* as well.[53] Another Franciscan, Gerald of Piscario (Géraud du Pescher), whom John praises by name and whom he heard lecture at the Toulouse *studium*, also embraced such concepts and did so in a specifically alchemical context.[54] In a surviving lecture treating the origin and composition of metals, Gerald states clearly that mercury and sulphur "are materially present in every metal and therefore are extracted from each of them," thus asserting the survival and subsequent reseparability of components from a composite substance. His various explanations of how differing "compaction of parts" (*compactio parcium*) results in the differing densities of the various metals and how a good or bad "cointimation" of

[50] See Appendix 1; John's usage of this term will be further analyzed in the forthcoming critical edition of *De famulatu*. For a lucid treatment of both particulate matter theory and the plurality of forms in alchemy, see Newman, *Atoms and Alchemy*, 23–58.

[51] On medieval "atomism," see Bernhard Pabst, *Atomtheorien des lateinischen Mittelalters* (Darmstadt: Wissenschaftliche Buchgesellschaft, 1994).

[52] Aristotle, *On Generation and Corruption* I.10, 327b30–328a17.

[53] Newman, *Summa*, and *Atoms and Alchemy*, 26–44.

[54] *LO*, 186 and note; on Gerald, who like so many Franciscans of John's time wrote also on the antichrist and prophecies, see "Géraud du Pescher, frère mineur," *Histoire littéraire de la France* 36 (1927): 614–17, 656, and Sylvain Piron, "Les *studia* franciscains de Provence et d'Aquitaine," 303–58 in *Philosophy and Theology in the Studia of the Religious Orders and at Papal and Royal Courts* (Turnhout: Brepols, 2012), on 330–31.

mercury and sulphur yields metals more or less porous and thus more or less dense and more or less resistant to corrosion all explicitly depend upon a particulate conception of matter.[55]

Unfortunately, John does not explain his reasoning explicitly in either the *Liber lucis* or the *De famulatu*; therefore, one cannot be certain about which of these two conceptions—plurality or particles—he had in mind, or indeed the degree to which he thought through all the implications of either. John's knowledge of the *Summa*, his experience with Gerald, and his own mention of "atoms" would seem to point more toward particles as an explanation. A potential objection, however, would be that the snow-white sublimate displays very different properties than the red tincture or the silvery mercury that combined to form it, implying that a new substantial *forma mixti* has been generated, thus pointing toward a plurality of forms—an impression strengthened by John's multiple references to disparate components "hiding beneath" the visible appearance of a composite. It is possibly to this issue that John alludes when he exclaims, "there is here the greatest miracle, for although that invisible spirit has the nature of the tincture of redness, it nevertheless congeals the mercury during sublimation into the whiteness of snow" (ll. 47–50). Thus one needs to be cautious, at least at present, in attributing one or the other conceptual background to him.

What is more certain is that both the plurality of forms and particulate conceptions of matter were linked specifically with *Franciscan* thought and theology in John's day. They were both thus readily available and undoubtedly well known to John. Moreover, the consequences that these two ideas have for practical alchemy may help explain the strikingly large number of Franciscan alchemists of the thirteenth and fourteenth centuries.[56]

2.2.6 John of Rupescissa and Assaying

Yet a third explanation is possible for what lies behind John's operational principles—one that shifts the focus from the disputative academic culture the friar knew in Toulouse to the realm of practical experience. In this case, a little-noticed aspect of John's biography is crucial. A section of John's *Commentary on the Oracle of Cyril* indicates that he was very familiar with, and seems to have

55 Gerald of Piscario, "Circa 4^m^ metheorum," in BNF, MS Lat. 4367, fols. 156v–58r. On 157r: "dicta duo sunt in quolibet metallo materialiter, et a quolibet ideo extrahuntur." Gerald uses a peculiar technical vocabulary (e.g. *partes contransite*) for describing the composition of metals which may be unique to him or perhaps more widespread among Franciscans, for further analysis of Gerald's ideas and terminology, see Lawrence M. Principe, "Friars Minor and *Opus Magnum*." F. M. DeLorme, "L''Ars faciendi sermones' de Géraud du Pescher," *Antonianum* 19 (1944):169–98 contains further information on the contents of this manuscript.

56 For a fuller survey of Franciscan alchemists and possible causes for their alchemical interests and theories, see Principe, "Friars Minor and *Opus Magnum*" and references therein. The surprising number of Franciscan alchemists was first pointed out by Thorndike, HMES, 3:347 and later scholars have written about "Franciscan alchemy," for example Antoine Calvet, *L'alchimie au Moyen Âge: XIIe-XVe siècles* (Paris: Vrin, 2018), 125–35; Michela Pereira, "I francescani e l'alchimia," *Convivium Assisiense* 10 (2008): 117–57; and Chiara Crisciani, "Giovanni di Rupescissa: sapere, alchimia e profezia," in *I Francescani e le scienze* (Spoleto: Centro Italiano di Studi sull'alto Medioevo, 2012), 239–79.

practiced, precious metal assaying at some point in his life. When explaining the term *paragolicum* ("a black stone which has the power of testing metals," i.e., a touchstone) that appears in the *Oracle*, John gives an extended digression about the touchstone and assaying.[57] John first notes "that when gold is pure by itself without any sort of admixture, it is said to be of twenty-four carats in the science of the minters of coins and in the science of alchemy." He then explains the system of carats and lists exhaustively the various ratios that provide gold of different carats.[58] Then, switching into the first person, John explains in minute detail how

> I will make twenty-four mixtures of gold and silver in precise ratios according to the twenty-four carats, and I will fashion twenty-four little rods, each mixture by itself. The first little rod will be of one carat, namely, having twenty-three parts of silver and one part of gold (by putting one ounce of gold into twenty-three ounces of silver) and I will inscribe "of one carat" upon that rod and keep it. Then I will make another of two carats, that is, by putting two ounces of gold upon twenty-two ounces of silver, and I will write "of two carats" upon that rod. In this manner I will make a rod of three carats, and of four, and so onwards until I will have twenty-four of them corresponding to the twenty-four carats according to the form of the science given above. I will then also have a very black touchstone, and I will anoint it with nut [oil] so that it is more black. And so, all these things having been thus arranged, if a piece of gold howsoever debased and mixed with other metals is brought to me, I will take it and touch the touchstone with that gold until the stone receives something by taking up some of that same gold. I will examine whether that gold which remains on the stone is of a red color, as is the rod of twenty-four carats that has no part of silver, and I will touch the stone with that rod next to the touch of the other gold that I want to test, and if the two marks are entirely of the same color, I will say that the gold is pure and is of twenty-four carats just like the rod of twenty-four carats because the marks on the stone of that rod and of the gold are wholly of the same color. If however I see that the mark of the gold is whiter than that of the twenty-four carat rod, I will view that gold as debased because it is whiter than pure gold and I will apply the debased rods, for example the one of eighteen carats, and will touch the stone with it alongside the mark of the gold to be tested, and if I find that these same marks are of the same color, I will say that that gold is of as many carats as the rod that makes the mark, because the gold makes a mark like the rod does, and both golds are like

57 One of John's lost theological works, *Book of the Touchstone*, once present in the papal library, also used the term: "liber paragolici Johannis de Rupescisa contra Franciscum de Montebelluna, apostatam et heresiarcham," see Maurice Faucon, *La librairie des papes d'Avignon*, 2 vols. (Paris: Thoren, 1887), 2:122 and Jeanne Bignami-Odier, *Études sur Jean de Roquetaillade* (Paris: Vrin, 1952), 190–91, and on the oracle commentary, 53–112.

58 BNF, MS Latin 2599, fols. 156v–57r: "est lapis niger qui habet virtutem probandi metalla et ut modum hujus probe intelligas debes scire quod aurum quando est per se purum sine mixtione quacumque in scientia trapesitarum id est campsorum nummulariorum et in scientia alquimie...dicitur habere xxiiii^or^ quadratos...." The introduction to *LO* (25 n.52) gives a dismissive assessment of John's mathematical skills based on this section but only as a result of using a faulty transcription in an earlier thesis that did not recognize that the copyist had trouble with Arabic numerals. On several occasions, the copyist confused the loop signifying the Arabic numeral 4 and the loop signifying the letter x, and in one case between the Arabic numerals 1 and 2. Once these obvious copyist's mistakes are recognized, John's calculations are correct.

> each other in color. If however the mark of the gold to be tested is whiter, I will apply another rod, perhaps of six carats, and will do as before until I find a rod that makes a mark entirely like the gold does, and then I will say that that gold is of as many carats as the rod.[59]

Given its context in the *Oracle* commentary, John's long excursus on carats and the touchstone is rather excessive. But given the level of detail and first-person narration, it reads like an explanation by someone with extensive and seemingly first-hand knowledge of assaying. If so, that artisanal experience might partly explain his confidence that compounded materials can be separated into their components, since separating gold and silver alloys was routine work for an assayer. Indeed, John describes how to carry out "the separation of gold and silver when they are mixed together" in a section of *De famulatu* that was omitted from the printed Latin edition.[60] Weight conservation may also be partly tied to assaying, since precious metals are valued by weight, and it is crucial to monitor weights in order to avoid unnoticed, and thus costly, losses. Nevertheless, the production and separation of metal alloys is significantly different on multiple levels from the production and separation of new "compounds" with strikingly new properties, since the former does not invoke the production of a new substantial form. Thus, assaying experience cannot on its own properly explain the origin of John's operational principles or his confidence in them. He would still have required a satisfying theoretical and explanatory background in the plurality of forms or in a particulate matter theory, unless he had no interest in foundational explanations, which would clearly conflict with the attitude he expresses consistently across both the *Liber lucis* and *De famulatu* and with his ten years of formal study in Toulouse. These theoretical and explanatory principles were ready

59 BNF, MS Latin 2599, fol. 157r–v: "ego faciam 24 mixtiones auri et argenti per partes distinctas secundum 24 quadratos et confingam 24 virgulas quamlibet per se. prima virgula erit unius quadrati scilicet habens 23 partes argenti et unam partem auri scilicet ponendo in 23 unciis argenti unam unciam auri et scribam super illam virgulam unius quadrati et servabo eam. Postea faciam aliam duorum quadratorum scilicet ponendo super 22 uncias argenti duas auri et scribam super illam virgam duorum quadratorum et per hunc modum conficiam virgam trium quadratorum et postea 4^{or} et sic donec habeam 24 secundum 24 quadratos juxta formam sciencie supradate. habebo autem postea lapidem paragolicum valde nigrum, et inungam [*here I correct the copyist's erroneous* "in virgam"] eum cum nuce ut sit magis niger. hiis itaque sic dispositis si portetur mihi aurum quamcumque sophisticatum et mixtum cum aliis metallis recipiam ipsum et tangam lapidem paragolicum cum illo auro donec lapis recipiat aliquid recipiendo de ipso consequenter aspiciam utrum aurum illud quod remansit in lapide sit rubei coloris ut est virga 24 quadratorum que nullam partem habet argenti et tangam cum illa virga lapidem juxta tactum alterius auri quod volo probare et si illi duo tactus sunt omnino ejusdem coloris dicam illud aurum esse purum et habere 24 quadratos sicut et virga habens 24 quadratos quia tactus illius virge et illius auri in lapide omnino sunt ejusdem coloris. si autem videam tactum auri albiorem quam tactum virge 24 quadratorum videbo illud aurum sophisticatum quia est albius auro puro et applicabo virgas sophisticatas puta virgam 18 quadratorum et cum ea tangam lapidem juxta tactum auri probandi et si inveniam quod ipsi tactus sint ejusdem coloris dicam illud aurum habere tot quadratos quot habet virga quia facit talem tactum sicut virga et utrumque aurum alteri in colore assimulatur. Si autem tactus auri probandi sit albior adhibebo aliam virgam forte 6 quadratorum et faciam ut prius donec inveniam virgam que faciat omnino tactum similem illi auro et tunc dicam illud aurum habere tot quadratos quot habet virga."

60 *De famulatu*, "Sciencia separacionis auri ab argento."

to hand in the foregoing alchemical, theological, and philosophical traditions to which John was surely exposed.[61]

The *Oracle* passage, however, points toward a much broader extent for John's knowledge and practical experience. Immediately after describing the use of the touchstone, John remarks that,

> I have treated other most fitting and highly scientific [*scientificos*] methods for testing gold and silver in the second book of my *Book on the Revelation of Secrets in the Weights of Things*. For I wrote that volume in three books, in the vernacular of the Aquitaine, at the request of a certain friend of mine, and therefore, I have here only briefly touched upon some of the science of testing metals.[62]

Thus what John writes in his *Oracle* commentary is but a small fraction of his knowledge of metals. The material he reprises there comes from only the second of the three books that comprised *The Weights of Things*, indicating that the complete text contained many more subjects than metal testing.[63] Unfortunately, that text, which must have been written before about 1347 when John began the *Oracle* commentary, is lost, and so one can only speculate on what else it might have contained.[64] Its composition in the vernacular implies a non-academic audience, and the statement that John wrote it at the request of a friend indicates that the friar enjoyed some level of notoriety for his knowledge of the subject. The emphasis given to *Weight* in the title suggests that this lost text might have provided some background to John's confidence in using weight conservation to reveal unseen dimensions of chemical transformations in the *Liber lucis*. Finally, the book's tantalizing title that invokes "the revelation of secrets" seems even to tie the practical nature of that text to John's broader endeavors in revealing other secrets, namely the secrets of the future hidden within Biblical and other prophecies, just as he uses weights in the *Liber lucis* to "reveal the secret" of the invisible sulphur hidden within his snow-white sublimate.

61 On an implicitly corpuscular mode of thinking in alchemy dating back to the Arabic traditions, see Newman, "Mercury and Sulphur."

62 BNF, MS Latin 2599, fol. 157v: "alios autem modos aptissimos et valde scientificos de probatione auri et argenti tradidi in libro revelationum secretorum in ponderibus rerum in 2° libro. Confeci enim volumen illud habens 3^{es} libros in vulgari aquitanico ad preces cujusdam amici mei et ideo breviter hic transivi de scientia probe metallorum."

63 The opening words to this section of the *Oracle* (fol. 157r) may suggest that it is a more or less direct Latin version of John's lost Occitan text. They seem directly to address the friend who asked for the treatise by beginning: "you want to see how all gold, howsoever much debased, will be tested with the stone here called *paragolicum*, I will make..." ("vis videre quomodo omne aurum quantumcumque sophisticatum probabitur cum lapide paragolico hic vocato, ego faciam...").

64 Questions remain of when, how, and why John gained knowledge of assaying; he provides no hints. It is difficult to imagine him engaging in such practice while studying at Toulouse or in the Franciscan *studium*. Might his family have been involved in goldsmithing or assaying, or perhaps was he assigned to receive and evaluate coinage and precious gifts offered to his friary at some point between 1337 and his imprisonment in 1344? Such suggestions, howsoever intriguing, remain mere conjectures.

2.3 Operation Two: Making a Quintessence

The second operation in John's preparation of the philosophers' stone involves the further elaboration of the white sublimate into a "quintessence." The use of this word ties the *Liber lucis* back to John's earlier *De famulatu*, particularly its first book, which is separately titled *On the Consideration of the Quintessence* (*De consideracione quinte essentie*). There John introduces the term *fifth essence* or *quintessence* as a metaphorical name for a range of substances prepared by alchemical processes into improved forms that he considers to be, as he says again here, "incorruptible, almost like the heavens" (l. 81). John argues in *De famulatu* that these "incorruptible" substances can potentially transfer their incorruptibility to human beings and thus fulfill a pharmaceutical function by fending off the corruption of disease, thereby extending life. John borrows the term from contemporaneous cosmology, which considered the region of the universe above the sphere of the moon (the supralunar world) to be composed of a "fifth element" or "quintessence." This region thus differs from the sublunar world (i.e., the Earth and its atmosphere), which is composed of the four elements—fire, air, earth, and water—whose constantly changing combinations and unbalanced interactions produce perpetual change through generation and corruption. In contrast, according to Aristotle, the heavenly substance (which the Stagyrite calls *aither*) is free from any heterogeneity or imbalance, which gives it a stable, unchanging, and thus incorruptible character that guarantees the eternity of the celestial realm. Exactly what John believed was the constitution of either the celestial substance or the terrestrial laboratory products that borrow its name analogically is a complicated issue that must be deferred to the close study and critical edition of *De famulatu*. At present, it suffices to underscore John's consistent use of the adverb *fere* ("almost") whenever he compares his prepared quintessences to the superlunar material; John did not believe he had some portion of the supralunar celestial matter in his flasks. For the present treatment of the *Liber lucis*, it suffices to note first the link to *De consideracione*, and second the homogeneity that characterizes the celestial substance, given that the second operation focuses on removing heterogeneities from the snow-white sublimate, thereby elevating it into a "quintessence" that is "incorruptible, *almost* like the heavens."

2.3.1 Distilling an Aquafort

John begins the second operation by preparing an aquafort—that is, a corrosive acid (in this case predominantly nitric acid) by heating a mixture of saltpeter and vitriol. The details of this operation are omitted from the printed editions and most of the later manuscripts, almost certainly because the general procedure for distilling aquaforts was well known by the start of the fifteenth century, which was not the case when John was writing in the mid-fourteenth. The earliest recipes for an aquafort comprised principally of nitric acid appear only in the first decades after 1300, thus fairly recently from John's perspective.[65]

John's instructions offer two points of interest. The first is his instruction to give only a gentle heat (*ignis suavis*) at the beginning so that no liquid distills out, a direction reinforced in a family of early manuscripts that add "indeed, most gentle, for a day and a night, that is, for twenty-four hours."[66] John's stipulation of an initial period of gentle heating initially seemed pointless, but the experimental reworking showed that it is important for obtaining a higher yield of aquafort and under more gentle conditions. During this "predistillation phase" (at 70–75 degrees C for 4–5 hours), the vitriol releases much of its water of crystallization, which, because of the low temperature, circulates in the flask rather than distilling away. This liberated water allows the potassium nitrate and iron sulphate to react, forming iron nitrate, turning the initially pale mixture dark rusty brown. When the temperature is raised, it is the iron nitrate produced during the period of gentle heating that decomposes to produce nitric acid.[67] If the salt mixture is heated strongly from the beginning, the water is expelled from the flask as it is liberated, precluding the complete formation of iron nitrate, and thus some proportion of the potassium nitrate is left unreacted. Whereas the thermal decomposition of iron nitrate provides nitric acid, that of potassium nitrate requires a higher temperature and gives no acid, only oxygen and potassium nitrite. (The reader may wish to compare the different results obtained in the first operation when a similar mixture containing saltpeter and vitriol was gently heated in an *open* dish, such that the water was immediately lost into the air, and thus very little iron nitrate was formed.) John's seemingly unnecessary direction thus once again indicates a level of optimization that must have been achieved through practical experience.

After the initial heating at low temperature, John orders that the fire be increased "until the head of the alembic begins to turn red" (ll. 89–90). This description seems to imply a heat high enough to produce a dull red glow in the apparatus. Indeed, before carrying out the process, I was worried about the high temperature and consequent damage to the apparatus. However, the alembic head began to redden long before a high temperature could be reached, not from heat but rather from the production of nitrogen dioxide, a red-brown gas, as the iron nitrates began their decomposition (Figure 2.5).[68] The fact that John was able to

[65] Vladimír Karpenko, "Some Notes on the Early History of Nitric Acid: 1300–1700," *Bulletin of the History of Chemistry* 34 (2009): 105–16.

[66] P[1] and Po: *ymo suavissimum per unum diem et noctem scilicet per viginti quatuor horas.*

[67] Ferrous nitrate is formed initially but is unstable; it is rapidly oxidized under the given conditions to ferric nitrate, producing iron oxide as a side-product which is responsible for the rusty brown coloration. In my trials, 200 g of mixed salts provided 50–55 mL of aquafort. For the reactions, see Gus van Weert and Yuxing Shang, "Iron Control in Nitrate Hydrometallurgy by (Auto)decomposition of Iron (II) Nitrate," *Hydrometallurgy* 33 (1993): 255–71; K. Wieczorek-Ciurowa and A. J. Kozak, "The Thermal Decomposition of $Fe(NO_3)_3 \cdot 9H_2O$," *Journal of Thermal Analysis and Calorimetry* 58 (1999): 647–51; and P. Melnikov, et al., "Thermal Decomposition Mechanism of Iron(III)nitrate," *Journal of Thermal Analysis and Calorimetry* 115 (2014): 145–51.

[68] The decomposition, and thus the production of nitric acid, begins at the relatively low temperature of about 110 degrees C and is complete by 140 degrees.

FIGURE 2.5 The "reddened head" of the alembic during the distillation of the aquafort, John's sign to increase the temperature. Author's laboratory.

observe this coloration in the head, which he used as the observable sign to increase the heat further, indicates that he must have been using a *glass* apparatus, rather than the more common and less expensive earthenware apparatus of the period. Moreover, his ability correctly to see the red color implies that John's alembic was made of a more or less colorless glass, that is to say, made of a higher quality material than the more common green or brown (often quite dark) glass. Reworking the process thus not only provided the correct understanding of the "reddened head" but also gave otherwise unobtainable information about John's material conditions in the mid-fourteenth century.

The aquafort obtained from this procedure was a liquid of an attractive pale blue-green color that emitted red-brown fumes.[69] Following John's instructions, this liquid was poured over a quantity of the white sublimate, the flask sealed and kept at a gentle warmth (40–50 degrees C) for twelve hours.[70] At the end of this time, only about half of the sublimate had dissolved, a result that John describes, and so his next instruction was followed, namely, adding "a seventh part relative to the water" of sal ammoniac (ammonium chloride) in order to complete the dissolution. The liquid instantly turned from pale yellow to orange, and the rest of the sublimate dissolved within thirty minutes.[71] It remained unclear to me for a long time why John considered the addition of sal ammoniac to be the "greatest secret" in the art and advised readers to "apply your heart to it" (ll. 94–95). A plausible explanation appeared only later when the replication of the process came to an impasse.

John next distills the aquafort away from the solution and heats the solid residue, whereupon the "spirit of mercury, that is, the quintessence whiter than snow, will ascend up the walls" of the flask, leaving behind "burnt and black dregs" (ll. 97–100). In the reworking, this occurred just as described: the snow-white sublimate arose quickly, leaving a small amount of dark non-volatile residue.[72] Again, following John's instructions, the sublimate was treated a second time with aquafort, distilled, and resublimed, and then a third time. On the final reiteration, no "dregs" were found after the sublimation, as John promised.

Although from a modern chemical perspective, this second operation might be summed up as a "purification of mercuric chloride through resublimation," such a summation would miss the operation's significance to John. In his typical fashion, John expounds upon the process, explaining that "in this operation you have done two things" (l. 102). The first is the removal of whatever was "earthy, corruptible, and imperfect." Significantly, these three adjectives are the direct opposites of the

69 The blue-green color is due to the presence of nitrogen oxides dissolved in the acid, whose decompositions generate part of the red-brown fumes.

70 I used 20 g of sublimate with 60 mL of aquafort, a ratio of 1 to 3; John does not provide any proportions. A later copyist (Ve_a) inserted the same ratio of 1 to 3.

71 Mercuric chloride's solubility in aqueous media is significantly increased in the presence of alkali chlorides and hydrochloric acid (the latter produced by the reaction of the aquafort with the sal ammoniac); *Merck Index*, s.v. "mercuric chloride." I would conjecture that John came to this discovery of how to force the dissolution of the sublimate thanks to his experience with gold and silver refining: gold does not dissolve in aquafort (nitric acid); however, if sal ammoniac is added (thus converting the nitric acid into aqua regia), it does dissolve. John indicates his knowledge of this fact in the penultimate section of the first book of *De famulatu*, which is not present in the printed editions. One surprising result—not mentioned by John—is that the sal ammoniac "disappears" in the process; its addition does not increase the weight of the sublimate. This is because under the conditions of the reaction, the ammonium ion is oxidized to nitrous oxide, a gas which passes away invisibly, see A. D. Kelmers, L. Maya, D. N. Browning, and W. Davis, "The Chloride-Catalyzed Decomposition of Ammonium Nitrate in Nitric Acid Media at 100°C," *Journal of Inorganic Nuclear Chemistry* 41 (1979): 1583–88.

72 This residue is iron oxide from ferrous materials carried upwards mechanically along with the sublimate when it was first produced.

properties of the superlunar quintessence—namely, celestial, incorruptible, and perfect. In terms of the separated dregs, once again John expresses surprise at the result, calling it "very marvellous" that something black could be removed from something white, in much the same way he marveled previously that a red tincture turns silvery mercury white. At each repetition, the crystals of sublimate become longer, larger, and more gleaming, and as a result, the material is fluffier and less compact, and these features, combined with its much more rapid ascent upwards in sublimation (less than two hours rather than twenty-four) and at a lower temperature may have suggested its increasingly "celestial and perfect" character. While John does not explicitly mention having made such an observation, he may refer to it obliquely when, later in the *Liber lucis*, he explains why the material of the stone is allegorically called an animal: "because it grows in subliming" (ll. 168–69).[73]

The second result that John cites is the greater incorporation of the "spirit of vitriol" (i.e., the invisible sulphur or tincture of redness) into the white sublimate. He claims that the aquafort contains more of this tincture of redness, presumably based on the use of Roman vitriol in its preparation, but surely more powerfully because he *saw* the red fumes produced during the distillation, and could not have missed the red fumes that issue constantly from the pale blue-green "water of vitriol." It has been suggested that John's repeated mention of red colors stems from a linkage with Christ's blood. While inventive, and while it is possible that a Franciscan *might* think of Christ's blood in connection with other red things, the connection is unnecessary and puts the cart before the horse.[74] John did not need to resort to metaphor and allusion for his reference to red colors; he actually *saw* red materials, as the experimental reworking demonstrates. Such redness was furthermore central to John's alchemical goal, directed toward producing the *redness* of gold by means of the *red* philosophers' stone. Thus, observable aspects of the practical process and established alchemical theories suffice to account for John's repeated references to red color.

2.3.2 John of Rupescissa and Arnald of Villanova

John justifies his aquafort procedure by quoting from the alchemical authority he cites far more than any other, namely, Arnald of Villanova, or rather, pseudo-Arnald of Villanova. The identity of the author or authors behind the many alchemical works attributed to Arnald of Villanova in the fourteenth century remains uncertain. But John had no reason to question the authenticity of these alchemical writings as products of the real Arnald of Villanova (1240–1311), an

73 Given the context of its use in the *Liber lucis*, one could also read "animal" as referring to the virgin's milk, but that liquid material is *distilled* not sublimed, and so the Deckname *animal* seems to refer instead to the snow-white sublimate before it is converted into the virgin's milk.

74 DeVun, *Prophecy*, 116–23. There are also some misreadings of the text in this section (e.g., conflating the white sublimate with the philosophers' stone, and the quotations from pseudo-Arnald with John's own words) and elsewhere, caused partly by use of a very corrupt manuscript of the *Liber lucis*.

author he knew well. The real Arnald was a Catalan physician deeply influenced by Franciscan thought and spirituality who, like John, wrote on the antichrist, interpreted prophecies, advocated for poverty and church reform, occasionally fell afoul of ecclesiastical authorities, and lived for a time in Avignon at the papal court.[75] John and Arnald also shared an interest in the *Oracle of Cyril* and the eschatological writings of Joachim of Fiore, and it is clear that John was already familiar with Arnald's *De cymbalis ecclesie*, from which he adopted his means of calculating the antichrist's arrival, by 1340.[76] It is thus easy to imagine John picking up the alchemical works of pseudo-Arnald alongside the prophetic and eschatological ones of the real Arnald with whom he was already familiar and shared so many interests and outlooks. Indeed, the pseudo-Arnald texts maintain some of the authentic Arnald's interest in eschatology and Joachism while also twinning these topics with alchemy—a combination that was surely attractive to John of Rupescissa.[77] John's indebtedness to pseudo-Arnald reappears repeatedly; the friar quotes him as an authority several times in the *Liber lucis*, in most cases to deploy an allegorical statement as justification for a practical laboratory operation.

In the present instance, John quotes a line that he then interprets as Arnald's obscure stipulation of the twelve-hour treatment of the sublimate with aquafort: "let it drink as much as it can drink for twelve hours" (ll. 107–8). John claims that he drew this line from the same Arnaldian text that he used in Operation 1 for the image of the grain of wheat dying in the ground, a text that he there calls a "tractatus parabolicus" that was "written using the words of the Gospel and of Christ's crucifixion as a trope" (ll. 53–55). Working from John's brief description of this text—which represents one of the earliest datable citations of a work by pseudo-Arnald—Antoine Calvet has identified this "tractatus parabolicus" with a short text attributed to Arnald that in its extant manuscripts is titled *Exempla* or *Liber propheciarum*.[78] Yet while part of this text correlates with John's description of a work that applies metaphors drawn from the life of Christ to alchemy, it

[75] On Arnald, see Joseph Ziegler, *Medicine and Religion c. 1300: The Case of Arnau de Vilanova* (Oxford: Clarendon, 1998); Francesco Santi, "La vision de la fin du temps chez Arnaud de Villeneuve: Contenu théologique et expérience mystique," *Cahiers de Fanjeaux* 27 (1992): 107–27; and Robert E. Lerner, "The Pope and the Doctor," *The Yale Review* 78 (1988): 62–79.

[76] Lerner, "Historical Introduction," in *LSE*, 28 and 57–58, and Gian Luca Potestà, "L'anno del antichristo: Il calcolo di Arnaldo di Villanova nella letteratura teologica e profetica del XIV secolo," *Rivista di storia del Cristianismo* 4 (2007): 431–64.

[77] Antoine Calvet, "Alchimie et joachimisme dans les *alchemica* pseudo-Arnaldiens," in *Alchemie et philosophie à la Renaissance*, eds. Jean-Claude Margolin et Sylvain Matton (Paris: Vrin, 1993), 93–107.

[78] For the text and an analysis of the work, see Calvet, *Oeuvres alchimiques*, 234–49 and 525–45; Calvet here reprises and updates his earlier treatment in Calvet, "Le *Tractatus parabolicus* du pseudo-Arnaud de Villeneuve, présentation, édition et traduction," *Chrysopoeia* 5 (1997): 145–71. Only four manuscripts of the text are currently known; to the three cited by Calvet can be added Torino, Biblioteca Nazionale Universitaria, E.II.23, fols. 13r–14v. While Calvet and a few other scholars following his lead use John's words *tractatus parabolicus* as the *title* of the pseudo-Arnald treatise, John may have intended these words only as a description of its *style* rather than as its title. No manuscript and no other contemporaneous reference uses *tractatus parabolicus* as a title; thus, I have refrained from using it as such in this study and from rendering John's words as a title in my translation. I refer instead to this pseudo-Arnald text under the title *Exempla* as it is found in the Venice manuscript; see also Thorndike, *HMES* 3:77.

does not contain the quotations that John claims to have taken from it.[79] Moreover, allusions to Christ occupy only about one-quarter of that pseudo-Arnald treatise; the majority deal with Old Testament prophets (thus justifying the indicated title *Liber propheciarum*), a content that does not match John's description of the "tractatus parabolicus" from which he was quoting. Calvet has reasonably proposed that John was citing from memory, and misremembered either the words or the correct textual source, or that he used a fuller version of the text that has not come down to us from the fourteenth century.[80] Phrases that resemble some of John's quotations do appear in other pseudo-Arnald texts—in the *De secretis naturae*, the *Quaestiones ad Bonifacium VIII*, and the major work of the pseudo-Arnald corpus, the *Rosarius philosophorum*.

While this situation might strengthen Calvet's hypothesis that John misremembered his exact source, it also suggests an alternate hypothesis that John used a "second-generation" compilation of pseudo-Arnaldian imagery gathered from multiple parts of the corpus and perhaps augmented with further metaphorical expressions. Several such brief compilations do exist, and in fact, a very short and terse text of this sort appears in the mid-fifteenth-century Darmstadt manuscript that also contains **D**.[81] While this compilation does not fully correspond to John's description of a text centered on imagery from Christ's passion, it does contain a line very close to "let it drink as much as it can drink for twelve hours," an instruction that cannot otherwise be found in any known work of pseudo-Arnald. This short text instructs in very similar words that the material for the stone must be "satiated with drinking of the most limpid water for at least twelve hours." In fact, *five* of the phrases John attributes to the "metaphorical treatise" by Arnald occur in this brief text, as well as several other turns of phrase that John uses elsewhere in the *Liber lucis* without citing Arnald. This brief text thus provides a greater number of correlations with the *Liber lucis* than any other pseudo-Arnaldian text. This similarity presumably explains why the copyist of the Darmstadt manuscript decided to juxtapose this particular pseudo-Arnald text with the *Liber lucis*. Following the example of that copyist, it is worth including the entire text here since it has not otherwise been published:

> Master Arnald presents the art of the stone's preparation obscurely in these words: The stone of the philosophers, which coming out of the earth is perfected or exalted in the air, having been satiated by drinking of the most limpid water for at least twelve hours, visibly swells on all sides. Having been placed in dry air moderately hot, it spends time in the oven; being deprived of extraneous vapor, it obtains a solidification of its parts, and having been diminished and emptied of superfluous humor, it is left fit for grinding. These things having been done, the virgin's milk is expressed from its

79 See the text printed in Calvet, *Oeuvres alchimiques*, 532–45.

80 Calvet, *Oeuvres alchimiques*, 236–37.

81 Darmstadt, Universitäts- und Landesbibliothek Darmstadt, Hs 2226, Faszikel III, fol. 50r–v.

> prior parts. And placed at once in the philosophers' egg it germinates, and withstands contrition and decoction for a long time until by nurturing and altering it will have been stripped of the variations of all colors. Thereupon it rejoices with a whiteness equal to that of snow. And thereafter it withstands the pains of an increasing fire without fear of harm, until having been drawn through crimsons and made glorious with purples, it comes forth from the tomb with regal power.[82]

Whatever the immediate source for John's quotations, it is clear that pseudo-Arnald's ideas and writings were of significant importance to the Franciscan friar. And this significance may have had as much to do with the spurious attribution to Arnald of Villanova, a person with whom John shared so much in common, as with the textual content. The obscurity and terseness of the brief pseudo-Arnald text presented here, its need for interpretation, and the breadth of its potential meanings recall the character of the prophetic texts (like the *Oracle of Cyril*) that engaged the attention of both John of Rupescissa and the real Arnald of Villanova.

2.4 Operation Three: Brief (In)digestion

The prologue and first two operations represent one-third of the *Liber lucis*. Readers who reached this point, perhaps impressed by John's careful instructions, demonstrations, and descriptions of observables, may be jarred by the brevity of Operation 3. In just seven lines, John stipulates that the material prepared in Operation 2 must now be heated gently in a special oven for eight days or more. While John subjoins a detailed description of the oven at the end of the treatise, he says little about this step whose purpose is to render the sublimate capable of being "dissolved into a water." That next transformation begins Operation 4, in which the roasted, or "digested," sublimate is ground, placed in a flask, and sealed within a vessel containing warm water until it turns to a liquid. This liquefaction corresponds to what early modern practitioners called *dissolutio per deliquium*, that is, the dissolution of a solid into a liquid through the spontaneous absorption of atmospheric humidity—what modern chemists (preserving the root of the earlier term) call deliquescence. John promotes this deliquescence by enclosing the sublimate in the humid air over a warm water bath. This liquefaction must happen or, as John warns rather omninously, "surely the work will not be brought to completion" (l. 121).

[82] Darmstadt, Hs 2226, Faszikel III, fol. 50r–v, "Magister arnaldus obscure tradidit artem de preparacione lapidis sub hiis verbis. Lapis philosophorum qui *de terra egrediens in aere* perficitur, seu exaltatur, *limpidissime aque potu saciatus ad minus horum 12* undique visibiliter tumet, moratur in stupha positus in aere sicco mediocriter calido vapore deparcus extraneo solidicacionem suarum parcium adipiscitur et extenuatus ac inanitus ab humore superfluo remanet aptus contricione, quibus peractis, ex prioribus ejus partibus *virgineum lac exprimitur*. Ipseque confestim repositus in ovo philosophorum pullulat contricionem et decoctionem tamdiu sustinet donec fovendo et alterando omnium colorum varietatibus fuerit denudatus. Inde cum compari niveo candore letatur. Et ex tunc *sine metu periculi penas sustinet ignis crescentis*, donec per sanguineos tractus purpureosque clarificatus, *egrediatur de monumento cum regia potestate*." I have here italicized phrases that reappear in very similar form in the *Liber lucis*.

The combination of gentle roasting and dissolution into a liquid comprise a crucial step in John's process because the resulting "water" must be distilled to yield the *lac virginis*, or virgin's milk, and which is "the purest material of the stone perfected by the final preparation" (ll. 127–28) and ready to be turned into the philosophers' stone. Many later practitioners clearly saw, or encountered practically, the challenges involved in the *Liber lucis*' tersely-written instructions at this point, and tried to address the problem in multiple ways. Since somewhat more profuse instructions for the process appear in *De famulatu*, several practitioners spliced parts of that text, or references to it, into their copies of the *Liber lucis* (e.g., ı′, **Fl**, **J**, **Mp**^{2}). Others excerpted these parallel sections and copied them out for comparison alongside the *Liber lucis* (e.g., **O**, **L**^{5}, **Mp**^{1}, **Ha**$_{d}$). Working from the other direction, one copyist wrote out the entire *Liber lucis* in the margins surrounding the parallel material in his manuscript of *De famulatu* (**Br**; see Figure 3.1); another did the same using just a partial copy of the *Liber lucis* (**Km**). All of these manuscripts record examples of practitioners actually deploying the alchemical principle that "one book opens another" by comparing the two texts in order to gain further information.

At the same point in another fifteenth-century copy of *De famulatu*, the owner inserted a full-page drawing of the "hot water bath or moist stove by which the spirit sublimed and calcined as above is resolved into water," along with a lengthy account of how to construct it.[83] Even more strikingly, the material added to the German translation of the *Liber lucis* published as *Thesaurus mundi* presents a detailed account of how practitioners working at the monastery of Santa Giustina in Padua were stymied at this point because the digested sublimate refused to liquefy, and so gave up on the process. This account, reportedly gathered from the fifteenth-century alchemist and priest Johannes von Teschen, recounts that the practitioners then obtained advice from a fellow alchemist, identifiable as Johannes Hinderbach (1418–1486), who was a student in Padua in the 1440s and later prince-archbishop of Trento. Hinderbach, who had himself followed John's method, advised the frustrated alchemists to heat the sublimate for a longer period, which they did for an additional twelve days. After that time it liquefied properly and was thereafter carried on to a (reportedly) successful conclusion.[84]

Other practitioners who ran into trouble at this same point in the process inserted further advice from their own struggles. One fifteenth-century practitioner wrote parallel lines from the *Liber lucis* (ll. 127–29) into the margin of his copy of *De famulatu*, marked which sections corresponded with the third and fourth

[83] Cambrai, Le Labo Cambrai, MS 920, fols. 180cv–180dr and 181v–82r.

[84] Kassel, Universitätsbibliothek Kassel - Landesbibliothek und Murhardsche Bibliothek der Stadt Kassel, 4° Ms. chem. 33[6, fol. 12r: "...meister peter yredt in dem virten arbeitt das sich dey matery nichtt wolte solvirn in langeren zitt / do les er vander arbeitt da sagt im min her dey orsagen / wan disser geist was noch nicht genugsam gedawert / und satze dey materi wider ihn dey dingstancy [*i.e.*, *digestion*] auff xii thagen in den Trucken Stuben der philosophen do solvirt er sich di materin." See also the description of **Ka**^{3}$_{d}$ in Section 6.1.1.2 and of the *Thesaurus* in Section 3.4. A further analysis of this account and the background to *Thesaurus* is in preparation.

operations, and, in a move similar to that reportedly suggested by Hinderbach, scribbled next to the prescription for eight days of digestion: "the truth is sixteen days."[85] Germanic exemplar γ″ (produced earlier than 1443) contained at least five glosses of a practical nature, three of which focus on this crucial roasting-liquefying step. One gloss recommends using a wide vessel for heating the sublimate so that—quoting from *De famulatu*—"hot air may touch the naked material." A second turns to Geber's *Summa* for guidance, juxtaposing his words with quotations from *De famulatu*. The most intriguing insertion describes how to construct a specially-designed water bath for the liquefaction, and ends with what sounds like a later practitioner's contented addition: "I also made that bath and I found what I sought."[86] In all these cases, the manuscripts bear witness to the many later alchemists who endeavored to follow John's procedure, encountered frustration at the same point in the process, and then labored to find solutions when their experimental results did not meet their expectations or John's descriptions.

But it was not only late medieval and early modern alchemists who encountered difficulties at the start of Operation 4. Around this same point in the process, my own reworking reached an impasse, and from the perspective of modern chemical knowledge, it remains unclear how one could surmount the obstacle. This difficulty requires an explanation because it relates directly to the potential conflict between the thesis of the current historiography of alchemy, which is backed by well-established evidence that alchemical authors based their writings on actual laboratory operations, and the apparent impossibility of achieving the final goal of gold-making that the same texts describe.

2.5 Operation Four: Expressing the Virgin's Milk

Operation 4 represents a hinge in John's experimental reporting where the workability of the process breaks down, and this crucial point is reflected in changes to the style and tone of the text. Operation 3 gently heated, or "digested," the mercury sublimate over several days in the "dry stove"; this operation is unproblematic, if of questionable effect.[87] John next exposes this dried material to a warm, humid atmosphere. At this point, my own reworking was stymied in the same way as were the endeavors of the monastic experimenters in Padua in the mid-fifteenth century, but with the added disadvantage of not having the future prince-archbishop of Trento to call upon for helpful advice. In short, the powder, exposed to a humid atmosphere at 45–50 degrees C (the "humid and warm oven" as it is called in *De famulatu*) refused to dissolve itself into a liquid, or even to become noticeably moist, even after three weeks.

85 Cesena, Biblioteca Malatestiana, MS Malatestiano S.XXIX.27, fols. 1r–82v, on fol. 56v: "veritas est 16 dierum naturalium."

86 Z, fol. 117r: "Ego autem illud balneum feci et inveni quid quesivi." For the text of these glosses, see Section 6.2.6.

87 It should be noted that this operation is also adumbrated in the short text of the Darmstadt manuscript: "Having been placed in dry air moderately hot, it spends time in the oven...."

2.5.1 Resolving Problems in Experimental Reconstructions

There are several routes to explore in order to resolve this issue. The first is that the experimental reworking made a wrong turn at some earlier point in the process. Despite the careful practical details given in Operations 1 and 2, and a promise of forthrightness, John did make a statement that raised doubts about the correct procedure. In Operation 2, after dissolving the sublimate in aquafort, he instructs readers to "distill off the aquafort, and keep it" (ll. 96-97), but never explains why this distillate should be kept rather than discarded. Is it to be used at some later point? He never says so. One possible interpretation (which assumes that there is intentionally omitted information) is that the distilled off aquafort, rather than fresh aquafort, is to be reused for the subsequent retreatments of the sublimate. At least one early practitioner was also puzzled at this point and, in order to resolve the ambiguity, inserted the word *nova* (new) into his copy (**Go**^{I}) so that it stipulates dissolving the sublimate "in new aquafort"; the copyist of the exemplar used to prepare **Po** and **P**^{I} ($\pi^{\mathrm{I}\prime}$) did the same. More significantly, the practitioner who produced **Go**^{I} later returned to his copy, crossed out the added word *nova*, and interlined instead "in the first aquafort," a change that was presumably the result of initial failure followed by further practical experimentation. There is in fact good precedent for reusing a distillate to treat the residue from which it was distilled; it is the process known as *cohobation*. In this operation, no longer part of modern chemical practice, the practitioner pours the distillate back over the distillation residue, distills it off again, and repeats the process some number of times. John uses this process himself for another procedure in *De famulatu*.[88] When this change is made to the reconstruction, the final product obtained is different. Since every repetition depletes the nitric acid in the aquafort, the sal ammoniac added in the second and third repetitions is not completely decomposed as it was in the first, and therefore persists into the final product. The product obtained from this modified procedure was a double salt of mercuric chloride and ammonium chloride. Exposed to the same warm, humid atmosphere within the *caldarium*, after two days the powder slumped into a moist mass, and after another four days dissolved completely into a crystal-clear liquid.

This modified interpretation of the procedure extends the successful replication of John of Rupescissa's results into Operation 4, and explains his puzzling instruction to "keep" the distilled off aquafort. Nevertheless, it cannot be determined conclusively whether this modification is a historically accurate interpretation (i.e., what John himself did or intended to be done) or a lucky coincidence. When interpreting alchemical processes, there can sometimes be more than one route to the same result, and the "success" of an experimental reconstruction is generally based on whether or not the results match up with the textual description. In this case, there is another possible interpretation of John's words that also makes the

88 *De famulatu*, "Non pretermittam."

process work and resolves a different puzzling statement that he makes. In Operation 2, when the addition of sal ammoniac caused the undissolved sublimate to dissolve in the aquafort, John named this material as the "greatest secret in the art" and abjured readers to "apply your heart to it" (ll. 94–95) as if hinting that there was more to be understood here. Possibly he intended his readers to reason that if the sal ammoniac causes an insoluble sublimate to dissolve in aquafort, it would likewise cause the same sublimate to dissolve into a water in the humid oven. This interpretation would have the practitioner use fresh aquafort at each iteration but then add more sal ammoniac to the dry sublimate before subjecting it to the humid oven. Put into practice, this method also worked, although it took longer for the mixed sublimate to dissolve itself entirely into a clear fluid.

In either case, the considerable satisfaction these interpretations provided was short-lived. The next step, the "expression of the virgin's milk," the most crucial step in the whole procedure, is simply logically (as well as chemically) impossible as written. John reports that when the liquefied sublimate is heated in an alembic, it yields a clear distillate, the virgin's milk, leaving behind a "gross combustible material." But since the only liquid present in the dissolved sublimate is the water that it absorbed in the "humid and warm stove," water is the only liquid that could be distilled off (unless a chemical reaction between the water and the sublimate has occurred) and the residue would be the sublimate simply dried out once again rather than a "gross combustible material." The implausibility of the result John describes increases with the claim that this "gross combustible material" can be separated into two oily distillates, leaving behind a black residue. The mercuric salt—which is of a very simple composition with or without added ammonium chloride—simply *cannot* generate so many different products. In practice, the liquefied material placed in an alembic and heated by a water bath as described, slowly provided only a colorless distillate of water, leaving behind the previously dissolved salt unchanged rather than a "gross combustible material."

At this point, two other explanations must be explored. The first is that John omitted a major ingredient, some substance that renders the material being distilled in Operation 4 sufficiently complex that it could produce at least four different thermal decomposition products, more than one of which—judging from the description given—would seem to be organic (i.e., carbon-based). John did omit a crucial ingredient in Operation 1, so it is possible that he did so again later. But while the missing ingredient in Operation 1 could be identified from the parallel process in *De famulatu*, that text gives no help for Operation 4, but instead becomes suddenly laconic at the distillation of the virgin's milk.

2.5.2 From Experience to Expectation

The other possible explanation, which I consider the more likely and the only one that provides historical insight more broadly, not only about John of Rupescissa but also about alchemy and its practices, is that Operation 4 is an inflection point where *experience* gives way to *expectation*. That is to say, up to this point, the *Liber lucis* has been reporting John's practical experiences, as is evident not only from the replicability of the process but especially from the minutely accurate descriptions he provides of both methods and results. From this point onwards, however, the *Liber lucis* instead extrapolates what would be *expected* to happen on the way toward the philosophers' stone. I previously proposed a similar explanation, based on what I termed "chymical optimism," for the late sixteenth-century route to the philosophers' stone described enigmatically by the author writing under the pseudonym of Basil Valentine. In that case, I suggested that an alchemical practitioner and author, encouraged by laboratory results and their coherence with theory and/or interpretations of earlier authoritative sources, shifted nearly seamlessly from what the practitioner had himself *seen* happen to what *should* happen according to theoretical expectations and the testimony of authoritative sources.[89] This hypothesis may explain much of chrysopoetic authorship and preserves intact the thesis that alchemical texts do contain descriptions of practical processes carried out by their authors. Leaving the laboratory and returning to the library, a close reading of the remainder of the text supports this explanation for the structure of the *Liber lucis*.

John's style reflects a shift from expounded past experience to extrapolated future results. Most obviously, the carefully detailed descriptions, the demonstrations, and the proofs that characterize the earlier operations vanish starting with Operation 4. Furthermore, while the first operations are striking in their originality, providing ideas, procedures, and results that are not found in earlier alchemical texts, the later operations rely increasingly on descriptions of the maturing stone, its changing appearance, and its use that had become established *topoi* by the mid-fourteenth century. It is surely no coincidence that the process in *De famulatu* that parallels the *Liber lucis* comes to an end at the same point. In that text, the close details of practical operations and the construction of the *furnus secretissimus philosophorum* stop just as the distillation of the *lac virginis* is mentioned.[90] Additionally, while John is authoritative in the early operations about the procedures that should be followed, permitting his readers only the most minor of potential alternatives, after Operation 4 he instead offers multiple parallel pathways, as if yet unsure which of them would give the desired results. Although John does cite pseudo-Arnald at several points throughout the *Liber lucis*, its later sections show a much greater interest in tying claims and methods to authoritative sources, as if, without the practical *demonstrationes ad sensum* of the earlier

[89] Principe, *Secrets*, 158. On the authorship of "Basil Valentine," see Lawrence M. Principe, "The Development of the Basil Valentine Corpus and Biography: Pseudepigraphic Corpora and Paracelsian Ideas," *Early Science and Medicine* 24 (2019): 549–572; reprinted in *Pseudo-Paracelsus: Forgery and Early Modern Alchemy, Medicine, and Natural Philosophy*, ed. Didier Kahn and Hiro Hirai (Leiden: Brill, 2021), 137–59.

[90] See Appendix 1, "Omnis spiritus."

sections, it became necessary to use authoritative precedent instead as support for the instructions offered. Some examples will serve to illustrate these observations.

The clear distillate produced in Operation 4 should transform into the white and red philosophers' stones by a straightforward process of gentle heating in a sealed flask, as described in Operations 5 and 6. It is therefore the most critical substance in the whole process. John immediately links this distillate to a series of authoritative texts by endowing it with multiple names found in those earlier sources. The first two, "May dew" (*ros madii*) and "water of life" (*aqua vite*), occur side by side in the *Quaestiones* of John's favored authority pseudo-Arnald.[91] These two names, along with the third name in John's list, "tear of the eye" (*lacrima oculi*), occur all together in the *Declaratio lapidis physici* where they describe a "water" so necessary for making the stone that "there is nothing in this world that could stand in its place."[92] The name that John will hereafter use the most frequently comes fourth in the list: "virgin's milk" (*lac virginis*), a term that dates back to Islamic alchemy (*laban al-'aḍrā*), for example, in al-Rāzī.[93] It also appears in the first Arabic alchemical text to enter Latin Europe, the *De compositione alchimiae* of Morienus, translated in the mid-twelfth century.[94] It is used in the section of Ibn-Sīnā's eleventh-century *Kitāb al-Shifā'* that circulated in Latin Europe appended to Aristotle's *Meteors*. Pseudo-Aristotle cites it twice in the widely-circulated *Liber perfecti magisterii*.[95] Closer to John's own time, the term virgin's milk shows up in the *Pretiosa margarita novella*, written in 1330 by Petrus Bonus of Ferrara, and in the early fourteenth-century *Testamentum* by the "magister Testamenti," later attributed to Lull.[96] In these earlier cases, the name *lac virginis* sometimes refers rather obviously to common mercury, but sometimes to a mysterious substance of undisclosed origin and undefined composition,

91 On the *Quaestiones*, see Calvet, *Oeuvres alchimiques*, 53–60.

92 *Declaratio lapidis physici Avicennae filio suo Aboali*, in *BCC*, 1:633–36, on 635: "non est aliquid in mundo quod possit stare loco ejus." The *Declaratio* is attributed to Avicenna, but is actually an original Latin composition basing itself on the *De anima* of pseudo-Avicenna with borrowings from the *Turba philosophorum*, see Julius Ruska, "Die Alchemie des Avicenna," *Isis* 21 (1934): 14–51 on 45–48.

93 Gabriele Ferrario, *The Book of Alums and Salts of Pseudo-Rāzī, Sources of Alchemy and Chemistry 2, Ambix* 70 (2023): S20, S121, and S155; Julius Ruska, *Das Buch der Alaune und Salze* (Berlin: Verlag Chemie, 1935), 90, and *Al-Rāzī's Buch Geheimnis der Geheimnisse* (Berlin: Springer, 1937), 104, 113, 117, and 180. The Islamic origin of this term indicates that it is not an "example of Rupescissa's Christian code names" as stated in DeVun, *Prophecy*, 113.

94 For a new critical edition and study of this important text, see Marion Dapsens, *Les* Masā'il Khālid li-Maryānus al-rāhib *dans leurs versions arabe et latine: Éditions critiques et traductions* (Florence: SISMEL, 2024); an English edition is available in Morienus, *A Testament of Alchemy*, ed. and trans. by Lee Stavenhagen (Hanover, NH: Brandeis University Press, 1974), 36–37.

95 Elisa Rubino and Samuela Pagani, "Il *De mineralibus* di Avicenna tradotto da Alfredo di Shareshill," *Bulletin de philosophie médiévale* 58 (2016): 23–87, on 36; pseudo-Aristotle, *De perfecto magisterio*, *BCC* 1:652 and 658. Guillaume Delmeulle has prepared a critical edition and study of this text (with the corrected title *Liber perfecti magisterii*) and demonstrated that it is an original Latin composition rather than a translation from Arabic: *Aux prémices de l'alchimie latine: Édition critique, traduction et étude du* Liber perfecti magisterii (Ph.D. diss., Université catholique de Louvain, 2025). My thanks to him for confirming the presence of this term in the manuscript tradition that predates John of Rupescissa.

96 Petrus Bonus, *Margarita preciosa novella*, in *BCC* 2:64; Michela Pereira and Barbara Spaggiari, *Il Testamentum alchemico attribuito a Raimondo Lullo*, (Florence: SISMEL, 1999), 56 and 506.

critical for making a transmutatory elixir and often described as a clear distillate. John differs from these earlier authors by giving a procedure for preparing the virgin's milk from clearly-named ingredients. At the same time, his use of this name connects his procedure to earlier authorities, while his clarity regarding how and from what to prepare the material is new. John underscores the identity of his product with the virgin's milk of previous authors by claiming that it turns a heated metal sheet white (i.e., like silver), a statement traceable to the pseudo-Aristotle's *Liber perfecti magisterii*, which reports the same effect in similar language.[97] One can imagine readers familiar with the earlier literature being excited by the linkage of this crucial and otherwise mysterious substance to a more or less straightforward description of its preparation (even if they too encountered difficulty at the last step for making it).

Whereas John nowhere else in the *Liber lucis* cites an authority other than pseudo-Arnald, in Operation 4 he rattles off a roster of his predecessors and judges their worth by whether or not they prepared this same virgin's milk. According to him, it was "never recognized by Geber [and] unknown to Avicenna," while Hermes "touched upon it," Alphidius "prepared it."[98] John's favored pseudo-Arnald caps the list of successful and worthy alchemists, for he "taught it."[99] Some later copyists altered both the list of authorities and the judgments made upon them to suit their own conceptions. The fifteenth-century **Mp**^{I} adds (pseudo-)Lull as Arnald's equal, and **Mp**$^{I}{}_{a}$ rehabilitates Geber and Avicenna by citing them positively rather than negatively: "known to Geber and not unknown to Avicenna." The Cologne edition and **Ua**, which are dependent upon these two witnesses, perpetuate these modifications.[100]

John initially asserts that the virgin's milk is the only substance needed for making the philosophers' stone. Subjected to gentle heating in the "philosophical furnace" described in Operation 8, it solidifies, turns black, passes through multiple colors, and finally becomes brilliantly white, yielding the completed white stone able to "convert all imperfect metals into silver more true than if it were from the mine" (ll. 193–94). In Operation 6, part of this white stone is set aside for making

97 Pseudo-Aristotle, *De perfecto magisterio*, 652: "extinguas in ipsa laminas aeris separati et fiet argentum." It is worth pointing out that if the term *lac virginis* in pseudo-Aristotle refers to common mercury, then indeed plunging hot copper plates into it would turn them silvery due to the production of a surface amalgam.

98 If by "Geber" John had in mind the texts of the Jābir corpus such as the *Liber de septuaginta* (rather than the *Summa perfectionis* of the Latin Geber), his claim that the true virgin's milk was unknown to Geber and Avicenna may reflect the change in his fundamental theory of alchemy from organic-material based elixirs advocated by Jābir and pseudo-Avicenna to the mercury-based alchemy of "master Arnald" and other contemporaries; see Sections 2.2.1 and 2.9 on this crucial shift in John's thought.

99 This is probably a reference to the appearance of the term in the *Tractatus aureus*, or *Septem tractatus*, a text attributed to Hermes, possibly originally of Arabic composition. See Moureau, "*Min al-Kīmīyā'*," 109–10 and Michela Pereira, "*I Septem Tractatus Hermetis*," 651–79 in *Hermetism from Late Antiquity to Humanism*, ed. P. Lucentini, I. Parri, and V. Perrone Compagni (Turnout: Brepols, 2003).

100 The inclusion of pseudo-Lull in the list of names given in the Cologne edition once again illustrates the dangers of relying on early modern printed editions of medieval texts; John could not possibly have listed Lull because his name was not attached to alchemical texts until after the *Liber lucis* was written. Thus, if the Cologne reading is naively taken as authentic, an erroneous dating of either the *Liber lucis* or the emergence of the pseudo-Lull results.

silver, and the rest is sealed again in a flask, heated more strongly, whereupon it turns red, thereby completing the chrysopoetic red philosophers' stone.

2.5.3 Separation of the Four Elements: An Alternate Path

Despite this seemingly easy process, John nevertheless provides a more complicated alternative procedure. The "gross combustible" residue left behind after distilling the virgin's milk can be divided into its four elements by distillation. The virgin's milk, which has already been removed, represents the element "water." By redigesting and redissolving the residue and then heating it in an alembic in a water bath, the "air" distills over in the form of a shining (*resplendens*) yellow oil. Stronger heating in an ash bath then liberates the "fire" as a "very red oil," leaving the "earth" behind as a black residue (ll. 149–63). This process of separating the four Aristotelian elements closely parallels the generalized methodology that John provided earlier in *De famulatu* for separating the four elements from all organic materials (e.g., blood, plants, flesh). There he begins with any material "putrefied and reduced to a water," a description that matches what is done to the inorganic sublimate in the *Liber lucis*. In *De famulatu* as in *Liber lucis*, a clear, colorless liquid—the "water"—is distilled off first. After further treatment and heating in a water bath, the "air" distills as a "shining golden oil" (*oleum splendidum aureatum*). Transferred to the greater heat of an ash bath, the "fire" distills as "red oil."[101] The very obvious parallel between the two processes and the descriptions of their products is clear. The general outlines of this process may derive from pseudo-Arnald's *Epistola ad Jacobum de Toleto*, which deals specifically with the distillation of human blood, the same substance that John uses for his example in *De famulatu*, although the Arnaldian text uses the separated elements solely medicinally and not transmutationally.[102] A description of extracting the four elements for chrysopoetic purposes does appear in pseudo-Arnald's *Rosarius*, but that process differs substantially from John's.[103] In all these processes, however, there is surely a dependence ultimately upon Jabirian methods for isolating the four elements from any substance and then recombining them in specific proportions to produce various transmutational elixirs according to the principle of balance.[104] The *Liber lucis* similarly stipulates using the separated elements in a specific proportion for making the stone, according to a process John parallels with the use of the virgin's milk alone, saying that this is the way "some" (unnamed) alchemists produce the stone. The water, air, and fire are to be combined in the

[101] *De famulatu*, "Non pretermittam."

[102] On the *Epistola*, see Calvet, *Oeuvres alchimiques*, 91–100 and 572–79; Thorndike, *HMES*, 3:79–80; and Michela Pereira, "Arnoldo da Villanova e l'alchimia: Un'indagine preliminare," *Actes de la Trobada internacional d'estudis sobre Arnau de Vilanova*, ed. Josep Perarnau, 2 vols. (Institut d'Estudis Catalans: Barcelona, 1995), 2:95–174, on 164–71. The text of the *Epistola* was published immediately following *De famulatu* in the 1561 Basel edition.

[103] See Calvet, *Oeuvres alchimiques*, 310–18: "De divisione lapidis per quatuor elementa."

[104] On the principle of balance, see Principe, *Secrets*, 37–42.

proportion of 8:4:1, sealed in a flask and heated. John is once again unique among all these authorities in that he discards the "earth" as having no value; the others employ it along with the other elements.[105] John suggests heating three flasks all at once, two with the virgin's milk alone—one for the white stone and one for the red—and the third with the mixed elements. The reason given is "so that you may explore the subtleties of the various methods and see the greatest miracles created within things by the glorious God" (ll. 218–19), but it may at least equally be a matter of listing multiple options for future success while the friar remained unsure which pathway would actually work. John's audience of post-antichrist Franciscans might not be as interested in exploring "subtleties" as in following an unproblematic procedure, since their duty will be to make enough gold and silver to rebuild a devastated Christendom.

The extent to which a shift from experience to expectation helps explain alchemical texts more generally can be decided only on a case-by-case basis. Nevertheless, this explanation does resolve the potential tension between the claim that many chrysopoetic authors performed and recorded sophisticated experimentation and notable practical results and the recognition that their final goals, no matter how described, were not always achieved as written. Before anyone should hastily, and rather anachronistically, accuse such writers of dishonesty or dissimulation, it is worth mentioning that a similar procedure remains in use among chemists today. Synthetic organic chemists strive to produce naturally-occurring compounds more easily, more purely, or in greater quantity in their laboratories—a goal which in itself recalls late medieval claims about the power of alchemy both to imitate and to outdo the productions of nature.[106] These modern chemists generate laboratory protocols that require an often-lengthy sequence of steps to construct a product of the desired molecular structure. In some cases, after a synthetic scheme has been worked out and published, a subsequent chemist discovers a shorter or more efficient route to some intermediate compound in the sequence, for example, a pathway that requires only three steps to get to the same point in the sequence for which the earlier route required ten. The later chemist can then "correlate" his own method for producing that intermediate in just three steps with the previously published route and legitimately claim to have created a new synthesis of the natural product without ever carrying out the remaining steps. The first few steps are the new contribution, actually performed in the laboratory, while the remainder do not have to be carried out again because there is an already-published—hence authoritative—description of the rest of the process needed to arrive at the desired product.[107] In both the modern and the alchemical cases, a later experimenter

[105] Compare pseudo-Arnald's *Rosarius*, bk. 2, ch. 24, in Calvet, *Oeuvres alchimiques*, 338–41. There pseudo-Arnald uses different proportions of the elements for the white and red stones. I have not identified a source for the proportion recommended by John.

[106] William R. Newman, "Technology and Alchemical Debate in the Late Middle Ages," *Isis* 80 (1989): 423–45.

[107] One of my last publications in organic chemistry relied on just such a "correlation": Philip Magnus, Martin J. Slater, and Lawrence M. Principe, "Stereospecific Dicobalt Octacarbonyl Mediated Enyne Cyclization for the Synthesis of Methyl Deoxynorpentalenolactone H," *Journal of Organic Chemistry* 54 (1989): 5148–53.

carried out practical processes of a new character until arriving at a correlation point (real or optimistically presumed) with the processes described by an earlier author. The authority given to the earlier author—whether by modern peer-review (which is always dependable, of course) or venerable age and pedigree—insures that this correlated synthesis is as good as done without repeating the remaining steps described by the earlier author.[108]

2.6 Operation Seven and Spurious "Operation Nine": Using the Stones

Operation 7 describes various methods for using the white and red stones to produce silver and gold from the various base metals. One unusual feature here is John's inclusion of iron and steel as base metals to be transmuted; given their very high melting points—temperatures not fully reached in the Middle Ages—they are rarely included among the metals appropriate for transmutation. John once again offers alternative methods for performing the transmutations, citing what various (unnamed) authors have done. Apparently unsatisfied with the friar's instructions, the copyist of π^3 cribbed the chapter on how to carry out transmutation from pseudo-Arnald's *Rosarius philosophorum* and appended it to the end of the *Liber lucis*.[109] This added section describes using the completed transmuting agent on successive portions of mercury in order to produce a precious metal. This added text propagated thence into subsequent manuscript copies through two families of descent, and in divergent forms into *both* printed editions, a feature that further indicates the perils of relying on early modern printed editions rather than a critical edition. The extraneous character of this added section was quickly forgotten in the manuscript tradition as it was integrated seamlessly into the *Liber lucis*; some later witnesses label it as "Operation 9" or incorporate it smoothly into authentic Operation 7.[110]

[108] The modern chemist is, of course, expected to signal the correlation point explicitly, which is not done in the alchemical examples. In John's case, it perhaps does not seem inappropriate or out-of-character for a confident prophet of the coming antichrist equally confidently to predict the future course of an alchemical operation for which there existed authoritative sources.

[109] Pseudo-Arnald, *Rosarius philosophorum*, bk. 2, ch. 31, "Quia grave est mille millium partes simul fundere...," in Calvet, *Oeuvres alchimiques*, 352–54, and *BCC* 1:676.

[110] All witnesses containing this addition can be traced back to π^3, either by direct descent or by contamination. In **P**^{2}, the earliest known descendant of π^3, this pseudo-Arnaldian text *follows* the explicit and colophon for the *Liber lucis* and bears its own heading (fol. 68v: "Quomodo dicta medicina prohiciatur"), thus signaling that it is *not* part of John's text. But **W**^{2}, which was copied directly from **P**^{2}, omits the intervening explicit and colophon, thereby integrating the pseudo-Arnald text into John's *Liber lucis*. This integration occurred similarly in other transmissions, such as in the production of ι^2, all of whose Italian descendants (**L**^{3}, **Mp**^{1}, **Fn**^{2}) contain the added material fully integrated into the authentic text. Likewise, γ' was contaminated by π^3 or one of its descendants, and as a result, all of its descendants contain the added material, although with minor but consistent textual variants from the readings found in descendants of ι^2. These variant readings divide Operation 9 into two types: one reads "rosas (bene) odoriferas" and "ad omne diminutum," while the other reads "rosas utiles et odoriferas" and "ad omne metallum." Descendants of π^3 show the first type, while those originating from γ' show the second.

B, **Or**$_f$, **P**^{3}, and **Pr** contain more or less complete versions of this added section, while **W**^{7} contains a very abbreviated form; **Hb**^2_d and **S** combine it with Operation 7, having omitted authentic Operation 8, and the same is true in **Fr**$_a$; **Ma** likewise runs it into Operation 7 but retains Operation 8 (much abbreviated); **Mu** labels the pseudo-Arnald text as the "eighth part" such that John's own "eighth part" becomes the "ninth part." Since

2.7 Operation Eight: The Philosophical Furnace

The eighth and final section of the *Liber lucis* fulfills the promise made in Operation 3 to describe the construction of a philosophical oven or furnace. This section parallels material presented previously in *De famulatu*, although the *Liber lucis* augments the earlier text and revises some of the measurements it provides (see Appendix 1). As in the case of the other parallel section, some later practitioners inserted lines from *De famulatu* into the *Liber lucis* in order to produce an even fuller description of the construction. That said, the furnace chapter is, along with the prologue, the section most frequently omitted in the manuscript tradition. Gratarolo, the editor of the Basel edition, even though he worked from a manuscript that contained this section, decided to omit it from print because "I have put better [material on the subject] elsewhere above," presumably in reference to his inclusion earlier in the collection of the *Liber fornacum* (*Book of Furnaces*) attributed to Geber.[III] Other omissions may likewise be due to a presumption that furnace technology had improved since John of Rupescissa's day, although physical losses probably also played a role, since the first and last pages of a manuscript are the most likely parts to be damaged or lost during transmission. In just this way, **$\mathbf{Y}^{\mathbf{I}}$** and its descendants all display an awkwardly truncated Operation 8, presumably due to the loss of the final leaf from their earlier exemplar.

In Operation 5, John emphasizes that he designed this oven specifically so that "by a clever technique [*ingenium*] the flask may have an equal fire from all sides... so that the virgin's milk can in no way fly to any part" (ll. 176–79). In other words, the structure of the oven forces the virgin's milk to remain in place, fully subjected to the digesting heat, because there is no cooler part of the flask (as there would be if it were simply heated from below in an ordinary furnace) into which it could retreat by volatilization. Why exactly John thought this feature crucial is a matter for speculation. Two features, however, stand out in his instructions for building this "philosophical oven." First, contrary to what is often depicted or imagined of alchemical furnaces, it is very small, less than two feet (60cm) in height. It could easily sit on a tabletop or workbench, and thus a large workspace would not be needed. Second, the close descriptive detail that characterized the first operations and that vanished at Operation 4 now returns. The furnace chapter gives precise measurements and construction details, making it very easy for readers to follow the description and to replicate the furnace (both in the fourteenth century and

the Basel edition stems from γ′, it mentions the existence of this section although does not reproduce it because "enough [on this topic] can be found elsewhere," *De confectione* (1561), 230. It appears also in the abridged **$\mathbf{Z}_{\mathbf{a}}$**, which was condensed very probably from a member of subfamily γ′. The spurious section appears in other families through various contaminations: **Ge** has it as a later marginal addition taken from a descendant of γ′; **G** and **T** have a modified and shortened version; **$\mathbf{Mp}^{\mathbf{I}}_{\mathbf{a}}$** titles it "ninth operation"; **$\mathbf{Oth}^{\mathbf{2}}$**, interestingly, seems to recognize the origin of the material, heading it (uniquely) "in eodem capitulo et ultima parte de projectione et multiplicatione usque ad infinitum dicit magister Arnaldus..." **Ua**, since it is based on a compilation of **$\mathbf{Mp}^{\mathbf{I}}$** and **$\mathbf{Mp}^{\mathbf{I}}_{\mathbf{a}}$**, likewise includes this section and titles it the "ninth operation," and from **Ua** it passed into the Cologne edition.

[III] *De confectione* (1561), 230; the *Liber fornacum* is in the first volume of Gratarolo's collection *Verae alchemiae* at 193–201. On the furnace chapter in the Cologne edition, see Section 2.7.

now). The same is true of the rather complicated assemblage of tripods and dishes for enclosing the "philosophical egg" that contains the developing philosophers' stone. The purpose of this arrangement is to ensure the "equal heating from all sides" by having the flask "raised up in the air and not touching the walls of the dishes" (ll. 322–23), thus avoiding hot spots through contact with any surface warmer than the air surrounding the flask.[112]

The text explicitly notes the inclusion of illustrative drawings of three parts of the furnace: the base, the iron plate, and the glass cups or goblets. Several early manuscripts contain the first two images in very similar forms. The third is somewhat more problematic; it is present in many fewer manuscripts, and it is not entirely clear from the text whether it was supposed to show the shape of the goblets or their positioning on the iron plate. If the latter, it appears as such in only two witnesses (**P**^{1} and **Br**); if the former, then in none. The absence of the third drawing is especially obvious in **Go**^{1}, where the first two images are drawn in boxes within the body of the text immediately following the words that introduce them, while the third box is left blank. If the third illustration was intended to show the shape of the goblets, it must have been lost early in the transmission, perhaps through abrasion or other physical damage, or never executed. I have included the extant illustrations in the English translation.[113] Manuscript illustrations are of course problematic because many copyists, especially those with mediocre draftsman's skills, simply omit them or leave blank spaces to be filled in later, perhaps by a more skilled hand.[114] In the case of the *Liber lucis*, the detailed textual description of the "philosophical oven" often invited later practitioners to elaborate upon or replace the very simple original drawings. Some endeavored to combine the individual parts depicted with the textual description to produce illustrations of the complete furnace with its various parts and functions labeled. One such depiction is shared by several Italian witnesses (Figure 2.6). There is no evidence that this image descends from a prototype authentic to John; it is presumably an addition by a later copyist that passed down through the manuscript tradition.[115] Some later copyists devised extremely imaginative or even impossible forms that bear little resemblance to John's narrative description (Figure 2.7). While no illustration appears in the Basel edition, the Cologne edition provides two engravings of the furnace, with added text titled "Description and Shape of the Physical Furnace." Neither these images nor

[112] This renewed level of detail does not necessarily imply that John built such a furnace—one can be precise about a design without actually making it a reality—although it is certainly possible that he did build and use such an oven. This situation is unlike his descriptions of the properties and appearances of chemical products, which could not have been predicted without carrying out the processes.

[113] Since the autograph has long since vanished, we cannot know what the original drawings looked like, or even if John actually drew the promised illustrations. However, since the text explicitly states *which* items were illustrated and *where* they were placed in the text one can, with a reasonable level of confidence, choose from among wide variety of drawings present in extant manuscripts the ones that correspond best with the claims of the text.

[114] E.g., **G**, **Go**^{1}, **Le**$^{3}_{cz}$, **Ko**^{1}, **Po**, and **W**^{6} all leave one or more such blank spaces that were never filled in.

[115] Another drawing of the complete furnace with a similar perspective appears with **Z** (fol. 118v) but more stylized and with considerably more descriptive labeling. I have shown elsewhere (Principe, "Missing Link," 195) that the copyist of Ua almost certainly used the Montpellier MS containing **Mp**$^{1}_{a}$ directly.

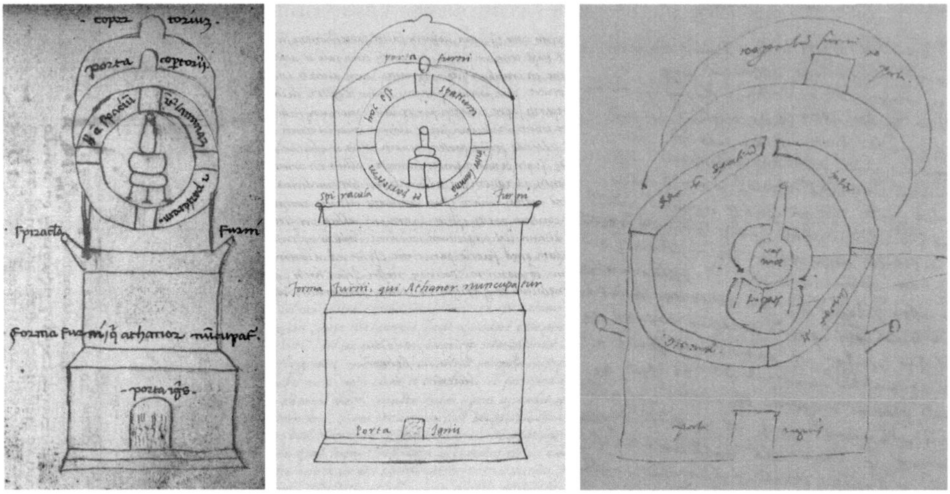

FIGURE 2.6 Three related illustrations of the completed furnace that appear in **Mp^1_a** (1459), **Ua** (ca. 1550), and **L^5** (1563).

the accompanying text has any known counterpart in the manuscript tradition. In fact, neither the images nor the text corresponds with what John himself describes in Operation 8. These engravings were reused in the 1606 edition of Andreas Libavius' *Alchemia*, where they are presented as John of Rupescissa's own designs.[116]

2.8 Alchemical Secrecy and Divine Revelation: The "Credas" and the *Mira ingenia*

Two sections of text entirely or largely omitted from the printed editions deserve special comment. The first is a lengthy section of Operation 6 on alchemical secrecy (ll. 220–51); I refer to this section by its opening word "Credas." This material is often omitted already in the manuscript tradition, like other sections (such as the prologue) that later copyists excised because they did not directly provide practical directions for making the philosophers' stone. The copyist of **Fn^2**, for example, began transcribing the "Credas" but soon gave up, signaling his frustration by writing "and many other things similar to these, not pertinent to the matter at hand, follow."[117] Therefore, the early modern editors need not have been

[116] Principe, "Missing Link," on 195–97.

[117] Fn^2, fol. 158v: "et multa alia impertinentia ad rem sequuntur his similia." Unfortunately, even a great scholar like Thorndike referred to the "Credas" and similarly non-practical parts of the *Liber lucis* as "what might irreverently be called guff," *HMES* 3:368; on the various omissions of the "Credas," see 367 n.65. It is to rather unfortunate expressions such as these, made from the perspective of early twentieth-century historiography, that DeVun would seem to refer when she repeatedly claims (*Prophecy*, 109, 127, 153, et passim) that historians of science ignore or dismiss the religious and theological content of John of Rupescissa's work (and of alchemy generally). Such criticism is not, however, justified for present-day historians of science. Amusingly, and not insignificantly, given what I note here about the editing out of the "Credas," prologue, and other theologically

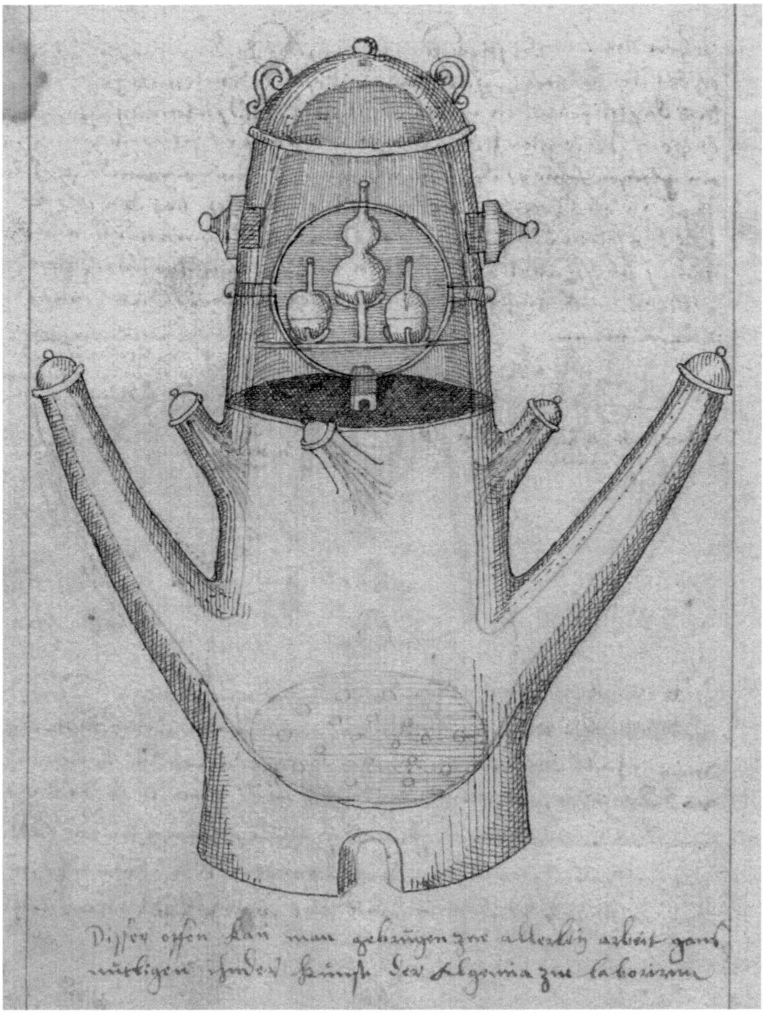

FIGURE 2.7 An extravagant interpretation of John of Rupescissa's philosophical furnace in **Ka**$^{3}_{d}$ (1576). The copyist, Wilhelm Vernuken (ca. 1540–1607), was a master-builder and sculptor employed at Kassel.

directly responsible for the omission from their published editions; this section may already have been missing from the manuscripts they used. This situation is demonstrably the case with the Cologne edition, which follows **Ua** nearly word for word, and that manuscript contains only about 20% of the "Credas."[118] Nevertheless, the "Credas," now restored in the critical edition, crucially resituates the *Liber lucis* within John's larger literary and theological program.

charged sections of the *Liber lucis* during its transmission in manuscript, it is clear that a great many late medieval and early modern alchemists *themselves* actively separated the theological from the practical in favor of the latter.

[118] Ua, fols. 53r–v. The Basel edition contains slightly less of the "Credas," *De confectione* (1561), 2:229.

The "Credas" begins by repeating the claim John made in the prologue that he has written more clearly than any previous author, a typical alchemical topos. He notes how the "great philosophers" to whom the secrets of the stone had previously been revealed invoked horrible curses upon those who would dare to reveal them openly. Why then has John risked putting stylus to parchment to produce the *Liber lucis*? He states that his actions do not incur such curses because he reveals this precious knowledge only "to the holy mystical body of Jesus Christ, the Roman Church" rather than indiscriminately to "wicked men" (ll. 224–26). His is thus a holy action, providing powerful knowledge only to the holy ones—meaning in particular the true Franciscans—to help them during and after the coming tribulation of the antichrist. John argues his case by incorporating a wealth of Biblical quotations in a manner typical of his prophetic writings, and concludes by putting his trust in divine power and wisdom, commending the *Liber lucis* "into the hand of the most holy Crucified One, our Lord Jesus Christ from whom I received this good" (ll. 241–42) who will protect it from wicked abusers. He specifies, once again, that his book is for the use of the "poor men of the Gospel" (*evangelici pauperes*), but warns his future brethren not to allow themselves to be seduced by the allure of wealth, and to use this privileged knowledge of transmutation only to "help the afflicted Church and the holy ones in the time of the tribulation" (ll. 250-51). John thus fulfills the qualified promise he made at the end of the first book of *De famulatu* to write about the secrets of transmutational alchemy specifically for his fellow faithful Franciscans.

The concluding line of the *Liber lucis*, entirely absent from both printed editions, provides one of the most concentrated expressions of the integral connections between alchemy and prophecy, between natural philosophy and theology, in John of Rupescissa's thought:

> Behold the marvelous techniques [*mira ingenia*] openly revealed in this time of present tribulation which in other generations were not known to the sons of men, as are now given to be known to the faithful poor ones, the sons of the holy Roman Church and the prelates of the same Church.... (ll. 324–27)

This final line clearly refers to intellectual and technological progress. The *mira ingenia* that John reveals in his book—his furnace design, his method for making the stone without gold, his techniques for revealing the hidden composition of his products, and presumably also his quintessences in *De famulatu*—are all *new*. They were previously unknown to the "sons of men." While concepts of progress are variable and not widespread in the Middle Ages, they do occur critically in one source of great relevance to John of Rupescissa, namely the progressive and melioristic vision of history proposed by Joachim of Fiore, the twelfth-century Calabrian theologian whose writings so deeply influenced Franciscan thought and eschatology, especially among the Spirituals. Both Joachim and Peter of John Olivi, who further developed Joachim's ideas, played fundamental formative roles in John's thought. Without delving into the complexities of Joachim's vision of history, it suffices to note that he argued for three overlapping developmental epochs of time (in his terminology, three *status*) related successively to the Three Persons of the Trinity: Father, Son, and Holy Spirit. According to this scheme, spiritual understanding would gradually increase among the people of God, reaching its summit in the epoch of the Holy Spirit and particularly during and after the time of the antichrist's appearance and downfall. Joachim predicted the creation of a new order of "spiritual men" (*viri spirituales*) who would play a critical role in this age of increasing understanding. The Franciscans, whose Order was founded shortly after Joachim's death, and especially the rigorists, saw themselves as the fulfillment of this expectation.[119]

The new revelations of John's *mira ingenia* follow the concept of progressively increasing understanding set forth by Joachim and Olivi. But whereas Joachim's *intelligentia spiritualis* is directed specifically toward allowing future Christians to understand more of the complexities and hidden meanings within Scripture, John here extends this view of increasing knowledge to include science and technology. In preparation for the arrival of the antichrist, "in this time of present tribulation" as the friar writes, God has revealed to John the new natural philosophical and technological discoveries that he in turn reveals to readers of the *Liber lucis*. John's sudden insight, the *intelligentia spiritualis* he received while chanting the *Te Deum* at Aurillac in 1340 that allowed him to understand the prophecies of Daniel, is fundamentally related not only to Joachim's own experiences of sudden understandings of Scripture and salvation history, but equally to the event mentioned at the end of the first book of the *De famulatu* when the transmutational secrets of the *Liber lucis* were revealed to John "through sublime theories and heavenly illuminations."[120]

But it would be a serious mistake to imagine that John was a passive receptacle of such understanding in either the prophetic or the alchemical case. He himself contradicts such an interpretation by twinning "sublime theories" acquired by

[119] The literature on Joachim's theory of history is extensive: a brief selection includes Marjorie Reeves, *The Influence of Prophecy in the Later Middle Ages: A Study of Joachimism* (Oxford: Clarendon Press, 1969), esp. 16–27 and 175–228; Bernard McGinn, *The Calabrian Abbot: Joachim of Fiore in the History of Western Thought* (New York: Macmillan, 1985), esp. 123–57; David Burr, "Bonaventure, Olivi, and Franciscan Eschatology," *Collectanea Franciscana* 53 (1983): 23–40 and "Olivi, Apocalyptic Expectation, and Visionary Experience," *Traditio* 41 (1985): 273–88; and especially with regard to John of Rupescissa, Lerner, "Historical Introduction," esp. 36–50. A very useful introduction is Bernard McGinn's exposition of Joachim's thought in *Apocalyptic Spirituality* (New York: Paulist Press, 1979), 97–112. For concepts of progress in the Middle Ages, see Alistair Crombie, "Some Attitudes to Scientific Progress: Ancient, Medieval, and Early Modern," *History of Science* 13 (1975): 213–30; George Molland, "Medieval Ideas of Scientific Progress," *Journal of the History of Ideas* 39 (1978): 561–77; Chiara Crisciani, "History, Novelty, and Progress in Scholastic Medicine," *Osiris* 6 (1990): 118–39; and J. R. Webb, "'Knowledge Will be Manifold': Daniel 12:4 and the Idea of Intellectual Progress in the Middle Ages," *Speculum* 89 (2014): 307–57.

[120] *De famulatu*, "Excusacio": *per sublimas theoricas et illuminaciones celestes*. On the *intelligencia spiritualis* in John, see Lerner, "Historical Introduction," in *LSE*, 36–50, esp. 49–50.

reading with the "heavenly illuminations" that together helped him understand the mysteries of chrysopoeia. Likewise, Joachim did not derive his system from his three (at least) "illuminations" alone: those moments of insight arrived amid intense study.[121] The *intelligentia spiritualis* acts to solve a problem that already exists, to unravel knots already deeply under consideration by the recipient. To think otherwise would be to adopt a view of divine intervention that is at odds with the bulk of orthodox medieval theology and spirituality. God co-operates with the recipient, rather than operating unilaterally. Geber expresses a similar view of God's cooperation with the alchemist's labor: "may the sublime God of natures, blessed and glorious, be praised, He who revealed the order of all the medicines [i.e., transmuting agents] to us with the experience that we diligently sought out by the soundness of this investigation and the application of our labor."[122] It would thus be wrong to conclude that since John received divine "illumination" he had no need of practical experimentation and was thus a sort of an "armchair" alchemist. Indeed, quite the opposite is the case, even in his prophetic writings. In those writings as well as in the *Liber lucis*, John repeatedly expresses his sentiment that "experience teaches," that is to say, experience gained through studious labor verifies or annuls the interpretations and predictions arrived at through "insights" alone.[123] It would clearly be absurd, not to mention dangerously heretical, for a recipient to believe that divine illuminations obviate the need for experience with the Holy Scriptures and their careful interpretation. In John's case, his alchemical endeavors clearly included reading a range of alchemical authorities, puzzling over their enigmas and contradictions, but depended critically, as the experimental reworking of this study shows, upon engaging in practical endeavors with materials and apparatus in a workshop laboratory.

This crucial line citing the *mira ingenia* contains yet more embedded meaning. John builds his final sentence for the *Liber lucis* around a word-for-word quotation from St. Paul's Letter to the Ephesians, using a verse that speaks of "the mystery of Christ, unknown to men in former ages but now revealed by the Spirit to the holy apostles and prophets."[124] John's choice of this particular verse implicitly makes multiple significant claims. First, it implies that his "marvelous techniques" of alchemy are "revealed by the Spirit." This celestial origin radically sanctifies such scientific and technological knowledge, and in doing so directly opposes the Franciscan Order's multiple prohibitions against alchemical study by its friars. Just as God inspired John to embrace radical poverty, to interpret and understand

[121] Reeves, *Prophecy*, 20–23.

[122] Newman, *Summa perfectionis*, 588. The same cooperative features of divine revelation in alchemy are pointed out in the context of the seventeenth-century chymists Joan Baptista Van Helmont and George Starkey in Newman and Principe, *Tried in the Fire*, 197–205.

[123] Cf. *Liber lucis*, ll. 15, 73, and 262; *LO*, 389; and *VM* (Tealdi, 222 and 224; Kaup, 148 and 155).

[124] Eph 3:5: "quod aliis generationibus non est agnitum filiis hominum, sicuti nunc revelatum est sanctis apostolis ejus et prophetis in Spiritu"; compare *Liber lucis*, ll. 324–26: "Vide mira ingenia patule revelata in tempore isto tribulacionis instantis quod aliis generacionibus non est agnitum filiis hominum sicut nunc datum est agnosci pauperibus fidelibus."

Scripture, and to preach about the coming antichrist much to the annoyance and frustration of his superiors, He also gave the friar insight to interpret, understand, and practice alchemy, an activity likewise unwelcome to his superiors. John's use of this verse even places the *mira ingenia* of alchemical and technical knowledge in parallel with the central mystery of Christianity and the gift of God to all people, powerfully underscoring the sanctified character of such knowledge. The verse simultaneously situates John and the future true Franciscans, the "faithful poor ones" for whom he is writing, in the position of the "holy apostles and prophets" of the Pauline text, sanctifying not only the knowledge but its possessors as well. Finally, in the immediate context of this Biblical verse, the *minimus apostolorum* speaks of his mission to the Gentiles and refers to himself as "Paul, a prisoner for Christ Jesus" (Eph 3:1). Thus, the friar's choice of this particular line sets up yet another implicit parallel, this time between St. Paul and himself, both of whom were writing from unjust imprisonments due to the disruptive yet inspired preaching that was required to fulfill the mission they both had been given—St. Paul to the Gentiles, John to the Roman Church.

When its full background is recognized, the last line of the *Liber lucis* encapsulates much about John of Rupescissa and his thought. It expresses his embrace of a Joachite vision of progressive history and expands that vision to include progress in technology and natural philosophy. It sanctifies such knowledge both by giving it a divine origin and by juxtaposing it with a central theme of Christianity, and it parallels John's dual alchemical and eschatological mission with St. Paul's mission to the Gentiles. For modern readers, it provides a concentrated expression of the interpenetration of John's alchemical, exegetical, and prophetic endeavors and underscores again how the *Liber lucis* could not have been written by any but a Franciscan. It is hard to imagine a more richly meaningful line that John could have produced for the conclusion of his *Liber lucis magisterii magni*.

2.9 Imprisonment and Practice: Where, When, and How Did John Practice Alchemy?

The text of the *Liber lucis* indicates that John was not an "armchair alchemist," but did in fact carry out the processes he describes, at least those of Operations 1–3, that is, up to (and probably including) the liquefaction of the sublimate in the humid oven at the start of Operation 4. The evidence for this conclusion comes primarily from his detailed descriptions of the visual aspects of the processes and products which have been confirmed by my experimental reworking. John could *not* have predicted these observables *a priori*, and they are not present in any other known texts from which he could have taken them. John's precise method for mixing mercury with the salts, his weight measurements, his description of the separation of the sublimate into its components, his choice of the strikingly appropriate term *marcidus* for the recovered mercury, his observation of the "black ferrous scum" that appears when the vinegar solution is heated (l. 70), his comparison of the separated "tincture of redness" to the color of a flower, his details about distilling aquafort and observation of the resulting red gas in the alembic head, his use of sal ammoniac to increase the solubility of the sublimate, as well as the several optimizations of the processes, all point unambiguously toward direct and repeated practical experience. Such observations and operational innovations could come about only in the course of practical experimentation.[125]

The question of timing for such experimentation is the first concern: *when* did John carry out the processes he describes in the *Liber lucis*? I have already noted the fundamental shift in John's understanding of alchemical theory, namely, his transition from the older organic-based elixirs characteristic of Arabic alchemy to the metal- and mineral-based chrysopoeia of recent Latin authors. This shift can be cleanly dated to the years between his *Oracle of Cyril* commentary, written predominantly between 1347 and 1349, in which he follows the organic alchemy of authors like Jābir, pseudo-Avicenna, and Roger Bacon, and his *De famulatu*, composed in 1352 (or very late 1351), in which he provides a brief version of the process more fully expounded in the *Liber lucis* in 1354 based in the metallic alchemy of more recent Latin authors like Geber and pseudo-Arnald.

Since the *Liber lucis* process is based on an alchemical theory John learned and adopted after 1347–49, the procedure could not have predated those years; it could not, for example, have been the result of alchemical activity in his youth before becoming a Franciscan or during the early years of his mendicant vocation. This conclusion certainly does not rule out the possibility that John was already interested in, and might even have pursued, transmutational alchemy at an early date—he clearly had experience with precious metal assaying and how to work with chemical substances by the time he wrote his lost *Liber revelationum secretorum in ponderibus rerum* cited in the Oracle commentary—it means only that the *specific* process described in the *Liber lucis* must have been worked out *after* 1349. By 1352, John had found time and opportunity to work out at least the outlines of the practical process cited in *De famulatu* based on that new perspective. These considerations thus place his conversion to the more modern Latin alchemy, and his experimental practices based upon it, in Avignon, where he arrived in August 1349.

This localization of John's alchemy, both theoretical and practical, in time and place (1349–54 at Avignon) presents an obvious question and a potential problem: could John have engaged in alchemical practice while incarcerated in the papal prison? While this possibility might at first seem outlandish, a closer examination of John's imprisonment shows otherwise. Multiple Rupescissa scholars have already pointed to his substantial authorship throughout the years of his confinement as evidence that his imprisonment could not have constantly been as

[125] It might be possible that John observed these processes being carried out by another practitioner rather than putting hand to flask himself, but in that case he would have been so close to the processes that he would qualify as a collaborator. Even if this is the case, it would not change the need to answer the historical questions that follow here about the time and place at which John performed (or witnessed) the processes of the *Liber lucis*.

deprived and dire as he describes.[126] He must, at least at times, have had access to books as well as ink, parchment, and space to work, read, and write. Indeed, he mentions "a certain place where I studied" and times "when I was seated in order to study" while confined in Avignon, even during what he describes as his worst persecution while in Avignon by the lunatic Simon Legat.[127] The official records of his confinement say very little—they are terse and rather ambiguous—and therefore the conditions of his imprisonment are almost entirely self-described. As such, they have both a contextual purpose and a self-fashioning value, and they draw upon Biblical topoi as much as upon actual personal experiences.[128] For example, John writes that when he was first imprisoned by Farinier at Figeac, he was thrown into a cell "in soft mud." It is certainly possible that John's cell had a wet dirt floor—he mentions that at the time "the whole friary was flooded"—but this description simultaneously ties John's experience to that of the prophet Jeremiah, who after prophesying the destruction of Jerusalem was cast "into a cistern where there was not water but mud, and Jeremiah sank into the mud" (*in lacum non erat aqua sed lutum descendit itaque Hieremias in caenum*).[129] Indeed, a little later in the same text John describes himself as a "type of Jeremiah," and in the *Vade mecum* he cites this very same account of Jeremiah in the cistern while explicitly linking that prophet's suffering and persecution with his own.[130] In other places, he identifies himself through his suffering with Daniel, and in the *Liber lucis* (as just mentioned) with the imprisoned St. Paul.

The fullest, most vivid, and most frequently cited description of John's sequential imprisonments over the course of twelve years from Figeac to Avignon appears in the ninth treatise of his *Liber ostensor*, written in 1356. Crucially, however, these accounts of his suffering occur within the context of his endeavor to prove that he is the *denunciator* predicted by Joachim of Fiore, who would foretell the arrival of the antichrist.[131] Several features were supposed to identify this "Announcer": he would be treated as foolish and few would believe him, he would be able to interpret Daniel, and most of all, he would undergo severe tribulations and punishments. For each descriptor, John argues that his own experiences match those predicted of the *denunciator*. Hence, describing his imprisonment and suffering in the most extreme terms serves to strengthen his claims about his own identity and thereby justify his entire prophetic program. This is not to say that

[126] To describe John's prodigious literary output, Robert Lerner coined the word "megalopolygraphy" in "John the Astonishing," *Oliviana* 3 (2009): http://oliviana.revues.org/335

[127] Robert E. Lerner, "Life, Works, and Fortune of the Prophet Immured," 25–39 in *VM* (Tealdi), on 35–37, and *LO*, 524; see also Elena Tealdi, "Political Propheticism: John of Rupescissa's Figure of the End Times Emperor and its Evolution," in *Cultures of Eschatology*, Vol 1: *Empires and Scriptural Authorities in Medieval Christian, Islamic and Buddhist Communities*, ed. Veronika Wieser, Vincent Eltschinger, and Johann Heiss (Berlin: DeGruyter, 2020), 441–63, on 443, note 10.

[128] See also Ciola, "Giovanni di Rupescissa: autobiografia, profezia e leggenda," 557–75.

[129] *LO*, 517 and Jer 38:6; cf. *LSE*, 137.

[130] *LO*, 537 and *VM* (Tealdi), 220–21 and (Kaup), 143–45.

[131] *LO*, 509–37; on the *denunciator*, see Gian Luca Potestà, "Il profeta degli Antichristi, del reparator e del millennio," 41–60, in *VM* (Tealdi).

John did not truly suffer at any point during his various confinements—he surely did—but only that it would have undercut his broader claims had he not emphasized or even exaggerated the worst periods of it and extended those times to the entirety of his varied confinements.[132]

Several remarks within his writings hint at quite a different condition than his own descriptions depict. In the *Vade mecum*, John recalls a visit in 1356 from the addressee of that work, the Franciscan physician Pierre Périer, and their lengthy conversation over dinner, even though John says this event took place *in carcere*.[133] It is hard to imagine such a meeting and meal occurring in the vile environment he describes in the *Liber ostensor*. More strikingly, in the *Liber ostensor* John recalls how, around Christmas 1350, he held a conversation about recent appointments of new cardinals with Hélie Talleyrand de Périgord (1301–1364), Protector Cardinal of the Franciscans, in his *camera paramenti* at the Petit-Palais in Avignon, that is, in the antechamber to the Cardinal's private quarters where visitors were received. The future Franciscan archbishop of Palermo, Roger de Palheriis (d. 1361), was also present.[134] John himself reports that this event took place *cum haberem arrestum per villam*, a somewhat obscure phrase since the exact legal meaning of *habere arrestum* is not clear. John however uses the same term *arrestum* again just a few pages later when he describes the period of time in 1345 between two bouts of imprisonment when, thanks to an agreement with Farinier, he was released from the muddy cell in Figeac and put "sub arresto" in the friary at Saint-Junien where for five months he "walked freely among the friars" before being imprisoned again. Thus, *arrestum* in John's usage would seem to describe a restriction considerably looser than confinement to a cell, seemingly a sort of house-arrest, which in turn suggests that his *arrestum per villam* of 1350 means that he was confined to the town of Avignon, but free to walk about on his own recognizance within that restriction, as he was when *sub arresto* within the Saint-Junien friary.[135] A similar terminological ambiguity clouds the term *relaxatus* that the official record of the papal prison uses to describe what was done with him a few months earlier on 30 October 1350, after the end of the trial that absolved him of heresy. In similar contexts, this term often means "released to secular authorities," generally for execution; but John obviously was not executed, yet neither did he leave the papal *carcer soldani*. It thus seems to describe a reduced level of confinement, and this "relaxation" of his confinement is presumably the *arrestum* that John states he was under when he visited Talleyrand less than two months later.[136]

[132] The worst periods seem to have been in several of the Franciscan prisons and possibly at Avignon in 1354 when Guillaume Court reexamined him for heresy.

[133] *VM* (Tealdi), 219; *VM* (Kaup), 140–41.

[134] *LO*, 181; in note 3, the editors of *Liber ostensor* interpret this passage to mean that John was in some sort of "prison ouverte."

[135] *LO*, 518; John's usage may be colored by the vernacular *arrester*, in the sense of staying in a place or restricted in movement.

[136] Vatican Archives, *Collectoriae* 50, fol. 91v; see *LO*, 3; Lerner, "Historical Introduction," 30 n.43, and "Life, Works, and Fortune," 33 n.23; and Gian Luca Potestà, "'Gli spiriti dei profeti sono soggetti ai profeti':

John claims in 1356 to have been imprisoned for seven years in the *carcer soldani*, that is, since his appearance before Clement VI in October 1349, and he provides extended descriptions of his tribulations there, but with few indications of what the physical structure of the place was actually like.[137] Fortunately, there are other accounts of the "Sultan's prison," also called the *carcer hospitii papalis Avinionensis* in official documents. Most importantly, it was not part of the papal complex, but was instead a large house in the town of Avignon rented by the papal authorities. The papal expense accounts preserve an inventory of its interior spaces dating from 1322: its lower part contained "seven rooms, a courtyard [or large hall], a well, a stable for two horses, a little dwelling below the stairs, [and] latrines near the door." Above this level, the house contained "three rooms, a kitchen, a supply-room with ambulatories and latrines, a large hall, [and] a little room with a dovecote."[138] The same large townhouse appears to have remained in use for the same purpose until well after John arrived there, since the rent for the *carcer soldani* was still being paid to the same landlord in 1355 as when the above description was recorded in 1322.[139]

Since the warden, or *soldanus*, who had direct oversight and responsibility for the locale, also lived in the house, it is quite possible that John could have had similar, reasonably comfortable (at least for a strict Franciscan) quarters there.[140] He was in the *carcer soldani* in 1349 when Guillaume Court commanded him to write the *Liber secretorum eventuum*, which John completed in just two months, and it would have made little sense for Court both to request a lengthy written document and at the same time keep him under conditions that would make writing difficult. Living conditions could only have improved after he was *relaxatus* in 1350. Since some of the interior spaces of the house were shared, John could at the same time also have been subjected to the troublesome (and occasionally murderous) Legat more or less as he describes. As for a place to perform the alchemical processes described in the *Liber lucis*, the upstairs kitchen would have readily sufficed since the only requirements for the process of the *Liber lucis* would be a source of heat

Da Giovanni di Rupescissa a Pietro Galatino," 47–59 in "*Per una severa maestra": Dono a Daniela Romagnoli*, ed. Vittorio Scotti Douglas (Fidenza: Mattioli, 2014), 47–59, on 48 n.4.

137 *LO*, 522.

138 Karl Heinrich Schäfer, *Die Ausgaben der apostolischen Kammer unter Johann XXII* (Paderborn: Schöningh Verlag, 1911), 411–14 and 608: "in parte inferiori 7 camere, curtis, puteus, stabulum pro 2 equis, domuncula sub scalis, latrine prope portam; in parte superiori 3 camere, coquina, gardaspensa cum deambulatoriis et latrinis, aula, camerula cum columbario." In 1365, when John was being tended in the Franciscan friary at the expense of Urban V, the records state that he was "once held captive in the palace at Avignon" (*olim detinebatur captus in palacio Avin.*) rather than in the *carcer soldanus*; however, this late record might be mistaken about John's earlier confinement or might signal that John was moved into quarters in the papal complex sometime after 1356. He is reported to have been seen there in the early 1360s by Heinrich Egher de Kalkar (see Section 1.1). Karl Heinrich Schäfer, *Die Ausgaben der apostolischen Kammer unter den Päpsten Urban V und Gregor XI* (Paderborn: Schöningh Verlag, 1937), 330–34.

139 Karl Heinrich Schäfer, *Die Ausgaben der apostolischen Kammer unter Benedikt XII, Klemens VI, und Innocenz VI* (Paderborn: Schöningh Verlag, 1914), 598.

140 Schäfer, *Ausgaben unter Johann XXII*, 608: "habitat in eo Gasbertus de Cagiarto Soldanus." When John arrived, the warden was Bernard of St. Margaret, 412.

like a hearth or stove and the requisite apparatus: a glass alembic and cucurbit as the only specialized equipment, along with completely quotidian culinary items like dishes, bowls, pans, bottles, funnels, goblets, and a mortar and pestle. All of his ingredients were readily available—the most exotic of them, mercury, could be obtained from an apothecary or gilder, while green vitriol was available from tanners and inkmakers, saltpeter from picklers, table salt and vinegar from the kitchen larder. The only drawback in using such a domestic space for his chrysopoetic endeavors would be the very real danger of poisoning himself or others with the highly toxic white sublimate that is central to the process, but John was clearly aware of that danger.[141]

This situation, considerably less dismal than John himself portrays, is further supported by a consideration of *why* John was kept confined in Avignon after he was acquitted of heresy; it was not for punitive purposes. As Lerner has argued, Clement IV's evident interest in the friar who appeared before him at the 1349 consistory can be explained with reference to the dire times during which John arrived.[142] Amid the unprecedented horror of the Black Death, famine, earthquakes, and war, John claimed to know the secrets of things yet to come and how the events of the present day fit into a broader apocalyptic vision. Thus, we can imagine that the ecclesiastical authorities, once convinced that John was an "an outstanding master of astronomy, theology, and philosophy" and not a heretic, considered it useful to keep him around as a source of knowledge and insight. This purpose aligns with what the notarial record of John's first days in Avignon states in its closing lines. The first part of the line records how he was treated after his initial interrogation: *acceptus de ca[r]cere et positus in custodiam ar[c]tam et certam*.[143] Here again, there are several possible nuances to the term *custodia*, but the anglophonic reflex to interpret it as "custody" (in the sense of imprisonment) seems less appropriate here than "protection," or perhaps more technically, "protective custody."[144] The line should thus be read as John "was taken from prison and placed under close and sure protective custody," which presumably includes not only the protection of a valuable resource, but also protection of that resource from the Franciscan authorities keen to have a troublesome and embarrassing friar back under their control. Given Farinier's actions three years earlier after John was set at liberty at Toulouse by the inquisitor Jean de Moulins, this provision would not have been unwelcome to John, especially in the light of the four attempts to get him back during Court's investigations of 1349–50. John says as much himself, expressing relief not to be "handed over to the prisons of the Friars Minor as they

[141] John alludes to its toxicity in *Liber lucis*, l. 304.

[142] Robert E. Lerner, "Prophet and Pope: John of Rupescissa confronts Clement VI," *Franciscana* 23 (2021): 97–122.

[143] Wolfenbüttel, Cod. Guelf 876 Helmstedt, fol. 1v; Bignami-Odier, *Études*, 225.

[144] This is also the sense in which the early German translator read the text, rendering the phrase as "gesaczt...in hut"; *John of Rupescissa, Vade mecum in tribulatione Translated into Medieval Vernaculars*, ed. Robert E. Lerner and Pavlina Rychterová (Milan: Vita e Pensiero, 2019), 322.

requested, but to remain in the *carcer soldani* under the protection [*protectio*] of the apostolic See."[145] There was probably also a parallel concern on the part of the authorities that it would be safer to keep the friar close at hand in case his bold proclamations strayed beyond those of a *fantasticus* in ways that would disrupt order at an already troubled time.

The following clauses of the notarial record strengthen the reading offered above when they provide the reason *why* John was put under protective custody: *ut majora ab ipso experientur Cum vero ab ipso plura percipientur*, that is, "so that greater things will be found out by him, since indeed many things will be understood by him."[146] Indeed, this is what in fact happened in the following years, as various cardinals, prelates, and others sought John out for his advice and insights. Moreover, the wide circulation of his writings that was made possible by his restriction to the papal city, thanks to the ceaseless comings and goings of diplomats, travelers, and visitors, fulfilled the notarial record's virtually prophetic last line about John's special understanding: *per totum divulgabitur mundum*, "it shall be spread throughout the whole world." Given this reassessment of John's existence in the *carcer soldani*, the notion of his practicing alchemy while in confinement is not at all implausible. In fact, his alchemical work—both the medicinal distillations of *De famulatu* and the chrysopoeia of *Liber lucis* might be considered to fall under the broader heading of the "greater things" that "will be found out by him."[147] To be sure, given the testimony of the now one hundred and six known witnesses of his *Liber lucis*, his original adaptations of and contributions to the medieval Latin alchemical tradition were indeed "spread throughout the whole world."

[145] *LO*, 522; the *misericordia* John says he found "in the sight of Clement VI" should not be read ironically.

[146] Cod. Guelf 876 Helmstedt, fol. 1v; Bignami-Odier, *Études*, 225–26. In this case, the German translation renders the purpose clause as "umb das noch mer ding von in erfar."

[147] At this point, I defer all discussion or hypothesis about how much of the material in *De famulatu* may have come from John's own material practices; this topic will be taken up in the introduction to the critical edition of that work now in preparation.

3 THE TEXT'S TRANSMISSION IN MANUSCRIPT AND PRINT AND ITS USERS

3.1 The Manuscript Witnesses

The text of the *Liber lucis magisterii magni* survives today in a large number of manuscripts dating from the start of the fifteenth century to the end of the eighteenth. In the 1930s, Lynn Thorndike gave a useful census of the eleven manuscripts of the *Liber lucis* known to him, noting that they varied substantially in content. He further provided a comparison of the different versions of the prologue as given in several of these manuscripts. He also observed that the two sixteenth-century printed versions differed significantly from the manuscripts and that "the printed versions are not alike."[1] In 1981, Robert Halleux expanded Thorndike's inventory in the context of an article on John's alchemy; he listed fifty-five manuscript witnesses.[2] In preparing the present critical edition, I first inspected all of the manuscripts listed by Halleux and found that eleven of them were not, in fact, the *Liber lucis*. Some were later compositions that coopted portions of text from the *Liber lucis*, sometimes quoting no more than a line or two; others were simply misattributed in the manuscript catalogs upon which Halleux relied.[3] In addition to the 44 authentic manuscript witnesses remaining, I then located a further 62, thus more than doubling their total number to 106, a surprisingly large number that indicates how extremely popular and widespread the work really was over the course of at least three centuries (see the *conspectus siglorum* for a full listing, Section 4.1). More witnesses are certain to come to light in the future. Since so many witnesses are anonymous, misattributed, or appear under alternate titles, it is virtually certain that many presently unrecognized or uncatalogued copies survive in various repositories. The extensive descriptions given in Section 6 of the textual variants that characterize the various families of witnesses will help other scholars quickly situate newly discovered witnesses in the lines of transmission shown in the *stemma codicum*.

Of the currently known witnesses, 75 are Latin (67 are complete, 8 are either extracts or fragments), 6 are epitomes, and 25 are vernacular translations.[4]

1 Thorndike, *HMES*, 3:365–69 and 734–40, and 4:645.

2 Halleux, "Rupescissa," 282–84.

3 I have listed and briefly described all of the manuscripts I removed from Halleux's list, as well as other misidentifications of the *Liber lucis*, in Appendix 2.

4 For more on the translations, see below, Section 4.1.3. Here "extract" means a witness *intended* to contain only a fraction of the text (**Km**, $\mathbf{L^1}$, $\mathbf{Oth^1}$, $\mathbf{W^8}$, and $\mathbf{X^{3b}}$), while "fragment" refers to a surviving portion of what was presumably a full copy originally (**H** and $\mathbf{L^2}$), or one that the copyist did not complete (**O**). Many witnesses omit the prologue and/or eighth operation, and a few break off before the conclusion; these are considered "complete."

DOI: 10.1201/9781003666547-4

Of the Latin witnesses, 43 date from the fifteenth century, 26 from the sixteenth, 5 from the seventeenth, and one from the eighteenth. Only three of these Latin witnesses were copied directly from a printed edition (**E**, **Lu**, and **W**^{5}), while a fourth (**Go**^{2}) combines elements from the two disparate editions. No unambiguously fourteenth-century witnesses have been located. The two documents (**P**^{I} and **Oth**^{2}) that have previously been claimed to date from the fourteenth century were misdated. **P**^{I} is certainly one of the earliest witnesses, but it is probably early fifteenth century rather than late fourteenth, while **Oth**^{2} was substantially misdated—it is definitively datable by watermark evidence as no earlier than the 1520s.[5] The two earliest explicitly dated witnesses (**B** and **Ko**^{2}) were produced in 1423 and 1428, respectively, and it is not possible to determine definitively which of the other early fifteenth-century witnesses (**P**^{I}, **P**^{2}, **L**^{3}, and **Y**^{I}) is the oldest, although **P**^{I} is certainly the most complete and reliable of them since the other three show substantial lacunae.

There are several plausible reasons for the lack of fourteenth-century witnesses. As a predominantly practical text, the earliest copies may simply have been "used to death." Moreover, since the text is quite short, it would have existed initially as a very thin gathering of only three to five sheets, and such small items are much more likely to be damaged or lost than large ones. It would have taken time for the text (and/or author) to gain sufficient notoriety to be copied into collections of alchemical texts, which, being larger in format as well as inherently more valuable, are more likely to have survived. Further, most copies of the *Liber lucis* are rather scrappy productions, that is, they were produced on paper by non-professional copyists, and often rather hastily—only 7% of the known witnesses are on parchment, and only a fraction of those were produced by a professional copyist. As a point of contrast, about a third of the roughly 250 known witnesses of *De famulatu* were produced professionally on vellum, presumably for physicians who could afford such luxury, and show less evidence of use. As relatively cheaper and less attractive documents, witnesses of the *Liber lucis* were less likely to be preserved. It is also worth mentioning the claim found in the colophon of several witnesses descending from γ' (produced prior to 1423) that the work "lay hid for many years, and would that it may remain hidden forever from evil greedy men and especially tyrants."[6] It might possibly be the case that the work began to circulate seriously only some years after its composition in 1354, but such claims of "long-hidden manuscripts" are not infrequently encountered in alchemical literature.

The six witnesses that preserve abbreviated Latin versions of the *Liber lucis* represent four different abridgments dating from the fifteenth century.[7] The most common version (transmitted by **Z**$_{a}$, **Hb**$_{a}$, and **Le**$^{I}{}_{a}$) was produced before 1468, probably in Bohemia. It omits all the theoretical considerations, demonstrations,

5 See the full descriptions of these two witnesses below in Sections 6.1.1.1 and 6.5.1.

6 See B, **Hb**$^{2}{}_{d}$, and S; Section 6.2.1.

7 These witnesses are described more fully in Section 6.10.

references to authorities, explanations of allegories, and theological/prophetic material, as well as Operations 7 and 8, evidently in an attempt to reduce the text into the form of a recipe. Two witnesses of this abridgment traveled alongside an otherwise unknown chrysopoetic recipe that is attributed to John, although its authenticity is open to question.[8] A second and roughly contemporaneous abbreviated version (**Fr**$_a$) occurs appended to a witness of *De famulatu*. Like the previous abridgment, it too descends from a Germanic family. A somewhat later abridgment (**Ve**$_a$) retains some of the theoretical and explanatory material and links John to the two most prominent pseudonymous authors of the fourteenth century by presenting itself as John's own abridgment of "Raymond [Lull]'s *Liber lucis*" and calling Arnald of Villanova "my [i.e., John's] master." A fourth abridgment (**Mp**$^{I}{}_{a}$), which dates from before 1459 and is of Italian (probably Southern) origin, is of special importance because it was one of the sources used to produce the hybrid text published as the 1579 Cologne edition (see below, Section 3.3). Its colophon remarks that "the sense has not been altered, but I have abbreviated it by setting aside much prolixity." In all these abridgments, the intent was to streamline the *Liber lucis*, apparently to make it more immediately useful in directing laboratory operations.

Of the twenty-five vernacular witnesses, four are fifteenth-century, thirteen are sixteenth-century, seven are seventeenth-century, and one is late eighteenth-century. Four of the late ones were prepared from printed editions (**L**$^{7}{}_{e}$, **Psg**$_s$, **Ua**$^{2}{}_{e}$, and **Zc**$_i$). Translations into German are both the earliest and the most abundant; fourteen manuscript witnesses represent eleven independent translations into German. The earliest of them, a South German production (**Bg**$_d$), dates from 1442 or 1443 and is also the earliest known translation into any vernacular. **Ha**$_d$ was produced just slightly later in the 1450s. Another German translation, produced before 1559 and shown in the stemma as θ_d, provides the text of the *Liber lucis* that was incorporated into a composite work titled *Thesaurus mundi, oder Schatz der Welt*, eventually published in Halle in 1621 (see Section 3.4). Yet another German translation, of which there is no known manuscript evidence, was published in 1610 in Joachim Tancke's collection of alchemical treatises *Proptuarium alchemiae*, but, as noted in Section 1.3, is attributed to "Pope Innocent."

A Catalan translation was made prior to 1472, when a surviving copy (**Lh**$_{cat}$) of it was produced by a Fleming, Arnald of Brussels, living in Naples. Yet another translation into a Romance language was made at some time prior to 1459, but has not survived; its existence can be deduced from the survival of its back-translation into Latin (**Mp**^{I}), and it is indicated in the stemma as τ.[9] There are also five known English translations of John's text, the first of which (**X**$^{6}{}_{e}$) dates from the third

[8] This "Practica" attributed to John is treated in Appendix 2.

[9] See below, Section 3.3. T may also be a back-translation from an unknown Italian translation.

quarter of the fifteenth century and is thus nearly as early as the earliest German translations. There are two translations into Italian—**Fn**^3_i and **Zc**$_i$—both from around 1600.[10] A single French version (**Or**$_f$) survives from 1518, as well as one into Czech (**Le**$^3_{cz}$, ca. 1582). A late translation into Castilian (**Psg**$_s$), based on the printed Basel version, was made in second half of the seventeenth century. Thus, the *Liber lucis* was rendered into a total of seven vernaculars: Castilian, Catalan, Czech, English, French, German, and Italian.[11]

3.2 Manuscript Families and Geographical Dispersion

The large number of recovered witnesses and their substantial textual variations allow for the construction of a well-populated and quite confident *stemma codicum* (Section 4.5). Prior to the production of the earliest known witnesses—that is, during the roughly fifty years after the book's composition in 1354—the text of the *Liber lucis* split into several very well-defined families, each substantially identifiable with a geographical region into which the text had circulated.

The earliest known manuscripts are either Provençal in origin, correlating with the original authorship of the work in Avignon, or Northern Italian. These witnesses and their later descendants are traceable by means of shared textual variants to a lost hyparchetype π standing at the head of a "Provençal-North Italian family." The family contains the largest number of reliable witnesses and five of the seven witnesses that date from the first third of the fifteenth century. This populous family also gave rise to three descendants that spawned their own distinct families (ι^2, φ, and ε^1) linked to other geographical regions (Italy, France, and England, respectively), which further indicate the primary role played by the Provençal-North Italian family in the wider diffusion of John's text.[12]

It is rather puzzling that despite the two early Provençal witnesses (**P**^{1} and **P**^{2}), only two of the remaining 104 witnesses were produced in France, suggesting that John's work was not widely circulated, not well received, or provoked little interest there. One of these, **Ko**^{2}, was produced in Paris in 1428 from the lost Parisian exemplar φ and is the work of a German arts student at the University of Paris named Walter Vallrevere, who brought his copy back to Germany with him. The sole French translation of the *Liber lucis*, copied or produced in Sens by the Celestine monk and humanist Pierre Bureteau in 1516 (**Or**$_f$), depends upon an exemplar from Germanic family γ' with additions from an Italian line of descent.

[10] This reckoning does not count two modern Italian translations, one made from the Basel version in the late twentieth century and published in Giovanni da Rupescissa, *Sulla preparazione della vera pietra dei filosofi*, ed. Paolo Cortesi (Rome: Atanor, s.d.), 37–60, and one based on **Oth**^{2} by Marcella Peruzzi and published in Giovanni da Rupescissa, *Il libro della luce*, ed. Andrea Aromatico (Venice: Marsilio, 1997), 121–59.

[11] *LO*, 24 cites a Swedish translation of the *Liber lucis*; however, this is an error, presumably based on a conflation with *De famulatu*, which exists in a 1522 Swedish translation preserved at Linköping, Stadsbibliotek-Stiftsbiblioteket, MS J8, fols. 91r–119r.

[12] Note that I have chosen, contrary to usual practice, the Greek sigla for the lost early hyparchetypes to signify the geographical regions where their descendants were most widely diffused, and thus the probable locale of their production: π for Provence, ι for Italy, γ for Germany, φ for France, and ε for England.

A mid-fifteenth-century manuscript produced in France preserves brief quotations from Operation 4, and these also come from an exemplar in Germanic family γ'.[13] The Provençal-North Italian line thus died out in (what is now) France early in the fifteenth century; it continued only in Italy and in a peculiar but significant instance in Germany (**Y**[I] and its descendants). It is hard to give a satisfying reason for this striking paucity of French witnesses; it seems unlikely to reflect no more than an especially low survival rate of manuscripts in France. A possible explanation is that the text traveled predominantly along the diplomatic routes to and from the Holy See eastward of Avignon, that is, into Italy and the Holy Roman Empire.

In terms of the number of extant copies, Germanic family γ is the largest, signaling a high and long-term level of interest in the work in Germanic lands. This family is also the most distinctive in terms of its numerous textual variants. In addition to the unique incipit mentioned above (Section 1.4), most members of family γ insert a clause into the prologue stating that the tribulations "of both the mystical [*misticus*] and real and true antichrist are sufficiently close at hand and stand nearly at the gates, unless God—for the consolation of the good and the correction of the wicked—should wish, since he is able, mercifully to defer the time." Besides providing an explanation for the antichrist's failure to appear on schedule in 1366, this insertion's reference to *two* antichrists, one "misticus" and the other "realis," indicates that its author was familiar with John's prophetic works that cite exactly these two antichrists or with the works of the Franciscan theologian and rigorist Peter of John Olivi (1248–1298) from whom John adopted the concept.[14]

Further sets of textual variants divide this large Germanic family into three subfamilies (γ', γ'', and γ'''). **B** is the oldest extant member of subfamily γ' (1423) and was produced in the South Tyrol near Trento; its copyist may have been a Franciscan, given its opening invocation of "Jesus, Mary, and Francis." The 1561 Basel edition's text also descends through this subfamily. Subfamily γ' is further distinguished by the inclusion of one version of spurious Operation 9 (see Section 2.6), which does not appear in its sibling subfamilies. The members of subfamily γ'' clearly record the engagement of reader-practitioners with the text: they all include multiple glosses providing further instructions, sometimes written in the first person, about how to carry out the procedures described.[15] These glosses must date from before 1442–43, since they are included in the German translation **Bg**$_d$, which dates from those years, but must have been added *after* John's text made its way into Germanic lands, since one of them uses the Germanic measure

[13] Bordeaux, Bibliothèque Mériadeck, MS 531, fol. 164v; see Appendix 2.

[14] *LSE*, 48, 52–53 and 140. John identified the *antichristus misticus*, a forerunner of the *antichristus maximus* (here *verus et realis*), with a combination of Emperor Ludwig IV the Bavarian and antipope Nicholas V.

[15] The glosses are reproduced in Section 6.2.6. The glosses were also copied into Ge by a different hand at some point after that witness (a descendant of family γ''') was completed; this connection is indicated by the coarsely dashed line in the stemma.

of weight *Loth* (half an ounce). The origins of the witnesses in this subfamily cluster tightly in northern Bavaria and Bohemia, as shown by the dialect of the German translation $\mathbf{Bg}_d$, the residence of Eggolsheim for its later copyist (ν, source for $\mathbf{Mu}^2{}_d$), the originally Bohemian ownership of **Z** and $\mathbf{Le}^{I}$, the production of $\mathbf{Wb}^2{}_d$ in Nuremberg, and the Czech translation $\mathbf{Le}^3{}_{cz}$.

The origins of the surviving witnesses of subfamily γ''' are geographically more dispersed—including one in a clearly Italian hand—but nevertheless cluster in southeastern parts of the Germanic lands: the South Tyrol, Bohemia, southeastern Saxony, Bavaria, and Vienna. **Ge** is a special case in that its main text, clearly part of subfamily γ''', was later corrected and augmented with readings taken from witnesses belonging to subfamilies γ' and γ'', thus reuniting all the descendant subfamilies of γ.

Two additional lines of descent, γ^2 and γ^3, further propagated the *Liber lucis* through Germany. The first is represented by only three witnesses, all fifteenth century, although a now-lost copy must have made its way to England by the 1450s, since the Oxford copyist of $\mathbf{X}^{I}$ drew upon it to supply the text missing from his defective English exemplar that descended through family ε^2. With only three known witnesses, a designation of the family's geographical reach remains provisional, but all three originate in either Northern Germany or the Netherlands. As for family γ^3, all but one of its five known members (three in Latin and two in German) originated in Northern Germany, the exception being an extract from John's warning about secrecy, the "Credas," used as an admonitory introduction to a North Italian collection of alchemical treatises ($\mathbf{Oth}^{I}$). A fourth line of descent in Germany, of particular interest and importance, stems from $\mathbf{Y}^{I}$ and will be discussed in Section 3.4.

Three lines of descent dispersed the *Liber lucis* through Italy. The most reliable witnesses arrived as part of the Provençal-North Italian family π. Considerably less reliable versions descended through family ι. The hyparchetype ι seems to have been badly damaged or otherwise illegible midway through Operation 5, which produced copyists' problems (or surrender in the case of **O**) in all of its descendants at the very same point in the text (*vitrea* at l. 117). It may also have been damaged at the start since none of its descendants contain the prologue. An attempt was made to repair the damage, resulting in ι', which actually made matters worse because it displaced, jumbled, and then repeated substantial amounts of text. Its descendants ($\mathbf{Oth}^2$, $\mathbf{L}^5$, **Pd**), which cluster in North Central Italy (Padua and around Gubbio), are therefore extremely corrupt. A third Italian line of descent is through ι^2, itself a descendant of π. Its oldest known witness ($\mathbf{L}^3$) was copied probably in the Veneto around 1410, while later descendants originated further south in Italy. This third Italian family played two significant roles in the propagation of the text, even though ι^2 also seems to have been damaged at the start of the text since its descendants lack part or all of the prologue and was adulterated at the end with spurious Operation 9 thanks to its descent from π^3.

First, it was the source for a vernacular translation (designated τ), possibly into Italian, that was then back-translated into Latin before 1459, possibly in Naples, to give one of the modified versions of the text (**Mp**^{1}) that would form the basis for the printed Cologne version of the *Liber lucis* (see Section 3.3). Second, this family also provided, through ε^{2}, a major line of transmission for John's work into England, although in a mutilated state.

John's text met quite a different fate in England than it did on the Continent. While its introduction to England appears to have been somewhat delayed relative to its diffusion into Germany and Italy, the interest in the *Liber lucis* upon its arrival was intense, yielding a considerable number of witnesses within a short period of time. Most English witnesses date narrowly to the mid-fifteenth century, which is contemporaneous with the importation and subsequent excitement surrounding other Continental alchemical texts, particularly the works of pseudo-Lull.[16] Yet this English interest was also short-lived; a century intervenes between the last known fifteenth-century witness, **Y**^{2} (curiously enough, copied in London by a visiting Italian in 1480 and then brought back to Italy), and the earliest known sixteenth-century witnesses: **X**^{2}, a scrappy and highly lacunose witness dated 1593, and the roughly contemporaneous **L**^{6}. The narrow mid-fifteenth-century spike of interest in the *Liber lucis* in England contrasts with its reception in Germany and Italy, where its high popularity continued unabated, particularly in Germany, for two centuries.

The two routes by which the text arrived in England provided divergent results. The descendants of ε^{1}, which branched off from the Provençal-North Italian family π, transmit a generally good and complete text, although with a modified prologue beginning "I, brother John of Rupescissa…." In contrast, the descendants of ε^{2}, which represent the majority of surviving English witnesses, transmit a very defective text. This poor state is due in part to losses inherited from their evidently damaged ancestor ι^{2}, but also to numerous intentional alterations made subsequently. None of these latter English witnesses attributes the text to John; many leave it anonymous, while some attribute it to Roger Bacon, Arnald of Villanova, or others. English copyists repeatedly endeavored to streamline the text by removing sections that seemed "irrelevant" to the practical process (for example, John's exhortations and warnings to his fellow *viri evangelici*, his demonstrations and proofs, and the references to alchemical authorities) in addition to many other lesser edits throughout. Nevertheless, new materials were also spliced in at $\varepsilon^{2\prime}$, most notably an alchemical text in verse, a genre of alchemical writing particularly popular in fifteenth-century England. Some English copyists were aware of the degraded state of their local copies and endeavored to amend them. In the 1450s, John Cokkes, a priest and physician at Oxford and copyist of **X**^{1}, restored the prologue and furnace chapter missing from his exemplar from family $\varepsilon^{2\prime}$ using

16 On the influx of Continental alchemy into mid-century England, see Rampling, *Experimental Fire*.

a copy from German family γ^2. The compiler of the late sixteenth-century **L**^{6} worked from at least three exemplars, one from each of the three English families ε^1, ε^2, and $\varepsilon^{2\prime}$, and chose, on an operation-by-operation basis, what he considered the best version to use in his own copy (to his credit he took most of his operations from the descendant of ε^1, which transmitted the most complete and authentic text).

Only one Iberian witness of the *Liber lucis* has been identified, a sixteenth-century copy now in Salamanca (**Sa**). It cannot be linked convincingly to any known lines of descent. The same is true of the single Catalan translation, which implies circulation and a certain popularity for John's work in Catalonia, but it is not known where that translation was produced since our only witness of it is a copy produced in Naples in 1472 (**Lh**$_{\text{cat}}$).

Judging then from the surviving witnesses, the *Liber lucis* found its greatest circulation in Northern Italy, Germany, and, in modified form, in England. In comparison with *De famulatu*, for which we have Scandinavian, Balkan, and multiple Iberian witnesses among the over 250 now identified, the *Liber lucis* seems to have been somewhat less widely distributed. In only thirteen (19%) of the sixty-seven complete Latin witnesses does the *Liber lucis* appear in the same alchemical collection as a complete *De famulatu*, and in only three cases (**Ccc**, **Ka**^{1}, and **Wb**) do the two texts appear immediately adjacent to one another, indicating that they did not customarily travel together.[17] Nevertheless, four other manuscript collections actually combine the two works, either by attaching the *Liber lucis* to *De famulatu* as its "third book" (**Sa** and **X**^{1}) or splicing it in between the two books of *De famulatu* and labeling it as the "second book of the quintessence" (**L**^{2} and **W**^{6}). Inserting the *Liber lucis* after the end of Book One of *De famulatu* makes sense if the copyists who did so recognized the text as the fulfillment of John's promise made at the end of Book One to treat transmutational alchemy in a future book. Interestingly, the insertion of the *Liber lucis* as the second book of *De famulatu* was done independently in the two witnesses where it occurs, since its text descends through different families. Two other manuscripts, however, contain all or parts of the *Liber lucis* copied out into the margins of preexisting copies of *De famulatu*. In the case of **Br**, the full *Liber lucis* begins roughly where the parallel material in *De famulatu* starts (Figure 3.1). In the case of **Km**, the copyist was interested specifically in the metallic and mineral alchemy of *De famulatu* and so excerpted those sections and later added parallel text from the *Liber lucis* into the margins surrounding the relevant material.[18] It was more common, however, for copyists to excerpt the sections out of *De famulatu* that parallel the *Liber lucis*, often placing them nearby as potential sources of insight onto the processes described; seventeen such witnesses exist. In more minor cases,

[17] B, Br, Ccc, Ct, Cu^2, Ka^1, Ko^1, L^4, Pd, Ua, W^1, Wb, and X^3.

[18] The excerpt from *De famulatu* runs from "Creavit sublimis deus" to the end of Book One.

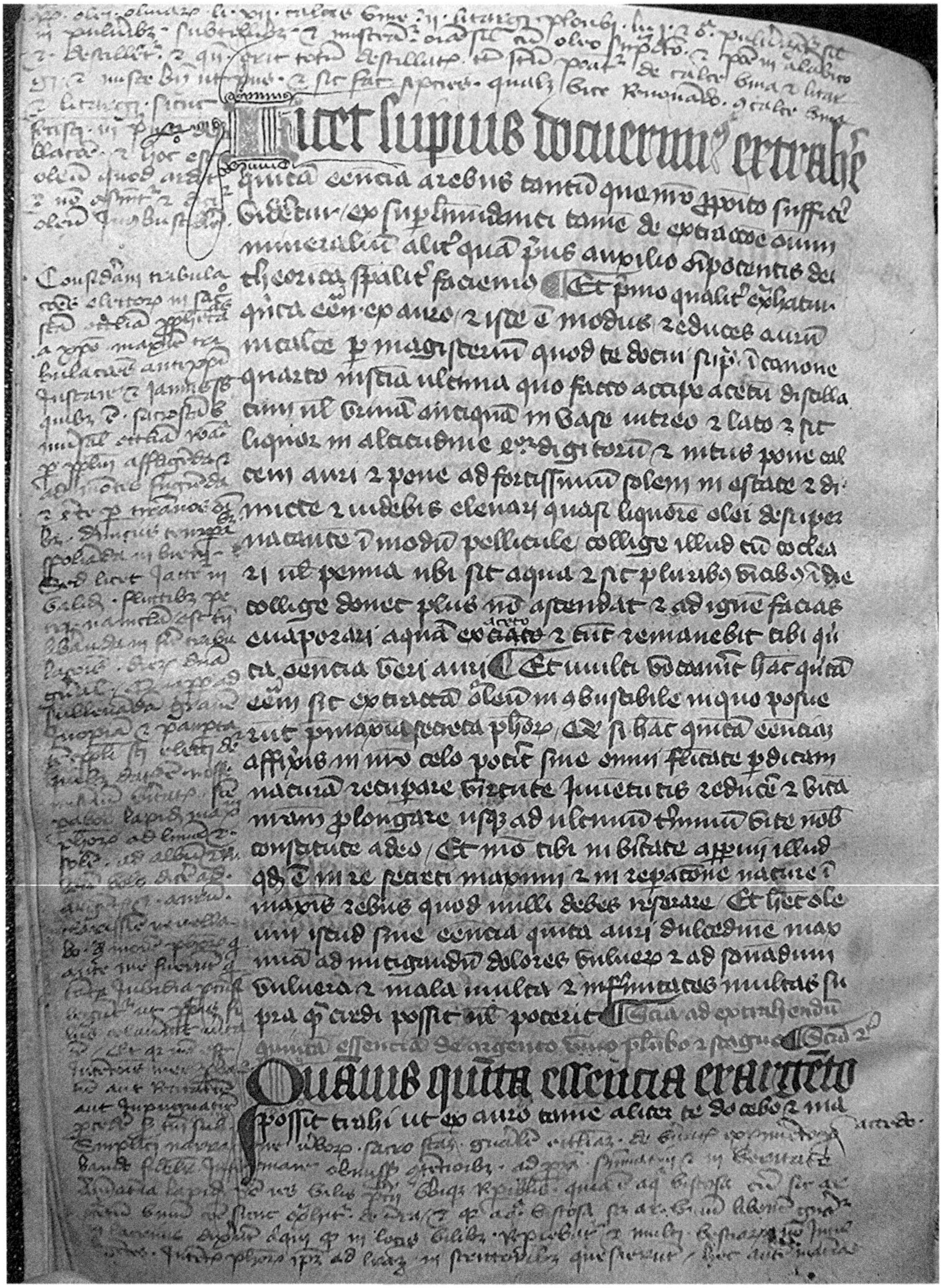

FIGURE 3.1 The start of the entire *Liber lucis* copied into the margins of an earlier witness of *De famulatu*; **Br**, fol. 24v. Courtesy of the Burgerbibliothek Bern.

as noted in Section 2.4, copyists spliced excerpts of relevant material from *De famulatu* directly into their copies of the *Liber lucis* or added marginal references to those other materials.

3.3 The Printed Versions: The Basel *De confectione* (1561) and Cologne *Liber lucis* (1579)

John of Rupescissa's *Liber lucis magisterii magni* remained solely in manuscript for over two hundred years after its composition in 1354. In the second half of the sixteenth century, however, it was printed in two different versions and under two different titles. In 1561, it appeared at Basel as *De confectione veri lapidis philosophorum* (*On the Making of the True Philosophers' Stone*) in the alchemical collection *Verae alchemiae artisque metallicae*, edited by the Italian physician Guglielmo Gratarolo (1516–1568).[19] The collection was reprinted at Basel in 1572. This "Basel version" of the text was later republished in other collections: first in the four-volume edition of Lazar Zetzner's *Theatrum chemicum* (1613) and its later six-volume version (1659–61), and then again in Jean-Jacques Manget's *Biblotheca chemica curiosa* (1702).[20] A second publication of John's chrysopoetic treatise appeared at Cologne in 1579, this time under the title *Liber lucis*, as part of a thin collection of alchemical works compiled by the physician Daniel Broekhuizen.[21] This "Cologne version" was also reprinted several times—in multiple later editions of the Broekhuizen collection, in the *Theatrum chemicum* (which reproduces the entire Broekhuizen collection), and in the *Bibliotheca chemica curiosa*; in both of these latter collections, the Cologne version immediately follows the Basel version.[22]

The Basel and Cologne versions differ significantly from one another. The Basel version contains a unique incipit, no prologue, seven operations, and three additional sections entitled *annotatio*, *de sale*, and *de elixir*. Based on its textual variants, its origin lies clearly in the Germanic family γ′. It shows all the characteristics of that group of witnesses and is especially close to the subgrouping that includes **B**. Relative to the critical edition, the Basel version contains significant omissions ranging from minor to very substantial, such that nearly one-half of John's original text is missing. Most notably, the prologue and most of the "Credas" are omitted, Operation 7 is cut to a small fraction of its usual length, and Operations 8 and 9 are merely cited as "two remaining chapters" and summarized in a single sentence with the excuse that the same or better material can be found elsewhere. The text ends with an added notice that "the master here makes no mention of fermentation," a complaint found in the marginalia of several later witnesses, and advises the practitioner to "be prudent...choose what is good...

[19] John of Rupescissa, *De confectione veri lapidis philosophorum*, in *Verae alchemiae artisque metallicae*, 2 vols. (Basel, 1561), 2:226–31. On Gratarolo, see Ian Maclean, "Heterodoxy in Natural Philosophy and Medicine: Pietro Pomponazzi, Guglielmo Gratarolo, Girolamo Cardano," in *Heterodoxy in Early Modern Science and Religion*, ed. John Brooke and Ian Maclean (Oxford: Oxford University Press, 2005), 1–29, esp. 17–19.

[20] *Theatrum chemicum*, ed. Lazar Zetzner, 4 vols. (Strasbourg, 1613), 3:179–89; 6 vols. (Strasbourg, 1659–61), 3:189–97; and *BCC* 2:80–83.

[21] John of Rupescissa, *Liber lucis*, in *Secreta alchimiae magnalia D. Thomae Aquinatis...accessit et Johannis de Rupescissa liber lucis...* (Cologne, 1579), 35–48.

[22] *Secreta alchimiae* (Leiden, 1598), 41–56; (Leiden, 1602), 41–56; *Theatrum chemicum*, ed. Zetzner, 4 vols. (Strasbourg, 1613), 3:278–89; 6 vols. (Strasbourg, 1659–61), 3:284–95; and *BCC* 2:84–87.

and apply a ferment as you know how to do."[23] The annotations which follow, presumably written by Gratarolo himself, amount to only a little less than half the length of John's own material and have no precedent in the known manuscript tradition. I have discovered, however, an early seventeenth-century German translation of this added material.[24]

In contrast to the Basel version, the Cologne version includes a shortened version of the prologue, nine (rather than seven) operations, and an additional furnace description with two woodcuts. Its origin is more complex and much more interesting than that of the Basel version. The two versions diverge sufficiently in wording while remaining mostly similar in gist that I initially suspected that John had written two versions of the text, but upon recovery and inspection of the surviving witnesses, a much more surprising explanation emerged. With the exception of five or six late manuscripts that were copied or translated from the printed Cologne edition (**E**, **L^7_e**, **Lu**, **Psg_S**, **Ua^2_e**, and possibly **W^5**), *only two* of the remaining one hundred manuscripts bear text that resembles that of the Cologne version. These two lone manuscripts, however, tell a remarkable story about the transformation of John's text and the twisted path of transmission it took on the way to its second publication.

The sole witness that contains text in very nearly the form of the Cologne edition is **Ua**, part of a mid-sixteenth-century Italian miscellany of alchemical treatises. In the few places where the Cologne edition gives a clearly incorrect or incomplete reading, **Ua** provides the correct or complete reading, thus establishing that its text is not dependent upon the printed edition.[25] The closeness of the two documents is further signaled by the fact that the text immediately preceding **Ua**, pseudo-Thomas Aquinas' *Tractatus in arte alchemiae* (fols. 44v–48v), also appears immediately preceding the *Liber lucis* in Broekhuizen's printed Cologne collection (24–36). The only significant difference is that the "description and form of the physical furnace" found in the Cologne edition (46–47) is absent from **Ua**.

The existence of this earlier manuscript of the Cologne text does not, however, explain the *origin* of the alternate text it contains. The answer is, however, beautifully provided by a mid-fifteenth-century southern Italian manuscript now in Montpellier. This document, Montpellier MS H 493, compiled in Naples in 1459, contains *two* versions of the *Liber lucis*: a full version (**Mp^I**) and an abbreviated version (**Mp^I_a**). The text of **Ua** draws upon *both* of these earlier versions and is, in fact, a messy pastiche of the two. **Ua** redistributes the prologue and nine

[23] *De confectione* (1561), 2:229–30; see above Section 2.2.

[24] Kassel, Universitätsbibliothek Kassel-Landesbibliothek und Murhardsche Bibliothek der Stadt Kassel, 4° Ms. chem. 48, fols. 13r–15v; see Appendix 2. In regard to the annotation providing readers with a sly hint about how to make the process succeed, at least in part, see above Section 2.2.2.

[25] This issue would be of concern only if the dating of Ua, now considered to be at least two decades earlier than the 1579 Cologne edition, is incorrect. The textual variants include the following: in caput I, Cologne leaves an apparently irresolvable abbreviation intact: "Et intentio mea est p^{re} bonis," while Ua reads "Et intentio mea est significare bonis"; in caput III, Cologne reads "Mercurius noster dissolvitur" and "invenias species nigras," while Ua reads more logically "mercurius non dissolvitur" and "invenias feces nigras."

operations given cumulatively by $\mathbf{Mp^I}$ and $\mathbf{Mp^I{}_a}$ into eleven "heads," and these headings continue into the Cologne edition. Some sections are taken fully from one or the other manuscript witness; other sections contain sentences from both interleaved on nearly a line-by-line basis. **Ua**'s close dependence upon the Montpellier manuscript is amply demonstrated by their nearly identical drawings of the furnace (see Figure 2.7). Thus, **Ua** and the Cologne text that is dependent upon it represent a *hybrid* of the two witnesses found in the Montpellier manuscript.[26]

But the story goes further; the two witnesses of the Montpellier manuscript have their own peculiar backgrounds that render the lineage of the Cologne version of John's text yet more tortured. The abbreviated version $\mathbf{Mp^I{}_a}$ descends through $\pi^{I'''}$ with minor contamination from π^3. The full version $\mathbf{Mp^I}$ descends through Italian family ι^2 but is a *back-translation into Latin* from an intermediate vernacular translation (labeled τ in the stemma) as demonstrated by its frequent use of cognates and alternative phrasings that cannot be explained as copyist's errors or random rephrasings.[27] During either the translation into the vernacular or the back-translation into Latin, several unique insertions were made: for example, the insertion of the Decknamen "terra foliorum" or "foliata" for the quintessence of mercury, the addition of "master Ramon Lull" to the list of alchemical authorities, and the claim that "some philosophers" call the invisible sulphur "Magnesia."[28] Thus, it is clear that the Cologne version transmits an extremely adulterated text resulting from the interleaving of an abridged version and a back-translation into Latin that descend through two different families, plus additional insertions.

3.4 Printed German Versions: *Tractat Bapsts Innocentii* (1610) and *Thesaurus mundi* (1621)

A previously unidentified publication of John of Rupescissa's *Liber lucis* in German translation appeared in 1610 under the title *Treatise of Pope Innocent on the Philosophers' Stone* (*Tractat Bapsts Innocentii Vom Stein der Philosophorum*). The text is part of Joachim Tancke's alchemical miscellany *Promptuarium alchemiae*.[29] Its possible preservation of some garbled connection to Innocent VI,

[26] These origins of the Cologne version are more fully treated in Lawrence M. Principe, "Bibliotheca Philosophica Hermetica MS 338: The 'Missing Link' in the Transmission of John of Rupescissa's *Liber lucis*," *Quaerendo* 53 (2023): 187–97. It should be noted that both **Ua** and the Montpellier MS are Italian in origin. A further curiosity of **Ua**/Cologne is that the heading for caput X promises not only details of the furnace construction but also "the calcination of mercury, Sol, and Luna." This heading is not present in either of the Montpellier witnesses, nor is the promised calcination actually treated in caput X of either **Ua** or Cologne. However, considerably later in the same Montpellier manuscript (fol. 267r), there is a rubric introducing a "modus calcinandi aurum et argentum secundum magistrum Johannem de Rupe Cissa." For some reason, the compiler of the text of **Ua** transferred this rubric (which is followed by material extracted from *De famulatu*) into his heading for caput X. This feature further argues for the immediate dependence of **Ua** upon the Montpellier MS.

[27] See Principe, "Missing Link," 193 for illustrative examples.

[28] On fols. 236v, 238r, and 239r of $\mathbf{Mp^I}$, and 36, 37, and 40 of the 1579 Cologne edition.

[29] *Promptuarium alchemiae* (Leipzig, 1610), 227-32; on Tancke and his publications, see Udo Benzenhöfer, "Joachim Tancke (1557–1609): Leben und Werk eines Leipziger Paracelsisten," *Salzburger Beiträge zur Paracelsusforschung* 25 (1987): 9–81.

the reigning pope in 1354 when John wrote the *Liber lucis*, was tentatively suggested in Section 1.3. The German text is fragmented and in places mistranslated, and its line of descent cannot be clearly ascertained due to its corrupt nature.[30]

A more complete and faithful German translation of John's text was published in Halle in 1621 under the title *Thesaurus mundi, oder Schatz der Welt*.[31] The origin and contents of this text are extremely interesting and require much more treatment and investigation than can be presented here.[32] *Thesaurus mundi* forms part of a collection of short alchemical tracts entitled *Heptas alchymica*, intended as a supplement to the two volumes of the just-mentioned *Promptuarium alchemiae* published by Joachim Tancke in 1610 and 1614. The editor of the collection signs himself as Theophilus Neandrus—either a pseudonym or possibly a Greek rendering of the German name Gottlieb Neumann—who remains unidentified at present. The only clue to his identity occurs in his self-description as "Tessanus Chymiatrus," that is, a chymical physician of Teschen, an important town in Silesia now straddling the border between Poland (Cieszyn) and the Czech Republic (Český Těšín).

Thesaurus mundi contains five sections, only one of which is the translated *Liber lucis*. It begins with practical advice for the reader/practitioner (294–97), which is followed by a "Prayer of John of Rupescissa" (297–98).[33] The translated *Liber lucis* comes next under the heading "here begins the discourse of John of Rupescissa's instructions for making the philosophers' stone" (*Hie faht an die Rede der unterweisung Johannis Rupecisse zu machen den Stein der Philosophen*, 298–316). An epilogue (317–18) then follows, answering the objection that John's process cannot produce a successful transmuting agent because it includes neither gold nor silver as a ferment (see above, Section 2.2). A proof of the efficacy of John's process (*Bewehrung dieser Kunst*, 318–20) comes last in the form of a transmutation history, that is, a detailed account of a successful transmutation using the method described in the *Liber lucis*.

30 E.g., *Verbrennung der Röhte* for *tinctura rubedinis* on 228, and *die Augen der Fische*(!) for *lacrima oculi* on 229. It contains the scribal error *Natur oder Materia* of l. 18 (on 227, from a minim problem reading the abbreviation *mā*), which is similar but not identical to a characteristic variant of γ (see Section 6.2), but the mistake is easy to make, and none of the other numerous variants of γ are present. It also contains the strange term *Suratrat* for the mercury sublimate, that is, *tartarus* spelled backwards.

31 *Thesaurus mundi*, 294–320 in *Heptas alchymica, das ist, Außerlesene philosophische Tractätlein von dem magisterio lapidis, so zum Complement des Promtuarii Alchymiae Herrn D. Joachimi Tanckii*, ed. Theophilus Neandrus (Halle, 1621). The *Thesaurus*, without publication information, is mentioned as Rupescissa's by both Pierre Borel, *Biblioteca chemica* (Paris, 1656), 194, and Giovanni Battista Nazari, *Della tramutatione metallica* (Brescia, 1599), 139. It also appears in the early modern list of alchemical manuscripts of Vatican, Barberini MS 273, fol. 294r, as cited by Thorndike, *HMES* 3:369, although Thorndike was unaware of the 1621 Halle publication. By odd coincidence, Halleux, "Rupescissa," 243 n.13, suggests that this publication, which he was unable to see, might contain *De famulatu* based upon the title *Thesaurus orbis* given to *De famulatu* in two of its witnesses: "Marburg B 12" (now MS 18) and BNF, MS fr. 2020, fol. 221v.

32 I am preparing a further study of this intriguing publication.

33 This "prayer" is adapted from the introductory section of *De famulatu*; however, the German is clearly not translated from the printed edition, which it predates.

The composite text of *Thesaurus mundi* can be traced in manuscripts back to at least the middle of the sixteenth century. I have identified three manuscript witnesses: **Ka**$^{3}_{d}$ (1576), **Ka**$^{4}_{d}$ (end of the sixteenth century), and **Go**$^{3}_{d}$ (early seventeenth century). Another manuscript, however, dated 1559, contains notes taken in Latin from a yet earlier (and presently unknown) witness, thereby providing a *terminus ad quem* of 1559 for the compilation of the *Thesaurus*.[34] There are notable differences between the printed version and the manuscripts; the manuscript materials are fuller and suffer from fewer obvious copyists' mistakes.[35]

The German translation of the *Liber lucis* incorporated into *Thesaurus* includes the prologue and all eight operations and can be confidently identified as a descendant of $\pi^{I'''}$; all of the characteristic textual variants of that subfamily are present (see Section 6.1.1.3). The translation's place in the stemma can, however, be further refined. Wherever there are differences between the known Latin witnesses of subfamily $\pi^{I'''}$, the German text follows **Y**^{I}, and the *Thesaurus mundi* translation breaks off at the same point where **Y**^{I} is truncated toward the end of Operation 8 (l. 319). In the stemma, I have indicated the German translation incorporated into *Thesaurus* using the siglum θ_{d}.

The connection of *Thesaurus mundi* with **Y**^{I} has some significant implications. First, **Y**^{I} is not only among the earliest witnesses known but is unusual for the Provençal-North Italian family π in that it is written in a German hand (**Ka**^{I} and **Be/D** are also in German hands, but the former is unambiguously a descendant of **Y**^{I}, and the latter two are closely related through their common ancestor $\pi^{I''}$). Thus, **Y**^{I} was yet another important and early vector for the transmission of the *Liber lucis* into Germanic lands. Second, the manuscript containing **Y**^{I}, Yale University Mellon MS 5, has been plausibly connected with the fifteenth-century priest and alchemist Johannes von Teschen or his immediate circle, owing to the strikingly large number of texts by him that it contains.[36] *Thesaurus mundi* refers explicitly to Johannes von Teschen when it provides its "proof of this art... as heard from the worthy and highly learned doctor Johannes von Teschen," who reportedly participated in the transmutational events that *Thesaurus* describes.[37] Significantly, all three manuscript witnesses relating to the *Thesaurus*, as well as the 1621 publication, appear side by side with a German prose version of Johannes von Teschen's *Lumen secretorum*, the original Latin verse version of

[34] Kassel, Universitätsbibliothek Kassel-Landesbibliothek und Murhardsche Bibliothek der Stadt Kassel, 4° MS chem. 29, fols. 53r–54v: "Ex Jo: de rupescissa: Thesaurus mundi."

[35] **Ka**$^{3}_{d}$, despite its Netherlandish dialect, is the most complete and reliable witness. Although the copyist, Wilhelm Vernuken, had some difficulty reading his exemplar, it nevertheless contains many fewer obvious transcription errors than the other manuscript witnesses (which probably descend from **Ka**$^{3}_{d}$) or the printed *Thesaurus* which does not appear to descend from **Ka**$^{3}_{d}$.

[36] Laurence C. Witten and Richard Pachella, *Alchemy and the Occult: A Catalogue of Books and Manuscripts from the Collection of Paul and Mary Mellon Given to the Yale University Library*, 4 vols. (New Haven: Yale University Library, 1968–77), 3:26–41.

[37] **Ka**$^{3}_{d}$, fol. 11v: "Als hirnach van dissen werdigen hogelarten docter Johan Techtschenens vernamen und gehortt wirdt."

which appears in Mellon MS 5. This fifteenth-century text by von Teschen is furthermore based, at least in part, on the method John of Rupescissa presents in the *Liber lucis*. These intriguing connections stretch yet further; the town of Teschen was home not only to the fifteenth-century alchemist but also to the editor of the collection *Heptas alchymica* that contains *Thesaurus*, a "chymical physician of Teschen."[38]

Further pursuit here of the origin of the various components of *Thesaurus mundi*, their compilation, and their relationship to Johannes von Teschen would run too far afield from this study's focus on the *Liber lucis*. Nevertheless, the intriguing connections highlighted by this cursory examination of the *Thesaurus* promise to tell us more about alchemical practices and the circulation of alchemical texts, although more historical documentation is required to flesh out the lines of influence and transmission. In the present context, it suffices to point to *Thesaurus mundi* as an illustration of the continuing practical engagement with John of Rupescissa's *Liber lucis magisterii magni* and the ways in which it continued to attract not only hopeful practitioners but also further textual elaborations for more than 250 years after its composition.

3.5 Copyists, Owners, and Users of the *Liber lucis*

As is evident from the surprisingly large number of surviving witnesses, the *Liber lucis* was a very popular and broadly disseminated text. The many times it was translated into multiple vernaculars and the frequent reprintings of both of its Latin editions further attest to its popularity and the continuing demand for it. Surviving marginalia, glosses, edits, additions, and other textual artifacts indicate, moreover, that many of its possessors were active practitioners as well, engaging with its text and working practically to complete the process that John proposed. But who were the copyists and owners of the many manuscript witnesses of the *Liber lucis*? Although in many cases their identities cannot be recovered, a little over one-quarter of the known witnesses actually do contain specific information—names, professions, residences, dates—about their copyists and/or their original or early owners. This information sketches a striking portrait of the diversity of people involved in alchemy and interested in John's particular approaches to the subject in the late Middle Ages and early modern period.

Four witnesses are connected with princely or noble ownership: a Tyrolian duke (**B**), a Bohemian nobleman ($\mathbf{Le}^{2}$), a Bavarian duke ($\mathbf{Mu}^{2}{}_{d}$), and a Hessian nobleman (ν). Many other witnesses were eventually subsumed into the library collections of noble practitioners, particularly that of the famous Moritz of Hessen-Kassel (1572–1632; **Ka**, $\mathbf{Ka}^{2}$, $\mathbf{Ka}^{3}{}_{d}$, $\mathbf{Ka}^{4}{}_{d}$, $\mathbf{Ka}^{5}{}_{d}$).[39] At least ten witnesses were produced by

[38] Regrettably, very little is known about Johannes von Teschen; see Joachim Telle, "Johannes von Teschen," in *Die deutsche Literatur des Mittelalters: Verfasserlexicon*, 4:774–76, and Thorndike, *HMES* 3:642–43.

[39] On Moritz and his alchemy, see Bruce T. Moran, *The Alchemical World of the German Court*, *Sudhoffs Archiv* 29 (Stuttgart: Steiner Verlag, 1991).

readers in religious orders: a German Benedictine alchemist and doctor of canon law ($\mathbf{W}^{7}$), a French Celestine monk ($\mathbf{Or}_{f}$), a cleric in the retinue of Cardinal Nicolas of Cusa (**V**), a Genovese friar ($\mathbf{W}^{3}$, $\mathbf{W}^{4}$, $\mathbf{Y}^{3}$, $\mathbf{Y}^{4}$, $\mathbf{Y}^{5}$), a German parish priest ($\mathbf{Ka}^{5}{}_{d}$), and the Oxonian priest-physician John Cokkes ($\mathbf{X}^{1}$). Several others are connected with physicians in various locales: a physician and professor of Greek at the University of Vienna (**Ge**), a Spanish physician and surgeon working in Italy ($\mathbf{Mp}^{2}$), the English physician and astrologer Simon Forman (1552–1611; $\mathbf{X}^{7}{}_{e}$), and the just-mentioned priest-physician John Cokkes at Oxford. A partly-overlapping group of witnesses was produced by others in academic settings: the copy made in 1428 by Walter Vallrevere while he was an arts student at the University of Paris ($\mathbf{Ko}^{2}$) and the one prepared by Peter Holzruter, an arts master also from Paris who became rector of a school in Konstanz (**Ma**). Finally, several artisans also produced and used copies of the *Liber lucis*: the Florentine goldsmith and empiric Lorenzo da Bisticci (β), the sculptor and architect Wilhelm Vernuken at the court in Kassel ($\mathbf{Ka}^{3}{}_{d}$), and an unnamed member of the Medici workshops ($\mathbf{Fn}^{3}{}_{i}$). Two other witnesses were the property of persons known as alchemical practitioners from other sources: Sebald Schwertzer (1552–1598; $\mathbf{Le}^{4}{}_{d}$) and Peter Moritz (1638–1702; $\mathbf{L}^{8}{}_{d}$), and yet another was owned and annotated by the Elizabethan mathematician and magus John Dee ($\mathbf{L}^{6}$). As mentioned in the previous section, the alchemical author and priest Johannes von Teschen may be connected with $\mathbf{Y}^{1}$.

Even while bearing in mind that such biographical information appears in only about 25% of the witnesses, this limited information still clearly indicates how alchemical interest and practice ranged widely across the various boundaries of social position and profession and of geographical and linguistic regions. Although John of Rupescissa stated that his intended audience for the *Liber lucis* was to be the *viri evangelici*, his fellow rigorist Franciscans of a post-antichrist world who could use its secrets to comfort and rebuild a despoiled Christendom, his chrysopoetic composition obviously reached a much broader readership. Those many readers from widely diverse backgrounds engaged deeply with the legacy John left to them in the *Liber lucis magisterii magni*: scrutinizing his words, endeavoring to follow his instructions, proposing new practical solutions to its difficulties, editing, annotating, criticizing, praising, glossing, translating, revising, and reworking the text from generation to generation over the course of three centuries and beyond.

Part 2

Edition and Translation

4 THE *LIBER LUCIS MAGISTERII MAGNI*

4.1 *Conspectus Siglorum* and *Stemma Codicum*

Latin Manuscripts

B Klagenfurt, Bischöfliche Bibliothek, MS XXIX d 24, fols. 202v–14v (1423)

Be Berlin, Staatsbibliothek zu Berlin, MS lat. oct. 472, fols. 96r–101r (2nd half 15th c.)

Br Bern, Burgerbibliothek, MS 480, fols. 24v–30r (1st half 15th c.)

Bo Bologna, Biblioteca Universitaria, MS 1496, busta 1, fols. 42r–47v (2nd half 16th c.)

Ccc Cambridge, Corpus Christi College, MS 99, pp. 249–54 (mid-15th c.)

Ct ____________. Trinity College, MS O.8.36., fols. 51r–55r (ca. 1500)

Cu^1 ____________. University Library, Dd.4.45, fols. 1r–4r (mid-15th c.)

Cu^2 ____________. University Library, Kk.6.30, fols. 62v–68v (mid-15th c.)

D Darmstadt, Universitäts- und Landesbibliothek Darmstadt, Hs 2226, Faszikel III, fols. 45r–50r (mid-15th c.)

E El Escorial, Real Biblioteca del Monasterio de San Lorenzo, Cod. Lat. H-1-9, fols. 181v–83v (18th c.)

Fl Florence, Biblioteca Laurenziana, Gaddi MS 174, fols. 10r–14v (2nd half 15th c.)

Fn^1 ____________. Biblioteca Nazionale Centrale di Firenze, Magliabechi Cl. XVI, cod. 110, fols. 14r–17v (1533)

Fn^2 ____________. Biblioteca Nazionale Centrale di Firenze, MS Palat. 885, fols. 154r–60v (end of 16th c.)

G Glasgow, Glasgow University Library, MS Ferguson 49, fols. 134r–42r (3rd q. 15th c.)

Ge Geneva, Bibliothèque de Genève, MS lat. 82, fols. 98v–104v (1546–60)

Go^1 Gotha, Forschungsbibliothek Gotha der Universität Erfurt, Chart. A 1017, fols. 76r–77r (ca. 1450)

Go^2 ____________. Chart. A 519, fols. 22r–34r (mid-1580s)

H San Marino, CA, Huntington Library, HU 1051, fols. 32r–v (frag., operation 7; mid-15th c.)

Hb Hamburg, Staats- und Universitätsbibliothek, Cod. Alch. 659, fols. 122v–29r (1468)

J Jena, Thüringer Universitäts- und Landesbibliothek, Ms. El. q. 21, fols. 151r–54v (mid-15th c.)

Ka^1 Kassel, Universitätsbibliothek Kassel-Landesbibliothek und Murhardsche Bibliothek der Stadt Kassel, 2° Ms. chem. 10, fols. 16r–19r (1462).

DOI: 10.1201/9781003666547-6

Ka^{2} ———. 4° Ms. chem. 10, fols. 25r–30r (1491)

Km Kremsmünster, Benediktinerstift, Codex Cremifanensis 72, fols. 51v–52r, 54r–v, and 227v (extr., 1430)

Ko^{1} Copenhagen, Kongelige Bibliotek, GKS 1712 kvart, fols. 111r–18v (2nd q. 15th c.)

Ko^{2} ———. GKS 236 folio, fols. 52r–54v (1428)

L^{1} London, British Library, Harley MS 3528, fols. 21v–22v (extr., Operations 5–7; mid-15th c.)

L^{2} ———. Sloane MS 75, fols. 114r–15v (frag., Operations 4–8; 3rd q. 15th c.)

L^{3} ———. Sloane MS 276, fols. 75r–77v (1st q. 15th c.)

L^{4} ———. Sloane MS 338, fols. 53v–56v (1461)

L^{5} ———. Sloane MS 2006, fols. 53r–62r (1563)

L^{6} ———. Stowe MS 1070, fols. 51v–56r (2nd half 16th c.)

Le^{1} Leiden, Leiden University Library, Vossianus Chym. Q. 2., fols. 123v–27v (ca. 1500)

Le^{2} ———. Vossianus Chym. Q. 25., fols. 2r–8v (1st q. 16th c.)

Lu Lübeck, Stadtbibliothek, MS. Med. Quarto 12, fol. 40r–53r (last q. 16th c., after 1579; lost)[1]

M Manchester, John Rylands Library, MS 65, fols. 164v–68v and 172r (mid-15th c.)

Ma Marburg, Universitätsbibliothek, MS 23, fols. 136v–39r (1496)

Mp^{1} Montpellier, Bibliothèque Universitaire Historique de Médecine, MS H 493, fols. 236v–42v (1459)

Mp^{2} ———. MS H 479, fols. 121v–26v (1558)

Mu Munich, Bayerische Staatsbibliothek, Clm 27438, fols. 10r–18r (early 17th c.)

O Ottobeuren, Benediktinerkloster Ottobeuren, MS O. 86, fols. 246v–48v (unfinished, Operations 1–5; ca. 1466)

Oth^{1} Philadelphia, Science History Institute, Othmer MS 2, fols. 1r–2v (extr.; 3rd q. 15th c.)

Oth^{2} ———. Othmer MS 14, fols. 23r–32v (1520s)

P^{1} Paris, Bibliothèque Nationale de France, MS NAF 4141, fols. 29r–38v (1st q. 15th c.)

P^{2} ———. MS Lat. 11202, fols. 61r–69r (1st q. 15th c.)

1 I have included a lost witness here because it was sufficiently well described that it can be fitted into the lines of transmission: see Section 6.11 for its description. Two other lost witnesses are not included in this conspectus since they disappeared before being fully catalogued. The first is Assisi, Sacro Convento, Fondo moderno MS 19, fols. 206–30; these folios are missing, but a description of each chapter is given in the table of contents (fols. 2v–3r, "Index Johannis de Rupescissa capitulorum lapidis philosophici") showing that this lost witness contained the ancillary sections added to the Basel edition but also Operations 8 and 9, which were omitted from that edition. The second is Bern, Burgerbibliothek, MS 215, which contained, along with John's *Liber secretorum eventuum*, four alchemical treatises, including "Liber magisterii magni"; the alchemical texts are all now missing. This would be the only known case where one of John's alchemical texts appeared in the same collection with one of his prophetic works (see *LSE*, 97–99 and Hermann Hagen, *Catalogus codicum Bernensium* (Bern, 1875), 261–62).

P^3 ___________. MS Lat. 7173, fols. 76r–91v (early 17th c.)
Pd Padua, Biblioteca Universitaria, MS 1380, fols. 157r–68r (ca. 1600)
Po Poppi, Biblioteca Communale Rilliana, MS 90, fols. 224v–32v (1470s–80s)
Pr Prague, Library of the Metropolitan Chapter of the Cathedral of St. Vitus, MS O. LXXIX, fols. 210r–25r (early 17th c.)
Rc Rome, Biblioteca Casanatense, MS 3567, fols. 274r–79r (late 15th c.)
S Schlierbach, Zisterzienkloster, MS 85, fols. 37r–48r (1560s)
Sa Salamanca, Biblioteca Histórica General de la Universidad de Salamanca, MS 2108, fols. 59v–65r (Operations 1–6 only; mid-16th c.)
T Torino, Biblioteca Nazionale Universitaria, MS E.II.23, fols. 17r–21r (mid-16th c.)
U Philadelphia, University of Pennsylvania, MS Codex 69, fols. 75v–80r (3rd q. 15th c.)
Ua Amsterdam, Universiteit van Amsterdam, Allard Pierson Depot, MS PH338, fols. 49r–55v (ca. 1550)
V Vatican City, Biblioteca Apostolica Vaticana, MS Palatinus latinus 1330, fols. 119v–25r (1463–64)
W^1 London, Wellcome Library, MS 117, pp. 261–74 (1462)
W^2 ___________. MS 707, fols. 1r–10v (1430s)
W^3 ___________. MS 384, fols. 291r–94r (1570s–80s)
W^4 ___________. MS 385, fols. 163r–67v (1570s–80s)
W^5 ___________. MS 512, pp. 49–60 (1660s)
W^6 ___________. MS 517, fols. 33r–36v (2nd q. 15th c.)
W^7 ___________. MS 526, fols. 388r–90r (1525–27)
W^8 ___________. MS 710, fol. 24v (extr., operation 8; late 15th c.)
Wb Wolfenbüttel, Herzog August Bibliothek, Cod. Guelf. 444 Helmstedt, fols. 241r–42v (1436–37)
X^1 Oxford, Bodleian Library, MS e Museo 155, pp. 497–507 (ca. 1450s)
X^2 ___________. Bodleian Library, Ashmole MS 1423, section 5, pp. 68–77 (1593)
X^3 ___________. Corpus Christi College, MS 185, fols. 121r–24v (mid-15th c.)
X^{3b} ___________. Corpus Christi College, MS 185, fols. 80r–v (extr., Operation 8 only; mid-15th c.)
X^4 ___________. Corpus Christi College, MS 226, fols. 27r–30r (mid-15th c.)
Y^1 New Haven, Yale University Beinecke Library, Mellon MS 5, fols. 115v–19r (1st q. 15th c.)
Y^2 ___________. Mellon MS 19, fols. 116v–20r (1480)
Y^3 ___________. Mellon MS 34, fols. 46v–51r (1570s–80s)
Y^4 ___________. Mellon MS 35, fols. 342v–45v (1570s–80s)
Y^5 ___________. Mellon MS 36, fols. 24r–25r and 180r–88v (1570s–80s)
Z Venice, Biblioteca Nazionale Marciana, Lat. fondo antico 323 (=1579), fols. 114v–19r (1470s–80s)

Abbreviated Latin Versions

Fr_a Florence, Biblioteca Riccardiana, MS 439, fols. 45v–47v (mid-15th c.)

Hb_a Hamburg, Staats- und Universitätsbibliothek, Cod. Alch. 659, fols. 119v–20v (1468)

Le^{1}_{a} Leiden, Leiden University Library, Vossianus Chym. Q. 2., fols. 122r–v (ca. 1500)

Mp^{1}_{a} Montpellier, Bibliothèque Universitaire Historique de Médecine, MS H 493, fols. 233r–36r (1459)

Ve_a Vienna, Österreichische Nationalbibliothek, Cod. Palatinus 5487, fols. 105r–6v (2nd half 15th c.)

Z_a Venice, Biblioteca Nazionale Marciana, Lat. fondo antico 323 (=1579), fols. 185v–86v (1470s–80s)

Vernacular Translations

Castilian:

Psg_s Paris, Bibliothèque Sainte-Geneviève, MS 3438, fols. 43r–49r (2nd half 17th c.)

Catalan:

Lh_{cat} Bethlehem, Pennsylvania, Lehigh University Library, MS 10, fols. 145r–49r (1472)

Czech:

Le^{3}_{cz} Leiden, Leiden University Library, Vossianus Chym. F. 3, fols. 199v–205r (1585)

English:

L^{7}_{e} London, British Library, Harley MS 4724, fols. 54v–61v (1st half 17th c.)

Ua^{2}_{e} Amsterdam, Universiteit van Amsterdam, Allard Pierson Depot, MS PH177, fols. 101r–106r (ca. 1800)

X^{5}_{e} Oxford, Bodleian Library, Ashmole MS 1424, fols. 24v–26r (1614–16)

X^{6}_{e} ———. Ashmole MS 1450, part 6, fols. 157r–76r (frag.; 3rd q. 15th c.)

X^{7}_{e} ———. Ashmole MS 1490, fols. 143v–47r (1585)

French:

Or_f Orléans, Bibliothèque municipale, MS 291, fols. 20r–25r (1516)

German:

Bg_d Bamberg, Staatsbibliothek, Msc. Nat. 6, fols. 6r–21r (1442–43)

Go^{3}_{d} Gotha, Forschungsbibliothek Gotha der Universität Erfurt, Chart. B 1233, fols. 75v–79r (1st q. 17th c.)

Ha_d Halle, Universitäts- und Landesbibliothek Sachsen-Anhalt, 1/C a/2, fols. 297r–308v (ca. 1450)

Hb^{2}_{d} Hamburg, Staats- und Universitätsbibliothek, Cod. Alch. 656, pp. 1–36 (1546)

Ka^{3}_{d} Kassel, Universitätsbibliothek Kassel - Landesbibliothek und Murhardsche Bibliothek der Stadt Kassel, 4° Ms. chem. $33_{[6}$, fols. 1r–12r (1576)

$Ka^4{}_d$ ______. 4° Ms. chem. 32, fols. 1r–14v (end of 16th c.)
$Ka^5{}_d$ ______. 4° Ms. chem. 73, fols. 27r–36v (1505)
$L^8{}_d$ London, British Library, Sloane MS 2701, fols. 142v–51r (1581)
$Le^4{}_d$ Leiden, Leiden University Library, Vossianus Chym. F. 6, fols. 397r–403v (last q. 16th c.)
$Le^5{}_d$ ______. Vossianus Chym. F. 27, fols. 239r–50r (last q. 16th c.)
$Mu^2{}_d$ Munich, Bayerische Staatsbibliothek, Cgm 3741, fols. 200r–5r and 264r (ca. 1550)
$Wb^2{}_d$ Wolfenbüttel, Herzog August Bibliothek, Cod. 185 Blankenburg, fols. 124v–30v (1629)
$Wb^3{}_d$ ______. Cod. 187 Blankenburg, fols. 164r–73v (1631)
$Wb^4{}_d$ ______. Cod. 306.1 Extravagantes, fols. 1r–7v (1st half 16th c.)

Italian:

$Fn^3{}_i$ Florence, Biblioteca Nazionale Centrale di Firenze, Magliabechi Cl. XVI, cod. 76, fols. 315v–21v (last q. 16th c.)
Zc_i Venice, Biblioteca del Museo Correr, MS Correr 106, fols. 125r–34r (ca. 1605)

Early Modern Printed Editions

Latin:

DC *De confectione veri lapidis philosophorum*, in *Verae alchemiae artisque metallicae*, ed. Guglielmo Gratarolo, 2 vols. (Basel, 1561), 2:226–31; in *Alchemiae quam vocant, artisque metallicae doctrina*, ed. Guglielmo Gratarolo, 2 vols. (Basel, 1572), 2:341–58; in *Theatrum chemicum*, ed. Lazar Zetzner, 4 vols. (Strasbourg, 1613), 3:179–89; in *Theatrum chemicum*, 6 vols. (Strasbourg, 1659–61), 3:189–97; and in *Bibliotheca chemica curiosa*, ed. Jean Jacques Manget, 2 vols. (Geneva, 1702), 2:80–83.

LL *Johannis de Rupescissa liber lucis*, in *Secreta alchimiae magnalia*, ed. Daniel Broekhuizen, (Cologne, 1579), 35–48; (Leiden, 1598), 41–56; (Leiden, 1602), 41–56; (Leiden, 1612), 41–56; in *Theatrum chemicum*, ed. Lazar Zetzner, 4 vols. (Strasbourg, 1613), 3:278–89; in *Theatrum chemicum*, 6 vols. (Strasbourg, 1659–61), 3:284–95; and in *Bibliotheca chemica curiosa*, ed. Jean Jacques Manget, 2 vols. (Geneva, 1702), 2:84–87.

German:

BI *Tractat Bapsts Innocentii, Vom Stein der Philosophorum*, 227–32 (partial and garbled) in Joachim Tancke, *Promptuarium alchemiae* (Leipzig, 1610).

TM *Thesaurus mundi*, in *Heptas alchymica*, ed. Theophilus Neander (Halle, 1621), 294–320.

Notes for the *Stemma codicum*

1. The Greek sigla (signifying lost witnesses) at the top of the stemma were chosen to indicate the geographical area where each family developed: γ for Germanic lands, ε for England, ι for Italy, π for Provence-Northern Italy, φ for France.

2. These sigla are *not* positioned relative to the timescale. Their dates may vary widely despite the positioning of their sigla at roughly the same level on the page. Their dates cannot be determined beyond the *terminus ante quem* provided by their earliest known descendant; e.g., γ^3 must have been produced earlier than 1436 when **Wb** was copied, but *how much* earlier cannot be determined, and thus the position given to γ^3 on the page carries *no* implication of its date of composition.

3. The Greek sigla that appear further down in the *stemma* represent lost intermediates whose existence is clearly evidenced: β is the copy owned by the Florentine goldsmith Bisticci, ζ is a hybrid of $\pi^{1}{'''}$ and π^2, θ_d is the German translation later published in *Thesaurus mundi*, λ is the common ancestor of $\mathbf{Hb}^2_d$ and **S**, ν is the copy owned by Philipp von Nassau, and τ is a vernacular translation. Their dates of composition (except for ν) cannot be determined save for the obvious fact that they are later than their ancestor and earlier than their descendants.

4. The lines of connection to α are highly provisional. No witnesses earlier than ca. 1400 survive.

5. The subscripts to sigla indicate the format or the vernacular language of the witness: a = abbreviated, cat = Catalan, cz = Czech, d = German, e = English, i = Italian, and s = Castilian.

6. Three types of lines connect sigla:
A *solid line* indicates a demonstrable line of descent. In almost all cases however, there are unknown intermediates between any two connected sigla; e.g., it is extremely unlikely that the late sixteenth-century $\mathbf{Le}^4_d$ was translated directly from ι, which dates from before 1466; *DeConf* was not published directly from γ′.

When a siglum is attached by *solid* lines to *two* earlier sigla, *major* contamination is indicated, that is, each ancestor contributed substantially to the text; e.g., **Ua**.

A *finely-dashed* line indicates *minor* contamination; e.g., $\mathbf{X}^1$, whose prologue and furnace chapter come from γ^2 while the rest comes from $\varepsilon^{2\prime}$.

A *coarsely-dashed* line indicates that another exemplar was used to *correct or augment* a witness *after* its initial production; e.g., **Ge** which bears marginal corrections taken from γ″ (also from γ′, not shown in the large stemma).

7. Not all contaminations are indicated in this overall stemma. Some lines indicating contamination would needlessly obscure the stemma; e.g., $\mathbf{Y}^1$ was corrected using a descendant of γ, but the coarsely-dashed line to indicate this connection would cut across the stemma in an intrusive manner. The origin of some contaminations cannot be clearly identified. All contaminations are mentioned, and their origins identified where possible, in the manuscript descriptions and displayed in the more detailed stemmata provided in Section 6.

8. Not all witnesses could be positioned in the stemma; about thirty have been omitted. Eight of them were copied from printed editions and omitted for that reason (see Section 6.10). Several others could not be fit convincingly into any of the established families or subfamilies (see Section 6.8). The rest were either too heavily edited or contaminated to be positioned with a suitable degree of confidence.

9. Underlined sigla indicate the witnesses that were used for the critical edition.

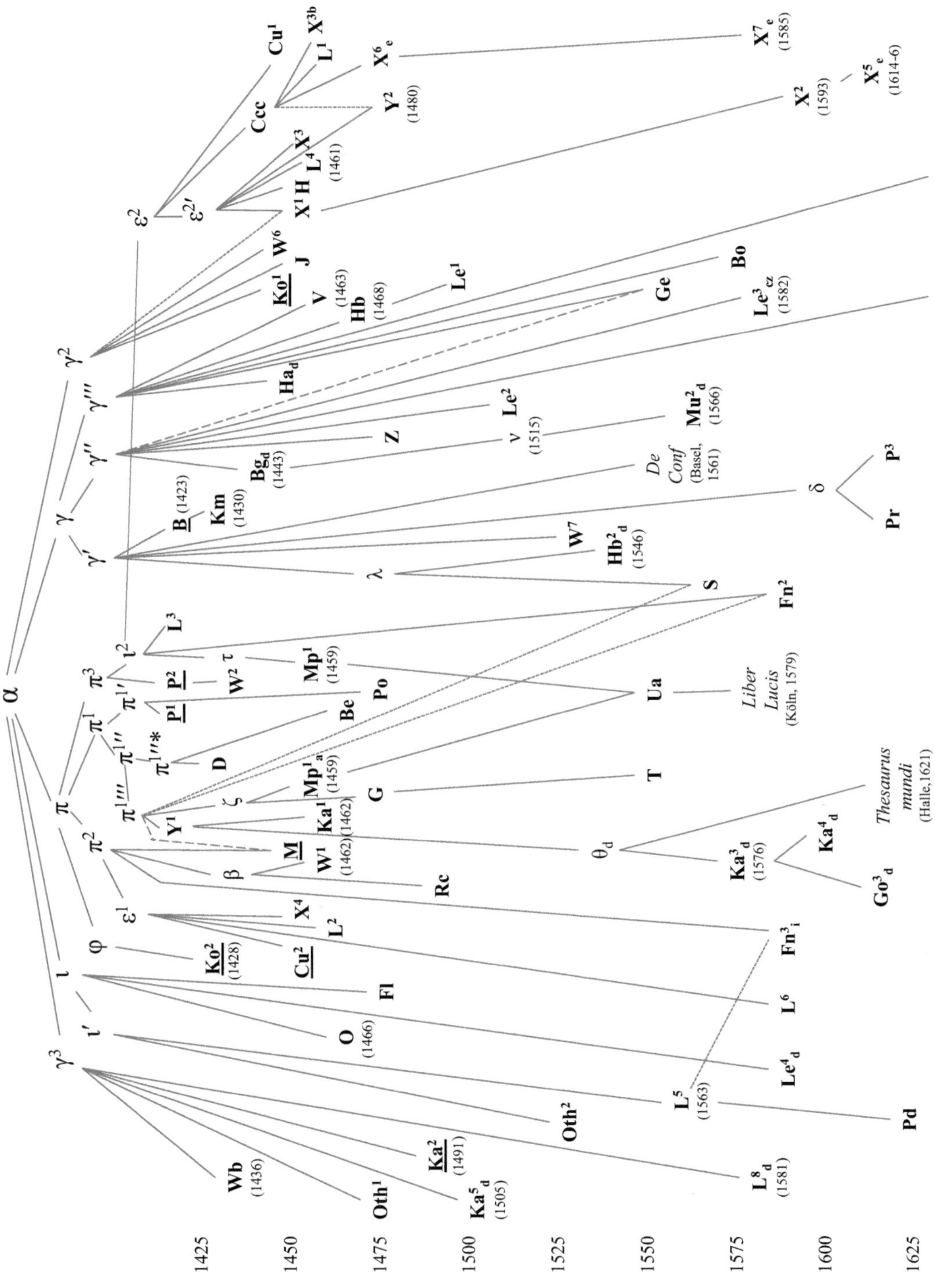
1425
1450
1475
1500
1525
1550
1575
1600
1625
Wb (1436)
Oth1
Ka5d (1505)
Ka2 (1491)
Oth2
L8d (1581)
L5 (1563)
Pd
O (1466)
Fl
Le4d
L6
Ko2 (1428)
Cu2
X4
L2
Fn3i
Rc
W1 (1462)
Ka1 (1462)
Y1
M
Mp1a (1459)
G
T
θd
Ka3d (1576)
Ka4d
Go3d
Thesaurus mundi (Halle,1621)
D
Be
Po
Ua
Liber Lucis (Köln, 1579)
P1
P2
W2
Mp1 (1459)
L3
Fn2
S
λ
W7
Hb2d (1546)
B (1423)
Km (1430)
De Conf (Basel, 1561)
δ
P3
Pr
Bgd (1443)
Z
Le2 (1515)
Mu2d (1566)
Had
Hb (1468)
Le1
Ge
Bo
Le3cz (1582)
Wb2d (1629)
Wb3d (1631)
V (1463)
Ko1
J
W6
X1 H
L4 (1461)
X3
Ccc
Cu1
L1
X3b
X6e
Y2 (1480)
X7e (1585)
X2 (1593)
X5e (1614-6)

4.2 Choice of Witnesses for the Critical Edition

The large number of extant witnesses had to be reduced to a manageable size for generating the critical edition. The principle employed was to choose the most complete and reliable witness from each of the families or subfamilies shown in the stemma, excluding those that descend from obviously damaged or edited hyparchetypes. Since the Provençal-North Italian family is so important, I chose five witnesses from that family. The two very early Provençal witnesses were chosen: $\mathbf{P^{I}}$ to represent π^{I} (via $\pi^{I\prime}$) and $\mathbf{P^{2}}$ to represent π^{3} even though it omits the prologue. **M**, due to its clarity and completeness, was included to represent subfamily π^{2}. $\mathbf{Ko^{2}}$, the only known descendant of φ, was likewise included given its early and well-documented origin in Paris in 1428 and the quality of its text. While German family γ contains the largest number of witnesses, I included only one in the critical edition. Subfamily γ′ is more reliable than subfamilies γ″ and γ‴; members of the former contain multiple interpolations and edits, while those of the latter have extensive lacunae and omissions and do not contain any authentic text that is not present in descendants of γ′. Therefore, I chose **B**, the earliest, most complete, and best-documented witness of subfamily γ′, to represent German family γ. I also included witnesses from each of the two much smaller German families, γ^{2} and γ^{3}. $\mathbf{Ko^{I}}$, an early and carefully prepared document that is much more complete than the highly edited **J** or the variously lacunose $\mathbf{W^{6}}$, stands in for γ^{2}. For family γ^{3}, I chose the later $\mathbf{Ka^{2}}$ over the earlier **Wb** since the latter omits operation 8 and contains a large number of clearly faulty readings.

I excluded all descendants of subfamilies $\pi^{I\prime\prime\prime}$ and ι^{2} due to their substantial lacunae and the fact that their readings could generally be supplied by witnesses stemming from π^{I} and π^{3} (respectively), which stand further back in the stemma. This decision also eliminates family ε^{2}, all of whose members omit the prologue, Operation 8, or both and contain other substantial lacunae and insertions. As a group, the English witnesses are the most corrupt, but I did decide to include $\mathbf{Cu^{2}}$ in the critical edition, the only complete and uncontaminated English witness, to represent ε^{I}. I also excluded the Italian family ι because of the apparent damage to ι that propagated through the family, and the even more severe corruption introduced into ι′ after a bungled attempt was made to fix it. Nevertheless, in one case, an uncorrupted portion of all the witnesses in family ι (and several outside that family) preserves a reading that I argue (see Section 4.3) is more authentic than that found in any of the witnesses included in the critical edition and has therefore been used to correct them. Throughout the making of the critical edition, when particular problems of choosing the most likely reading arose, I consulted witnesses not included in the edition to see if they offered help in resolving difficult issues. I did not make use of a "base text" as the final resort in the event of otherwise irresolvable doubts about the correct reading, since I could not demonstrate that any one witness out of the more than one hundred now extant could adequately serve that function universally. The most problematic choices between alternate readings are explained at length in Section 4.3.

The eight witnesses from which the critical edition was generated are therefore

B Klagenfurt, Bischöfliche Bibliothek, MS XXIX d 24, fols. 202v–14v (1423)
Cu² Cambridge, University Library, Kk. 6. 30, fols. 62v–68v (mid-15th c.)
Ka² Kassel, Universitätsbibliothek Kassel-Landesbibliothek und Murhardsche Bibliothek der Stadt Kassel, 4° Ms. chem. 10, fols. 25r–30r (1491)
Ko¹ Copenhagen, Kongelige Bibliotek, GKS 1712 kvart, fols. 111r–18v (2nd q. 15th c.)
Ko² Copenhagen, Kongelige Bibliotek, GKS 236 folio, fols. 52r–54v (1428)
M Manchester, John Rylands Library, MS 65, fols. 164v–68v (mid-15th c.)
P¹ Paris, Bibliothèque Nationale de France, NAF MS 4141, fols. 29r–38v (1st q. 15th c.)
P² Paris, Bibliothèque Nationale de France, Lat. MS 11202, fols. 61r–69r (1st q. 15th c.)

4.3 Difficult Readings and Editorial Interventions

1. Lines 20–21: "Et quia aqua viscosa, scilicet argentum vivum, generatur libenter in *cavernis*, dixerunt aliqui quod in locis vilibus reperiebatur."

Mercury's place of origin is the most variable word among the many witnesses of the *Liber lucis*. The word *cavernis* is transmitted by none of the witnesses chosen for the critical edition, which variously supply *latrinis*, *lacrimo*[?], *cloacis veteribus*, and *latrinis terre*. The reading *cavernis* is, however, given by all descendants of ι (excluded from the critical edition because of their corruption beginning in Operation 5), by **D** (a descendant of π) as *cavernis terre*, and by several later witnesses descending from γ (**Bo**, **U**, and **P³**), presumably as a correction. Other witnesses give *laminis terre* (**Po**), *locis vilibus lapideis* (ε¹), *lacis* (**Cu¹**), *lapide* (**L²**), and *sassis* (**Fn²**), all of which place the origin of mercury in a mineralogical context. On both logical and contextual grounds, I have chosen *cavernis* as the correct reading; its widespread corruption into *latrinis* is easily explained paleographically.

The idea that quicksilver is generated in latrines is somewhat strange. Authors before John nearly universally locate its origins underground in the Earth, beginning with the concept of Aristotle's twin exhalations in the *Meteorologica*, which was elaborated into the medieval mercury–sulphur theory of the metals. John himself states clearly at the beginning of this section that quicksilver is "extracted from the Earth" (*extrahitur de terra*). His definition of quicksilver as *aqua viscosa* (thick water) indicates his dependence upon Geber's *Summa perfectionis*, which likewise describes it as *aqua viscosa in visceribus terrae* (a thick water in the viscera of the Earth), clearly also giving it a subterranean origin.[2] Geber's exact phrase was also coopted into pseudo-Albert's late thirteenth-century *Semita recta*, which goes on to say that quicksilver is found "in the mountains and especially in old drains." The latter phrase, *in cloacis veteribus*, reappears verbatim in **M/Rc/W**^{I} (π^2), having been taken from the *Semita recta* by a copyist and inserted into the *Liber lucis* at this point, presumably to replace what that copyist considered doubtful reading. The variant *latrinis terre* (latrines of the Earth) given by **Be**, **P**^{I}, and **Ka**^{2}**/Wb** (γ^3) is itself largely nonsensical, but would make perfect sense as a corruption of *cavernis terre*, the reading given by **D**.

The corruption of *cavernis* into *latrinis* is an easy mistake, and given the frequency of this reading in several families it must have occurred early in the transmission. Initial letters c and l could easily be confused in Gothic scripts; a letter v with a vertical stroke (the usual abbreviation for *ver*) if slightly rotated could be read as a flagged t representing *tri*; and the last syllable (*-nis*) is identical in both words. Given a poorly legible manuscript, the subsequent mention of *stercoribus* (l. 23) could incline a copyist toward the related *latrinis*, as would the widespread notion—dating back to Arabic sources—that the starting material for the philosophers' stone is cheap, common, despised, and "trampled in the dungheap," a topos alluded to by John himself when he writes that quicksilver is of "low price and able to be found everywhere."[3] The reading *latrinis* also undercuts John's subsequent mockery of the *filii bestiarum* who seek the material "literally in excrements"; if quicksilver were in fact "generated freely in latrines," then seeking it in excrement would not be so foolish. John implies that their foolishness lies in not understanding the multiple meanings of *vilis* that "some authors" (*aliqui*) use to refer to the places where quicksilver is generated: low, common, vile, despicable (cf. *vilis precii* in the previous line). Hence, they seek the material "literally" (*ad literam*) in a "vile and despicable" place rather than metaphorically in a "low, common" place, such as the caverns of the Earth.

2. Lines 72–73: "in colore floris *gaudii* amenissimo supramodum"

The exact meaning of "in the exceedingly pleasant color of a flower of delight" is difficult to divine. Other witnesses (**W**^{I} and most descendants of ι) replace *gaudii* with *candidi* or *albi* (white), both of which are certainly wrong because, having performed the process as John describes, I obtained a bright orange-red material that correlates perfectly with his claim that this material is the separated "tincture of redness." Two witnesses (**Go**^{I} and the additions to **Ge**) read "auri" here,

2 Newman, *Summa*, 333 and 669, and "The Genesis of the *Summa perfectionis*," *Archives internationales d'histoire des sciences* 35 (1985): 240–302.

3 See for example, Morienus to Khalid, in *A Testament of Alchemy*, ed. Lee Stavenhagen (Hanover, NH: Brandeis University Press, 1974), 24 and 26: "in sterquiliniis calcatur." This author nevertheless ridicules those who "labor in the dungheap" to extract the stone or its starting material. It is something quite different to state that *quicksilver*, as both the liquid metal (Hg) and the basis of metal composition, is found in excrement or generated in latrines.

which is a contamination originating from an appeal to the parallel process in *De famulatu* which calls the residue *pulchrior omni auro* (more beautiful than all gold), which is not necessarily a reference to color. $\mathbf{P}^{\mathrm{I}}$ gives *floris giladii*, which is difficult to interpret; **Br** provides *floris glocis*, which is equally problematic; **Po**, which descends from the same hyparchetype as $\mathbf{P}^{\mathrm{I}}$, gives *floris granarii*, which is nonsensical (flower of a granary) but which might be a corruption of *floris granati* (a pomegranate flower)—such flowers are in fact very close in color to the substance I prepared. It seems reasonable that John, who displayed his botanical knowledge elsewhere, named here some flower with which he was familiar, perhaps using its Occitan name, which would have been unrecognizeable to most copyists. The same occurs in *De famulatu* where John gives "in the vernacular of the Aquitaine" [*in vulgari aquitanico*] the Occitan name *trescalam* for St. John's wort (*Hypericon perforatum*), which is variously corrupted in many of the extant witnesses.[4] It is not possible therefore to be certain about John's original word, although there are several reasonable possibilities—the *granati* (pomegranate) arguably suggested by **Po**; *gladioli* (gladiolus, variously *glaieul*, *glaug*, and *glaujol* in local vernaculars, possibly an origin for the *giladii* and *glocus* of $\mathbf{P}^{\mathrm{I}}$ and **Br**; **Ma** gives the vaguely similar *glaudiali*, but it is not a very reliable witness); and so forth. Yet one can be quite confident that the original word was the name of a red or orange-red flower. Given the apparent irretrievability of that original term, I have left the common reading *gaudii* in place in the edition, and signaled its unsuitability in a footnote.

3. Lines 84–91: "extrahe aquam fortem *hoc modo...et serva*."

This section on how to prepare the aquafort appears in several important early witnesses belonging to the Provençal-North Italian family π, specifically those descending through π^{I}: **M**, **Po**, $\mathbf{P}^{\mathrm{I}}$, $\mathbf{Y}^{\mathrm{I}}$, **G**, and **T**, as well as **D** and the later **S**, but not in a majority of the witnesses chosen for the edition. I have chosen to include it in the critical edition on the grounds that the practical details of a laboratory operation that was well understood by the fifteenth century (when the copies were made) are much more likely to be omitted than to be added by a copyist. For example, John's practical details for performing the sublimation in Operation 1 are omitted in many witnesses, sometimes (as in the descendants of γ''') replaced with "[do] as you know how" (*ut scis*). In fact, the extended instructions for making aquafort are omitted in exactly this way from the abbreviated version $\mathbf{Mp}^{\mathrm{I}}{}_{\mathbf{a}}$ which itself descends through π^{I}: "make an aquafort as you know how" (*fac aquam fortem ut scis*). **Be**, also a descendant of π, preserves the beginning and ending words of the instructions while omitting the rest, and thus reads "distill an aquafort from them in this way as is fitting, which water keep in its its own vessel" (*distilla ab eis aquam fortem hoc modo ut decet quam aquam serva in vase suo*).

4 E.g., in *De consideratione* (1561), 141, the word is corrupted into *transcalam*.

4. Lines 143–45: "Licet mercurius in se habeat tincturam ad rubeum intrinsecus absconsam sed per purgaciones, *quia nimis est terrea, diminutam* et ob hoc indiget sulphure purissimo vitrioli epotari ad plenum."

The reading *terrea diminutam* occurs in only a minority of the chosen witnesses (**Ko**^{I} and **Ko**^{2}). Both **P**^{I} and **P**^{2} give *terra diminuta*, which gives the meaning that "the *earth* is too diminished" by the purgations, which does not make sense since one of the stated purposes of the purgations is to remove the earthiness from the mercury. Both **M** and **Cu**^{2} insert the word *materia* earlier in the sentence, such that *diminuta* refers back to it, giving the meaning that the "material is diminished." But all of the chosen witnesses conclude that "on this account, it [i.e., the mercury] needs to drink up fully the most pure sulphur of the vitriol," indicating that the sulphur (the tincture of redness) must come from the vitriol even though the "mercury has in itself a tincture for the red hidden within." Why is the mercury's own tincture not sufficient? A logical explanation is given only by the readings of **Ko**^{I} and **Ko**^{2}, in which *terrea* is preserved and *diminutam* refers back to *tincturam*, which explains that the mercury has lost its own tincture of redness because during the purifications the *tincture* within the mercury is diminished because "it is too earthy."

5. Lines 167–70: "et dixerunt ipsum quidam philosophorum animal *cum sanguine suo quia crescit sublimando, sanguis quia habet animam* sanguinis, id est tincturam rubedinis, spiritus vitrioli supradicti."

In addition to numerous variants found throughout the entire sentence, the words italicized above appear in only a minority of the witnesses chosen for the edition (fully in **P**^{2}, partly in **B** and **Ka**^{2}), and even then the exact phrasings vary. They are lacking in all descendants of the important Provençal-North Italian family except those that descend through π^3.[5] Choosing either the *lectio difficilior* or the *lectio plurium codicum* would eliminate them. The question is whether the italicized words were *added* to John's text as an explanatory gloss or *omitted* from his original.

I decided to include them in the critical edition for the following reasons: John always explains the metaphorical terms he cites from other authors, for example, with his quotations from pseudo-Arnald in the *Liber lucis*, and likewise in his prophetical writings; thus it is likely that he did the same here. Furthermore, without this phrase (as in **Cu**^{2}, **Ko**^{2}, and **M**), the "tincture of redness" is equated with the "animal," rather than with its blood, which is less fitting. What is here being called "animal" is not itself red like blood, but instead holds the tincture of redness unseen within itself, like an animal holds its red blood hidden within; hence, it is clear that the blood, and not the animal, should be equated with the *tinctura*

[5] The families descending through π^3 also have some version of these words; they are fully present within ι^2 with variants in ε^2.

rubedinis, which occurs when this phrase in question is included. The widespread loss of these words could easily be the result of an eye-skip from *animal* to *anima* that occurred early in the transmission. The only *partial* omission of these words from descendants of γ^3, such as **Ka²** (which reads "animam cum suo sanguine, id est tinctura rubedinis") and elsewhere, could be the result of a different eye-skip from *sanguine* to *sanguinis*, plus the transformation of *animal* into *animam* from a misreading. **Pᴵ** likewise gives *anima* here for *animal*, while **Koᴵ** ungrammatically gives *alius*. The large number of variants throughout this sentence possibly indicates a legibility problem in the autograph or a very early witness.

6. Lines 240–44: "Commendo igitur hoc revelatum archanum in manu crucifixi sanctissimi domini nostri Jhesu Christi a quo recepi hoc bonum, et custodie angelorum sanctorum et evangelicis pauperibus qui sollicitudine divitiarum *seduci, deo servante,* non possunt."

The last part of this line is problematic. Both the *lectio difficilior* and the *lectio plurium codicum* (**Cu²**, **Ka²**, **Koᴵ**, **Ko²**, **M** and **Pᴵ**), provide the reading "I commend this revealed arcanum...to the poor men of the Gospel who, having been seduced by the desire for riches, cannot serve God" (*evangelicis pauperibus qui sollicitudine divitiarum* seducti deo servire *non possunt*). This reading does not make sense, nor does it fit the context since John has already stated at length that he wants the arcanum kept from those who would misuse it for purposes other than serving God and the Church. How would the *evangelici pauperes* remain true to their Franciscan calling if they have been seduced by the allure of wealth and thus "cannot serve God"? Such corrupted Franciscans would hardly align with the other two worthy custodians of the arcanum cited here in parallel: Christ and the holy angels. But several witnesses, all with some link to π^3 (**P²**, **W²**, **B**, and **L²**), give a reading that fits far better with the sense of this section: "to the poor men of the Gospel who, with God protecting them, cannot be seduced by the desire for riches" (*evangelicis pauperibus qui sollicitudine divitiarum* seduci deo servante *non possunt*).

4.4 Editorial Principles

In producing this critical edition, the following minor differences in readings have been ignored in the apparatus:

Orthographic variants (e.g., sulphur/sulfur; forficibus/forpicibus; alutel/aludel/aludellus; -tio/-cio; rotundo/rotondo; destruatur/dextruatur; yhesu/jhesu; inmittas/immittas)

Transpositions (e.g., ibi est/est ibi, sacrosancta romana ecclesia/ecclesia romana sacrosancta)

Roman vs. Arabic numerals vs. words for quantities and dates (all are given as Roman numerals)
Use of an adjective of the wrong gender, number, and/or case (e.g., medicine reservato)
Whether or not a non-Latinate noun is declined (e.g., elixir/elixiris, aludel/aludello), these have been normalized as indeclinable nouns
Errors in whether a verb is deponent or not (e.g., operari/operare)
Use of subjunctive vs. indicative (in the same tense, person, and number), unless there is a difference of meaning
An added or omitted *et* that does not alter the sense
Minor words with equivalent meaning: vel/aut, ante/antequam, post/postea, seu/sive, neque/nec, nil/nichil, igitur/ergo, prout/ut, and quia/quod when used in the sense of "because."

In terms of normalization: u/v and i/j are standardized (u/i used for vowels, v/j for consonants); capitalization is normalized according to modern usage, and punctuation and paragraph breaks are added according to sense and clarity; numerals used as abbreviations (e.g., 2° for *secundo*, 4^{m} for *quartum*) are spelled out; proper names are given in their standard Latin orthography (e.g., Arnaldus, Geber, Avicenna). Medieval spelling, as used in the majority of chosen witnesses, has been preferred (e.g., *sublimacio* not *sublimatio*, and *dyadema* not *diadema*), and the loss of the diphthongs *ae* and *oe* has been preserved (e.g., *celum* and *tincture*).

Italics are used in both the critical edition and its translation to indicate quotations from the Bible or from other sources, with reference information in the footnotes. There may well be additional quotations that I have failed to recognize.

5 CRITICAL EDITION AND ENGLISH TRANSLATION

DOI: 10.1201/9781003666547-7

Liber lucis magisterii magni

Consideravi tribulaciones electorum in sacrosancto evangelio prophetatas a christo, maxime tribulaciones temporum antichristi, *instare in januis*, quibus est sacrosancta universalis romana ecclesia haud dubium plurimum affligenda, et ad montes fuganda, et certe per tyrannos omnibus diviciis temporalibus spolianda in brevi. Sed licet *jactetur in validis fluctibus petri navicula, est tamen liberanda in fine tribulacionis dierum* domina generalis. Quapropter ad solvendam et sublevandam gravem inopiam et paupertatem futuram populi sancti et electi dei, *quibus datum est nosce misterium veritatis sine parabolis*, lapidem maximum philosophorum ad lunam et ad solem, ad album et ad rubeum, volo dicere ad argentum et aurum, clarissime revelabo contra morem philosophorum qui ante me fuerunt qui tanta invidia percussi leguntur ut propriis filiis celaverunt veritatem. Et quia non est intencionis mee probative aut recitative aut impugnative procedere, sed tantum sub simplici narracione

---

1 Consideravi tribulaciones] Consideravimus magnas tribulaciones P1; *om. prologum* P2; *antea add. rub.* Jhesus maria franciscus. Incipit lieber magisterii confeccionis veri lapidis philosophorum clare et absque quacumque palliacione editus per magistrum Johannem de Rupescissa ordinis minorum virum utique a deo illustratum quem libellum composuit ad sublimandas inopias pape et cleri et tempore future tribillacionis ecclesie eciam ad hoc divina premonitus revelacione B; *antea add. rub.* Incipit magisterium lucis. In nomine domini Amen. Cu2; *antea add.* Ego frater Johannes de Rupescissa Cu2; *antea add. rub.* Incipit rupssissa Ka2; *antea add. rub.* Incipit liber lucis fratris Johannis de rupecissa. Et primo incipit prologus M; *antea add. rub.* Incipit liber magistri Johannis de rupescissa qui dicitur liber lucis et tribulacionis P1 | sacrosancto evangelio] sacrosancta ecclesia B; evangelio Cu2; sacrosancta evangelia Ka2; sancto evangelio Ko1 **2** prophetatas] prophetas Ko2; per prophetas dictas et M maxime] *add.* et principaliter prophetatas M | temporum] *om.* B; tempore Ka2 P1 | instare] *om.* B **3** in januis] *om.* B; januis Cu2 M; in annis Ka2; *add.* tam mistici quam realis et veri satis scito et quasi in januis nisi deus pro consolacione bonorum et correccione malorum velit tempus cum possit misericorditer prolongare vel prorogare B | quibus] in quibus B; sub quibus Ka2 | est] *om.* B universalis] *om.* Ka2 P1 | haud dubium] sine dubio Ka2; hoc dubium Ko1; hanc dubium Ko2; *om.* P1; *add.* ut M **4** fuganda] fugienda B Ka2 P1 **5** temporalibus spolianda] *om.* Cu2 | in brevi] in successu temporis Ko1 | Sed] et Ko2 | jactetur] jacture michi P1 | in validis] in variis Ka2 **6** fluctibus] fructibus P1 | tribulacionis] *om.* Ko1 **7** generalis] generalibus P1; *add.* gentium M | Quapropter] quod propterea P1 | solvendam...sublevandam] sublimandam B; sublevandam Ka2 Ko1 | gravem] *om.* B **8** futuram] *om.* B; futuri Ka2 **9** nosce] noscere Ko2 | veritatis] sue veritatis P1 | sine] in P1 lapidem maximum] lapidis maxime B **10** ad[3]...rubeum] *om.* B P1 | volo dicere] *om.* B; id est P1 **11** ad...aurum] *om.* B | revelabo] relevando Ka2 | contra...philosophorum] De quo omnes philosophi Cu2; quod moris philosophorum non erat M **12** qui[2]...invidia] *om.* Cu2 | percussi leguntur] *om.* Cu2; perculsi fuerunt M | ut] *om.* Cu2; quod M P1; *add.* numquam aperte de hoc scripserunt quinymmo M **13** celaverunt] celarunt B | veritatem] *om.* B | Et] Sed M | probative...14 recitative] probacione aut recitacione Ko1; probative P1 **14** aut...procedere] aut impugnacione probare Ko1 sub] *om.* B

B Cu2 Ka2 Ko1 Ko2 M P1 P2 **1** Consideravi...17 accedo] *om. prologum* P2

The Book of the Light of the Great Magistery

I have considered how the tribulations of the elect prophesied by Christ in the holy Gospel,[1] especially the tribulations during the times of the antichrist, *stand at our gates.*[2] By these tribulations the holy universal Roman church shall without doubt be greatly afflicted, be made to *flee to the mountains,*[3] and surely be despoiled by tyrants in a short time of all her temporal wealth. But although the *little ship of Peter shall be tossed about by powerful waves, it shall nevertheless be liberated at the end of the days*[4] of tribulation as the universal ruler. For this reason, in order to remove and relieve the severe want and future poverty of the holy and elect people of God, to whom *it is given to know the mystery of truth without parables,*[5] I shall most clearly reveal the greatest stone of the philosophers for Luna and for Sol, for the white and for the red, that is to say, for silver and gold, contrary to the custom of the philosophers who came before me, who are found to be so stricken with such great envy that they hid the truth even from their own sons. And because it is not my intention to proceed in a probative, declamatory, or argumentative fashion, but instead to inform the holy universal Church faithfully with only a straightforward narration of words about

---

1 Especially Mt 24 and Mk 13.

2 Mt 24:33; Mk 13:29; John uses the same verse similarly in the *VM* (Tealdi, 217; Kaup, 134-36).

3 Mt 24:16; Mk 13:14; Lk 21:20; used also in *VM* (Tealdi, 252; Kaup, 239)

4 See Mt 8:23-27 regarding the image of the ship of Peter (a metaphor for the Church) tossed by violent waves but saved by Christ's intervention. The phrase that John uses here is taken from the prophecy *Alta cedrus Libani* (*The Tall Cedar of Lebanon*) or "Tripoli Prophecy": *navicula Petri iactabitur validis fluctibus, sed evadet et dominabitur in fine dierum*"; see the edition in Robert E. Lerner, *The Powers of Prophecy: The Cedar of Lebanon Vision from the Mongol Onslaught to the Dawn of the Enlightenment* (Ithaca, NY: Cornell University Press, 2009), 204. John uses the same words again in *LO*, 429 (*navicula Petri...jactabitur...in validis fluctibus*) and alludes to the same image in *VM* (Tealdi, 235; Kaup, 161n83 and 187).

5 Cf. Lk 8:10.

verborum sacrosanctam generalem ecclesiam de veritate experimento probando fideliter informare, omissis contencionibus, ad praxim summatim et breviter accedo.

Materia lapidis est res una, vilis precii, ubique reperibilis, quia est aqua viscosa, scilicet argentum vivum commune, sicut extrahitur de terra. Et quia aqua viscosa, scilicet argentum vivum, generatur libenter in cavernis, dixerunt aliqui quod in locis vilibus reperiebatur. Et multi filii bestiarum non intelligentes intentum philosophorum ipsam ad literam in stercoribus quesierunt. Hec autem materia preparatur aliquid amovendo, scilicet terrestreitatem quam habet, et aliquid apponendo, scilicet sulphur philosophorum quod non est sulphur vulgi sed est sulphur invisibile, tinctura rubedinis. Et ego nominabo tibi eum nomine suo, et certe est spiritus vitrioli romani. Et iste due preparaciones fiunt simul sic:

Accipe salis petre et vitrioli romani ana libras ii, et tere eas subtiliter, et in patella terrea movendo desicca ad ignem lentum aliquantulum ut

---

15 generalem] romanam Ko1 | de] *om.* B | veritate] *om.* B; veritatis M | experimento… 16 probando] doctrina et experimento B; experimento probandi Ka2; exprimendo Ko1; et experimento probandi probabili P1 **16** fideliter] et fideliter probando Ko1; fedelibus P1 | contencionibus] contemptionibus M P1 | ad] *om.* Ko2 P1 | praxim] propositum B; praticam Ka2 M; *om.* P1 | summatim] *om.* M P1 **17** et breviter] et in brevitate B Ka2; et breviare Ko2; breviter M **18** Materia] Natura seu materia B; *ante add. rub.* Materia lapidis Ka2; *ante add. rub.* Que sit materia lapidis nostri et qualiter preparatur M | res una] res B **19** viscosa] *add.* tantum Ko2 P1 | scilicet] tamen sit B | commune] *om.* P2 | de] a Ko1 | terra] minera terre B M **20** aqua… vivum] *om.* P1 | generatur libenter] generatur B Cu2; quod generatur libenter Ka2; *om.* P1; *incipit lacuna* P2 **21** cavernis *correxi secundum ι et D*] latrinis B Cu2 Ko1; latrinis terre Ka2 P1; lacrimo[?] Ko2; cloacis veteribus M; *add.* ideo Cu2 M | aliqui] antiqui phylosophi M | vilibus] ubilibet B | reperiebatur] reperitur M P1; *add.* dicentes Est omnium rerum vilius quod nos phylosophi veneramur M | Et] unde P1 | filii… 22 non] bestiales sic B **22** intentum] sensum Ko1; dicta P1 | ad literam] in latrinam Cu2; ad latrinam vel cloacam Ko1; ad latrinam Ko2; *om.* M; ad terram in latrinis P1 | in… 23 stercoribus] stercorum Cu2; in stercore P1 **23** quesierunt] quesiverunt Ka2 P1 | Hec autem] Et Cu2; Hec M; *finitur lacuna* P2 | materia preparatur] natura operatur B; preparatur P2 | aliquid] *om.* P1 | amovendo… 24 terrestreitatem] inde abmovendo scilicet terreitatem P2 **24** et aliquid] aliud et aliquid P1 | apponendo] *om.* P1 | scilicet[2]… 25 quod] *add.* non habet et illud quidem sulphur M **25** est[1]] *om.* Ko2 | est[2]] *om.* Ka2 | sulphur invisibile] *add.* philosophorum et M **26** tibi eum] id tibi Cu2; tibi ipsum Ka2; eum P1 | nomine suo] *add.* vero et proprio M **27** spiritus] species P1 | romani] scilicet romani B; viridis M; *add.* id est leonis Cu2 | iste] ille Ko1 **28** Accipe] Recipe B M P1; *add.* ergo in nomine domini nostri yhesu christi M | salis petre] salis communis vel salis petre P1; sal petra P2 | romani] viridis M | libras ii] libram i Cu2 M; *add.* de quolibet M | eas] *om.* B M P1 **29** patella] patena Ka2 Ko2 P2 | terrea] ferrea B; *om.* Cu2 M desicca] et sicca Cu2 | ad… aliquantulum] aliquantulum ad ignem Cu2 Ko2 M; cum igne aliquantulum Ko1 | ut] et fondatur et P1

B Cu2 Ka2 Ko1 Ko2 M P1 P2 **20** generatur libenter] *incipit lacuna* P2 **23** Hec autem] *finitur lacuna* P2

the truth by means of experience that will be demonstrated, I now, with all controversies laid aside, briefly and summarily come to the practical part.[6]

The material for the stone is a single thing, of low price, and able to be found everywhere, because it is a thick water, namely, common quicksilver as it is extracted from the Earth.[7] And because this thick water, namely quicksilver, is freely generated in caverns some have said that it was found in despicable places.[8] And many offspring of fools, not understanding the intent of the philosophers, have sought for the material literally in excrement. This material is prepared by removing something, namely the earthiness which it has, and by adding something, namely, the sulphur of the philosophers, which is not common sulphur but an invisible sulphur, the tincture of redness.[9] And I shall name it for you with its own proper name: it is for certain the spirit of Roman vitriol.[10] These two preparations are done at the same time thus:

Take saltpeter and Roman vitriol, two pounds of each, grind them finely and dry them a little by stirring them in an earthenware dish on a gentle fire so that

---

6 Cf. the similarly worded promise of brevity in *VM* (Tealdi, 218; Kaup, 136-38): "in hoc loco narrative, sine manifesta probatione, procedam."

7 This description of quicksilver as an *aqua viscosa* probably depends upon Geber, see Newman, *Summa perfectionis*, 333 and 669.

8 On the reading of this important line, see the discussion of my editorial intervention in Section 4.3.

9 On the earthiness of quicksilver in Geber, see Newman, *Summa perfectionis*, 462-65 and 721-23.

10 The Latin text "ego nominabo tibi eum nomine suo" may well be an allusion to the corresponding revelation of the alchemical starting material in the pseudo-Aristotelian *Secretum secretorum*--"ego nominabo ipsum nomine suo"; see Robert Steele, *Opera hactenus inedita Rogeri Baconi*, Volume 5: *Secretum Secretorum* (Oxford: Clarendon, 1920), 115 and 261. While "spirit of vitriol" in much later chymical texts refers to the distillate of green vitriol, namely, sulphuric acid, here John refers instead what he perceives to be an "invisible spirit" hidden within the vitriol, as he defines it a little later in the text.

aquea humiditas recedat. Deinde recipe libram i argenti vivi optimi de minera extracti et non de plumbo facti, involve in panno lineo et sparge ipsum super materiam predictam calidam aliquantulum quia seminabitur per poros panni. Et movebis aliquantulum cum baculo et imbibetur in materia predicta ut non appareat in ea. Pone totum in vase aludel ad sublimandum, primo cum lento igne donec humiditas per foramen capitis vasis recedat aliquantulum, quod cognosces cum gladio si positus super foramen non recipiat ex fumo ascendente humorem. Post claude foramen cum luto sapiencie. Sed tucius est ut ante materia in patella paululum desiccetur et in aludel non sit foramen quodcumque. Et fortifica ignem per xii horas et in fine da forciorem ut bene totus mercurius sublimetur. Postea extrahe ipsum album ut nix.

Et in hoc opere fecisti duo. Primo, purgasti mercurium a terrestreitate et nigredine et humiditate maxima qua vivus currit, et congelasti eum, et

30 aquea…recedat] aque humiditas recedat B Cu2; aqua et humiditas recedant Ka2 | recipe] accipe Cu2 P2; *om.* Ko1 Ko2 | argenti vivi] mercurii B | optimi] *om.* B P1 P2 **31** non de] non Cu2 Ko2; non in Ko1; *om.* P1 | plumbo] *om.* P1; *add.* nec de ferro vel per aliquod aliud artificium scilicet de herbis vel plantis extracti vel M | facti] *om.* Cu2 P1 | involve] wulve ipsum B; volve Ka2 P2; et dimitte transire M | in…lineo] in coreo leporis B; per pannum lineum vel corium M; in panno lini P1 | et sparge] spargendo M; et asperge P1 **32** ipsum] eum P2 | materiam…aliquantulum] dictam materiam calidam P1 | quia] cum baculo quia Cu2; et Ko1; *incipit lacuna* M; *om.* P2 | seminabitur] spergetur magis subtiliter P1; *add.* et exibit et mundificabitur transeundo B **33** panni] corei predicti B; dicti panni P1 | Et] *incipit lacuna* B | movebis] move P1 | aliquantulum] materiam aliquantulum Ka2 P2; *finitur lacuna* M | cum baculo] *add.* miscendo M | et imbibetur] ut imbibatur Ka2; imbibe M | in] cum Ka2 P1 **34** predicta] *finitur lacuna* B | in ea] *add.* et B Ko1; *add.* et parum move et quando fuerit bene mixta Ka2; *add.* Postea M; *add.* Et post pone dictam materiam cum infrigidata fuerit in mortario de petra et incorpora cum pistello ligni. Et postea P1; *add.* dictum argentum vivum P2 | totum] hoc totum M | vase aludel] vas vitreatum in alludel Ka2; cucurbita M **35** primo] *om.* Cu2 M; Et primo fac P1 **36** vasis] *om.* Ka2 P1 | recedat] exeat P1 | aliquantulum] *om.* M P1 | cognosces] cognoscitur Ko1 | gladio] lamina ferrea Cu2; cutello Ko1 M | si] *incipit lacuna* Ko1; *om.* P1 | positus] posita Cu2; fuerit positus Ka2 M; posito P1 | super…37 foramen[1]] supra foramen Ka2; *add.* quando P1 **37** ex] de Ka2 M P1 | ascendente] astante Cu2 | humorem] *finitur lacuna* Ko1 **38** est ut] *om.* B P1 | ante] *om.* Cu2; antea Ka2 | in patella] *om.* B Ka2 P2; in patena Ko2 | paululum] paulatim P1 **39** et] ut Cu2 Ka2 Ko2 P2 | non] nullum B | quodcumque] *om.* B; aliquod Ko1 P1 | Et fortifica] *add.* paulatim Ka2 **40** forciorem] ignem fortiorem B M P1 | bene] *om.* P1 | Postea] *om.* Cu2 Ka2 Ko2; Postea eo refrigerato M **41** extrahe] extrahes P2 | ipsum] eum Ka2; *add.* et erit P1 | album] *add.* illud sublimatum P1 | ut] sicut B P1; ut est Ko1 P2 **42** Et] *om.* B; *incipit lacuna* Cu2 M; Ecce enim quod P1 | in…opere] In isto autem opere B; *om.* P2 | purgasti] separasti P1 | mercurium] ipsum B; eum Ka2 Ko2 P2; argentum vivum P1 | terrestreitate] terreitate P1 **43** et nigredine] *om.* Ko1; *add.* maxima P1 | et[2]…qua] *om.* P1 | vivus] unde B; *om.* Ko1 P1; argentum vivum P2 | currit] *om.* P1 et[3]] Secundo P1 | congelasti] coagulasti B

B Cu2 Ka2 Ko1 Ko2 M P1 P2 **32** quia] *incipit lacuna* M **33** Et] *incipit lacuna* B aliquantulum] *finitur lacuna* M **34** predicta] *finitur lacuna* B **36** si] *incipit lacuna* Ko1 **37** humorem] *finitur lacuna* Ko1 **42** Et] *incipit lacuna* Cu2 M

the aqueous humidity departs. Then take one pound of the best quicksilver, extracted from the mine and not made from lead, and wrap it in a linen cloth and sprinkle it over the aforesaid material made a little warm, for it will be disseminated through the pores of the cloth. And you will stir a little with a stick, and the quicksilver will be drunk up in the aforesaid material so that it is not visible in it. Put everything into an aludel to be sublimed, at first with a gentle fire until the moisture departs a bit through the opening at the top of the vessel. You will know this using a sword, if when placed over the opening it does not pick up any moisture from the ascending fume. Then close the opening with the lute of wisdom.[11] But it is safer if the material be dried somewhat beforehand in a dish and there not be any opening in the aludel. Increase the fire for twelve hours, and at the end give a stronger fire so that all the mercury is well sublimed. Then take it out, white as snow.

In this work you have done two things. First, you have purged the mercury from earthiness and blackness, and from the very great humidity by the means of which it runs quick.[12] And you have congealed it and turned it white as snow,

[11] Lute of wisdom (*lutum sapiencie*) is a mixture of various ingredients including clay that was used for sealing small openings and for protecting soft glassware from being damaged or melted by the heat of the fire.

[12] I.e., "flows as a liquid." The idea here is that the "very great humidity" ordinarily present in quicksilver is what allows it have a highly-mobile liquid form. Once this "humidity" is removed, the mercury "congeals" into a solid, white sublimate.

reddidisti eum album ut nix et conteribilem et miscibilem et aptum ad digestionem in stupha sicca, et aptasti ad reducendum eum in aquam prout inferius apparebit. Secundo, immisisti in ipsum sulphur invisibile philosophorum quod est tinctura rubedinis et est spiritus vitrioli romani. Et est ibi miraculum maximum quia cum spiritus ille invisibilis habeat naturam tincture rubedinis tamen congelat mercurium in sublimacione in albedinem nivis. Et hoc est quod multi maximorum philosophorum scripserunt, quod lapis fit ex mercurio et sulphure quod non est sulphur vulgi sed sulphur philosophorum. Et de predicta preparacione, sublimacione, et operacione dixit magister Arnaldus de Villanova in tractatu parabolico de majori, edito sub tropo verborum evangelii et crucifixionis christi, illa verba: *Nisi granum frumenti cadens in terram mortuum fuerit, ipsum solum manet; si autem mortuum fuerit multum fructum affert.* Intelligens per granum frumenti in terra mortificandum argentum vivum mortificandum in terra salis petre et vitrioli romani et ibi mortificatum et cum igne sublimatum multum fructum affert quia lapidem magnum quem omnes philosophi quesierunt. Signum autem quod spiritus vitrioli incorporetur mercurio sublimato est quia licet mercurius sublimatus multas feces terreas dimittat in terra tamen si posuisti unam libram

---

44 reddidisti eum] reduxisti B; *om.* P1 | nix] *finitur lacuna* Cu2 | conteribilem] convertibilem Ko1; contractabilem P1 | miscibilem] *finitur lacuna* M **45** sicca] calida et sicca M | aptasti] aptum Cu2 M; *add.* eum B Ka2 | eum] *om.* B Ka2 P1; ipsum M **46** Secundo] *om.* M; tercio P1 | immisisti] post misisti B; immiscuisti Ka2 | in] infra P2 | ipsum] eum Ka2 P1 P2 **47** est[2]] *om.* B Cu2 | spiritus] species P1 | vitrioli romani] leonis id est vitrioli romani Cu2; vitrioli M **48** miraculum] mirabile Ko1 quia cum] et tamen P2 | spiritus…habeat] ille species invisibiles habeant P1; spiritus invisibilis habeat P2 **49** naturam tincture] tincturam B Ko1 | congelat] coagulat B; congellant P1 | mercurium] materiam P1 | in[1]] *om.* P2 **50** nivis] nimis B Ko2; ipsius P1 | maximorum] *om.* P1 **51** scripserunt] dicunt Cu2; dixerunt M | mercurio…sulphure] materia sulfuris P1 | quod[2]…52 sed] *om.* Ko1 **52** sulphur] *om.* Ko1 Ko2 M | philosophorum] *om.* Ko1 | preparacione] racione Ko1 **53** dixit] dicit Ka2 P2 | magister] *om.* P1 | de Villanova] *om.* Ko1 | in] *add.* quodam suo Cu2; *add.* ejus P1 **54** parabolico] *om.* Cu2; parabole M | de majori] *om.* Cu2 Ko1 | edito] edicto B; *om.* Cu2 | sub] *incipit lacuna* B; *om.* P1 | tropo] typo Ka2 Ko1; thipho P1; stirpo P2 | evangelii] evangelicorum M; euvangelium P1; angeli P2 **55** illa] ista Ko1 P2; hec P1 | verba] *finitur lacuna* B | frumenti] *om.* P1; *add.* etc. P2 | cadens] cadat Cu2; *incipit lacuna* P2 **56** fuerit[1]] *om.* B | ipsum…manet] ipsum semen manet Ka2; nullum fructum affert P1 **57** affert] *add.* etc. P1; *finitur lacuna* P2 | Intelligens] Intellegatur B; Intellige Ka2; Intellige ergo P1; Item intellige P2 | in…mortificandum] *om.* Ko1 P2; in terram mortificatum id est P1 **58** argentum vivum] mercurius B | mortificandum] mortificatum P1 terra] *add.* id est Ka2 | et[1]…romani] et vitrioli M; *om.* P1 **59** mortificatum] mortuum P2 | multum] *om.* Cu2 Ko1 Ko2 | quia] Quasi dicat P1; quia affert P2 **60** autem] *om.* Ko1 | spiritus] *om.* Cu2; species P1 **61** vitrioli] *om.* Cu2; *add.* romani Ka2 | incorporetur] *om.* Cu2 | mercurio] mercurius incorporetur in spiritu vitrioli Romani sic Cu2; argento vivo P1 | sublimato] sublimatus Cu2; *om.* Ko2 | quia] quod P2 | mercurius sublimatus] *om.* Ko1; argentum vivum sublimatum P1 **62** in terra] cum terra predicta P1 | posuisti] tu posuisti P1 | unam libram] *add.* de mercurio crudo Ko2 M

B Ka2 Ko1 Ko2 P1 P2 **44** nix] *finitur lacuna* Cu2 | miscibilem] *finitur lacuna* M **54** sub] *incipit lacuna* B **55** verba] *finitur lacuna* B | cadens] *incipit lacuna* P2 **57** affert] *finitur lacuna* P2

grindible and miscible and fit for digestion in the dry oven, and rendered it fit to be reduced into a water, as will appear below. Second, you have put into it the invisible sulphur of the philosophers which is the tincture of redness, and is the spirit of Roman vitriol. And there is the greatest miracle here, because although that invisible spirit has the nature of the tincture of redness, it nevertheless congeals the mercury during sublimation into the whiteness of snow. And this is what many of the greatest philosophers have written, that the stone is made from mercury and a sulphur that is not common sulphur but the sulphur of the philosophers. And in regard to the aforesaid preparation, sublimation, and operation, master Arnald of Villanova said these words in a metaphorical treatise concerning the greater [stone], written using the words of the Gospel and of Christ's crucifixion as a trope:[13] "Unless the grain of wheat falling into the earth has died, it remains a single grain; but if it has died, it brings forth much fruit."[14] Understanding the grain of wheat which is to die in the earth as the quicksilver which is to die in the earth of saltpeter and Roman vitriol, and there having been mortified and sublimed with fire, the quicksilver brings forth much fruit because it brings forth the great stone that all the philosophers seek.

The sign that the spirit of vitriol is indeed incorporated into the sublimed mercury is that even though the sublimed mercury leaves behind many earthy dregs in the earth, nevertheless if you have put in one pound you will get the

[13] For Antoine Calvet's identification of this reference with a pseudo-Arnald text called *Exempla* or *Liber propheciarum* in the known manuscripts, see Calvet, *Oeuvres alchimiques*, 234-49 and 525-45. The line quoted here by John, as well as most of those he cites further on, does not however occur in the text of any known manuscript of pseudo-Arnald's *Exempla*. I have chosen to translate John's words "tractatu parabolico" as a *description* of the text he is citing rather than as its *title*, since that title is not indicated in any currently-known manuscript of pseudo-Arnald. See the discussion in Section 2.3.2.

[14] Jn 12:24-25.

tantumdem rehabebis. Sed hoc esset impossibile nisi quia mercurius albior nive secum traxit purissimum spiritum vitrioli predicti quod est sulphur invisibile. Demonstratur autem spiritus invisibilis vitrioli ad sensum per ingenium sic:

Recipe acetum et in eo extingue ferrum quatuor vicibus. Deinde pone ibi mercurium predictum sublimatum tritum per unam noctem. In crastinum distilla acetum per filtrum bis vel ter donec sit clarum. Post mitte in patella terrea vitreata ad ignem lentum et amove cappam nigram ferream que elevabitur, et post purgacionem da ignem et evapora acetum. Et remanebit tibi spiritus vitrioli in fundo in colore floris gaudii amenissimo supramodum, si experto credatur. Et ita habebis ad partem mercurium marcidum effectum et ex alia parte sulphur invisibile supradictum. Quare michi est demonstratum quod argentum vivum per vitriolum sublimatum secum portat spiritum ejus qui sulphur invisibile lapidis, tinctura rubedinis appellatur. Nam de argento vivo sine vitriolo sublimato non poteris extrahere spiritum supradictum. Hec est operacio prima.

---

63 tantumdem] tantum M | rehabebis] habebis P1; *add.* de sublimato M | Sed…impossibile] quod impossibile esse videtur P1 | quia] *om.* Cu2 P1 | mercurius…64 nive] argentum vivum P1 **64** traxit] transcisset albiorem nive P1 | vitrioli] *add.* romani Ka2 | predicti] *om.* P1 | quod] quia spiritus P1 est] *om.* Ko2 **65** Demonstratur] *incipit lacuna* P2 | spiritus] species P1 | invisibilis] *add.* viridis leonis id est M | vitrioli] *om.* Cu2 Ko1 | sensum] modum P1 **67** Recipe acetum] *add.* vini Ko1 | in… extingue] extingue ibi B Ko1; in eum extingue ibi Ko2; intus extingue M | ferrum] frenum Ko2 quatuor vicibus] quater P1; *add.* vel pluries M | pone] impone M **68** mercurium] argentum vivum P1 | predictum sublimatum] sublimatum B | tritum] *om.* P1 **69** crastinum] crastino Ko1 | acetum] *om.* Ko1 | filtrum] linguas B Ko2; linguas filtri Cu2; linteas Ka2 | mitte] pone Ko1 P1 **70** patella terrea] ampulla B; patena terrea Ka2 Ko2 | vitreata] devitreata Ko1 M | ad…lentum] *om.* Ko1 cappam] ejus capam P1 | ferream] ferri P1 **71** elevabitur] elevatur Ko1 M | post purgacionem] bene purga postquam est purgatum P1 | evapora] fac evaporare P1 **72** tibi] ibi B; *om.* Ka2 Ko1 | spiritus] species P1 | vitrioli] *add.* romani Ka2; leonis M | in fundo] *add.* vasis Ka2 P1 | in[2]] de B Ka2 Ko1 Ko2 | gaudii] giladii P1 | amenissimo] amenissimus M **73** si…credatur] si experte operetur Cu2; *om.* P1 | habebis] habes Cu2 Ko1 | mercurium] materiam que tinctura rubedinis appellatur P1 **74** marcidum] mortuum B; mortificatum Ka2; mortificati Ko1; *om.* P1 | effectum] *om.* P1 | et…parte] *add.* habebis P1 | Quare] Quia M; Et ita P1 **75** michi] nobis Ka2; modo Ko1; tibi jam M | est] fuit P1 | quod] scilicet M | argentum vivum] mercurius B | vitriolum sublimatum] vitriolum sublimatus B; vitrioli sublimacionem Ko1; leonem sublimatum M; *add.* quod M **76** spiritum] speciem P1 | qui] quia B; quod Ko1; quid M | lapidis] *om.* Ka2 P1 **77** appellatur] est P1 | argento vivo] mercurio B vitriolo] *add.* romano Ko1; leonem M **78** supradictum] predictum B; *finitur lacuna* P2 | Hec…prima] Et est operacio prima B; et hoc de primo capitulo Cu2; *om.* Ko1 Ko2 M

B Cu2 Ka2 Ko1 Ko2 M P1 P2 **65** Demonstratur] *incipit lacuna* P2 **78** supradictum] *finitur lacuna* P2

same amount back.[15] But this result would be impossible unless the mercury, whiter than snow, has carried up with itself the purest spirit of the aforesaid vitriol which is the invisible sulphur. Likewise, by an ingenious technique the invisible spirit of vitriol is demonstrated to sense thus:

Take vinegar and extinguish iron in it four times. Then put the aforesaid sublimed mercury, ground, into it for a night. In the morning distill the vinegar through felt two or three times until it is clear.[16] Then put it into a glazed earthenware dish on a gentle fire and remove the black ferrous scum that will rise to the surface, and after the purification give it fire and evaporate the vinegar. And the spirit of vitriol will remain for you at the bottom in the exceedingly pleasant color of a flower of delight,[17] if you will believe the one who has experienced it.[18] And so you will have mercury, rendered languid, separated on the one side, and the abovementioned invisible sulphur on the other, whereby it has been demonstrated to me that quicksilver sublimed with vitriol carries with itself the spirit of the vitriol which is called the invisible sulphur of the stone, the tincture of redness.[19] For you cannot extract the abovementioned spirit from quicksilver that is sublimed without vitriol. And this is the first operation.

---

[15] See the discussion of this important line in Section 2.2.4.

[16] This usage of the word "distill" is no longer present in modern languages, in which "distill" now always implies a cycle of vaporization and recondensation. The root meaning of the Latin *distillare* however means simply to *drip down*, and thus until the early modern period it encompassed both the modern meaning of distillation as well as filtration either by gravity or more commonly by capillary action through a piece of cloth, which is the meaning here. In this latter case, a cloth is dipped halfway into the liquid, then draped over the edge of the vessel leaving one end of the cloth immersed in the liquid. The liquid ascends the cloth, up and over the edge of the vessel, and drips out from the other end into another container. The variant reading *per linteas* ("through linen cloths") refers to the same process, and the further variants *per linguas* ("through tongues") and *per linguas filtri* ("through tongues of felt") express this technique clearly.

[17] On this problematic reading, see Section 4.3.

[18] A reference to the famous line in *Aeneid* 11:283; John uses the same verse similarly in *De famulatu*, remedium 8.

[19] On the seemingly odd (but remarkably accurate) usage of the adjective "languid" here, see Section 2.2.4.

Secunda operacio hujus artis est ab illo mercurio sublimato, quod est album ut nix, extrahere spiritum ejus qui a quibusdam magnis philosophis quinta essencia clamitatur quia spiritus ille est incorruptibilis fere ut celum licet non sit fixus sed volatilis quia sic est necesse. Extrahitur autem hec quinta essencia, hoc est spiritus mercurii, ab ipso supra sublimato sic:

Recipe salis petre libram i et tantumdem vitrioli romani, et extrahe aquam fortem hoc modo: Recipe predictam materiam et tere eam subtiliter in patella terrea, post pone in vessica et apta desuper alembicum, et pone ad distillandum in cinere. Et primo da ignem suavem et non fortem ne materia ascendat ad caput alembici, et cum videris materiam non amplius ascendentem da tunc ignem forciorem et continua ignem donec caput alembici incipiat rubescere et tunc da ignem fortissimum donec tota aqua exierit, quam aquam recipe et serva. Et tere mercurium supra sublimatum et dissolve eum in illa aqua forti supra cineres calidos in amphora vitri clausa per xii horas. Quod si non dissolveretur pone intus de sale armoniaco trito septimam aque partem vel plus, et hoc est maximum archanum in arte; appone cor ad ipsum. Et post dissolucionem pone in urinali et desuper alembicum et mitte super ignem et distilla aquam fortem

---

79 Secunda] *antea add. rub.* Secunda operacio Cu2 Ka2; *antea add. rub.* Modus extractionis quinte essentie M | hujus] istius lapidis Ko1 | artis est] *add.* quomodo M | mercurio] argento vivo P1 sublimato] *om.* M | quod...80 album] albo M; qui est albus P1 **80** extrahere] extrahe B; abstrahere Cu2 Ka2; abstrahe Ko1 Ko2; extrahas P1; *add.* possis M | magnis] *om.* P1 **81** clamitatur] appellatur P1 | ille] *om.* M **82** licet] *om.* Cu2 | sed] *om.* B | quia] quare Ka2 | Extrahitur] Extrahatur B P1 autem] enim B; ergo P1 | hec] *om.* Cu2 M **83** mercurii] argenti vivi P1 | ipso] illo B; *add.* argento vivo P1 | supra] *om.* M P1 | sublimato] *add.* ut supra et hoc M | sic] ut sequitur P1 **84** romani] viridis M | et[2]] *add.* de eis M | extrahe...85 fortem] extrahatur aqua fortis P1 **85** modo] *incipit lacuna* B Cu2 Ka2 Ko1 Ko2 P2 | Recipe] Accipe P1 **86** vessica] *add.* vitrea P1 **87** da...suavem] *add.* ymo bene suavissimum per unum diem et unam noctem videlicet per viginti quatuor horas P1 **88** amplius] superius P1 **89** ascendentem] ascendere M | tunc...forciorem] *om.* P1 **90** rubescere] fieri rubeum P1; *add.* et continua dictum ignem donec nil destillabit vel parum P1 | aqua...91 exierit] aqua grossa exinde destillaverit P1 **91** serva] *finitur lacuna* B Cu2 Ka2 Ko1 Ko2 P2; *add.* Forma vero vessice supra nominate in qua distillatur aqua fortis predicta in fine libri hujus prout melius depigni potest M | Et] Postea M; Deinde P1 | mercurium] materiam P1 | supra] superius M **92** eum] eam P1 | illa] dicta P1 | supra] super Ka2 Ko1 M | amphora] forti anfora P1 | vitri] vitrea Cu2 Ka2 M **93** xii] xxiiii P1 | Quod] Et Ka2 P2 | dissolveretur] dissolvitur Ka2 | pone...94 armoniaco] *add.* sublimato M **94** trito] *om.* B | septimam...partem] vii partes aque B | vel plus] *om.* Ka2 | et] *add.* statim dissolvetur P2 **95** archanum] secretum Cu2 | in[1]] *add.* hac M | appone] appone ergo B; et pone Ka2; apponere Ko1; *om.* P1 | cor] cor tuum B Ka2 P2; *om.* P1 | ad ipsum] *om.* P1 dissolucionem] distillacionem Ko2 **96** urinali] cucurbita Cu2 | et[1]...alembicum] et supra alembicum Ka2; cum alambico supra P1 | mitte] pone P1 | et distilla] ad destillandum P1 | aquam fortem] *om.* P1; *add.* cum lentissimo igne M; *add.* cum parvo igne carbonum dictam aquam fortem P1

B Cu2 Ka2 Ko1 Ko2 M P1 P2 **85** modo] *incipit lacuna* B Cu2 Ka2 Ko1 Ko2 P2 **91** serva] *finitur lacuna* B Cu2 Ka2 Ko1 Ko2 P2

The second operation of this art is to draw out from that sublimed mercury, which is as white as snow, its spirit which is called its quintessence by certain great philosophers because that spirit is incorruptible, almost like the heavens, even though it is not fixed but volatile because it is necessary that it be so.[20] This quintessence, that is, the spirit of mercury, is drawn out in this manner from the same mercury sublimed above:

Take one pound of saltpeter and the same amount of Roman vitriol and extract an aquafort in this way: Take the aforesaid material and grind it finely in an earthen dish, then put it in a rounded flask, attach an alembic on top of it, and put it in ashes to be distilled.[21] Give it at first a gentle and not a strong heat so that the material does not ascend to the head of the alembic, and when you see the material ascending no further, then give a stronger fire and continue the fire until the head of alembic begins to turn red, and then give a very strong fire until all the water comes forth; take this water and keep it.[22] Then grind the above-sublimed mercury and dissolve it in that aquafort upon hot ashes in a sealed glass flask for twelve hours. If it should not dissolve, put therein ground sal ammoniac, a seventh part relative to the water or more, and this is the greatest secret in the art; apply your heart to it. And after the dissolution, put it in a urinal glass, and an alembic on top, and place it upon the fire and distill off the aquafort

---

[20] The term "quintessence" is very rarely used to name an alchemical substance prior to John, and never (to my knowledge) for the mercury sublimate. While this line might be a rather sly self-reference to John's own usage of this term for mercury sublimate in his earlier *De famulatu* (see Appendix 1), it may also refer to pseudo-Arnald's *Quaestiones...ad Bonifacium VIII*, alluding to the claim made there that the whole material of the stone is "converted into a spiritual material or substance, which is called quintessence among the philosophers," *BCC* 1:698-702, on 701. For John and pseudo-Arnald, see Section 2.3.2; for the difficult issue of the authorship and date of the *Quaestiones*, see Calvet, *Oeuvres alchimiques*, 53-60. The use and meaning of the term "quintessence" in Medieval alchemy will be further explored in my forthcoming critical edition of *De famulatu*.

[21] I.e., in an ash-bath, that is, the flask is partly buried in a bowl of sifted ashes. The ashes (sand could be substituted) distribute the heat more evenly and protect the flask from thermal shock and direct contact with the fire.

[22] This part of the text cannot be correctly understood without undertaking the experiment. The initial gentle heat is intended to cause the expelled water of crystallization from the vitriol to circulate in the flask, but not cause it to rise as high as the alembic head and distill into the receiver. During the experimental reworking, this expelled water was clearly visible condensing in the neck beneath the alembic head and running back down into the distilling flask onto the salts. This "predistillation" period actually causes a chemical reaction to take place that increases the yield of aquafort. Thereafter, the "stronger fire" drives the water further upwards and causes the head of the alembic "to turn red," but not from the heat as might well be assumed from the text alone, but rather from the production of nitrogen dioxide, a gas of a deep red-brown color. For more on this process, see Section 2.3.1.

quam recipe, et post totalem distillacionem continua ignem et ascendet per latera urinalis spiritus mercurii, hoc est quinta essencia, albior nive. In fine fortifica ignem ut tota ascendat ad plenum. Aperi vas et extrahe quintam essenciam et invenies in fundo vasis feces combustas et nigras a spiritu mercurii separatas.

Et in hac operacione fecisti duo. Primo, separasti illud quod in albo mercurio separato erat terreum, corruptibile, et imperfectum; totum enim illud aqua fortis combussit. Et est mirum valde quod de re alba ut nix possit extrahi res terrestris et nigra. Secundo, imbibisti infra spiritum mercurii amplius de spiritu vitrioli qui est in aqua forti. Et hoc est quod in predicto tractatu dicit magister Arnaldus de Villanova: *bibat quantum bibere poterit per xii horas* quia mercurius dissolutus in aqua vitrioli plus bibit de spiritu vitrioli, hoc est de sulphure invisibili, quam habuit et dissolucio fit in xii horis. Secundario dissolve illum spiritum mercurii in aqua forti et sublima ut prius et remanebunt feces nigre, sed non tot ut prius, et tercio itera opus et nichil de nigris fecibus invenies aut parum. Et tunc habes perfectam

97 quam] quod P1 | distillacionem] dissolucionem Cu2; *add.* cum nil plus destillabit amove capam et pone antonarium id est coperciolum cucurbite bene lutando et P1 | continua ignem] *add.* lentum M ascendet] ascendit Ko1 **98** urinalis] cucurbite Cu2; vasis P1 | mercurii] argenti vivi sublimati P1 | hoc est] hoc B; hec est Cu2 P1 P2; id est Ko1 | quinta essencia] *add.* ejus Ko1 **99** fortifica] nota fortifica P2 | tota] totum B Ko1 P1; tota quinta essentia M | ascendat…plenum] *add.* postea M | Aperi vas] aperies M | extrahe] extrahes M P2 **101** mercurii] argenti vivi P1 | separatas] *add.* quas serva et sublima et iterum extrahe de alia quinta essencia a dictis fecibus P1 **102** Et] Et nota quod M | Primo] Nam P1 | illud…albo] album et perfectum secundo separasti P1 **103** mercurio separato] mercurii non separatum Ka2; id quod P1; mercurii separato P2 | corruptibile] *om.* P1 | et imperfectum] *add.* totaliter P1 | totum…104 illud] *add.* scilicet terreum dicta P1 **104** mirum] mirandum M | ut nix] *om.* Cu2 M **105** Secundo] item P1; Secundario P2 | infra] in P1 | spiritum mercurii] ipsum mercurium Ko1; spiritum argenti vivi P1 **106** vitrioli] vitrioli romani Ka2; leonis M | est[1]] erat P2 in[2]…107 tractatu] *om.* Ko1; in dictis suis parabolicis M **107** dicit] dixit B M | magister] *om.* M | de Villanova] quod B; *om.* Ka2 Ko1 Ko2 P2 | bibat] Bibat Bibat Bibat B **108** per] in Ka2 | xii] xxiiii P1 mercurius] argentum vivum P1 | dissolutus] positus Cu2; *om.* Ko1 Ko2 M; dissolvitur P1; *add.* est B in…vitrioli] *add.* tunc M; *add.* et quanto plus bibit de aqua forti tanto P1 | plus] *om.* B Cu2 P2 bibit…spiritu] *om.* B P2 **109** vitrioli] *om.* B P2; leonis M | de] in P2 | sulphure invisibili] *add.* plus habet Ka2; *add.* omnem terrestritatem dimisit P2 | habuit] habet B P1; haberet Cu2 M; habebat Ka2; *om.* P2; *add.* de terra P1 | et] *add.* hec B; *add.* talis P1 | dissolucio] dissolutum Cu2; *add.* et imbibicio P1 | xii…110 horis] dictis xxiiii[or] horis et aliquando in pluribus P1 **110** Secundario] Secundo B P1 dissolve] *om.* P1 | illum] *om.* B Ka2; ipsum M P2; illud P1 | spiritum mercurii] argentum vivum P1; mercurium P2 | in…sublima] in aqua forti et fac Ko1; destilla et sublima in aqua forti nova P1 **111** prius[1]] primo P2 | feces nigre] in igne feces B; adhuc feces nigre Ka2; feces Ko1 | sed…tot] non tot B; *om.* P1 | tercio…opus] tercio reitera opus B; *om.* P1 **112** et] donec Cu2; et sic pluries itera donec M | invenies] reperies Ka2 | aut parum] *om.* Cu2 M | habes] habebis B Ka2 | perfectam] perfecte B; *om.* M

B Cu2 Ka2 Ko1 Ko2 M P1 P2

and keep it. After the distillation is complete, continue the fire, and the spirit of mercury, that is, the quintessence, will ascend up the walls of the urinal flask whiter than snow. At the end, increase the fire so that it all ascends fully. Open the flask and take out the quintessence, and you will find at the bottom of the flask burnt and black dregs separated from the spirit of mercury.

And in this operation you have done two things. First, you have separated whatever there was in the white separated mercury that was earthy, corruptible, and imperfect, for the aquafort has burned all of it. And it is very marvelous that an earthy and black thing can be drawn out from something as white as snow.[23] Second, you have imbibed into the spirit of mercury more of the spirit of vitriol that is in the aquafort. And this is what master Arnald of Villanova says in the aforesaid treatise, "let it drink as much as it can drink for twelve hours,"[24] because the mercury dissolved in the water of vitriol drinks up more of the spirit of vitriol, that is, of the invisible sulphur, than it had before, and the dissolution is done in twelve hours. Dissolve that spirit of mercury a second time in the aquafort and sublime as before, and there will remain behind black dregs, but not so many as before. And repeat the work a third time and you will find no black dregs or very few. And then you have the completed quintessence, the

[23] The dark "dregs" left behind are iron compounds from the vitriol that were carried upwards mechanically into the white sublimate during the initial sublimation in operation 1.

[24] See note 13; this line also does not appear in pseudo-Arnaldian text *Exempla* nor does it seem to be appropriate for that treatise; see Calvet, *Oeuvres alchimiques*, 236-37. A very similar line does however appear in the short pseudo-Arnald text found in the Darmstadt manuscript, see above in Section 2.3.2.

quintam essenciam, materiam lapidis secundario preparatam, incorruptibilem, volatilem non fixam, et albiorem nive.

Tercia operacio est ista. Recipe prefatum spiritum mercurii separatum seu quintam essenciam ejus et tere mundissime super lapidem et pone pulverem in capsea vitrea et mitte infra furnum philosophorum in stupha sicca ad digerendum prout quidam volunt nominare, licet alii dicant ad calcinandum, ad ignem de carbonibus lentum per viii dies vel ultra quem furnum inferius describam. Ibi enim decoquitur ut possit dissolvi in aquam quoniam aliter non dissolveretur nec opus perficeretur pro certo.

Quarta operacio est ut post digestionem predictam completam in stupha sicca, ponas spiritum in urinali clauso in balneo infra ollam aqua plenam donec totus spiritus dissolvatur in aquam. Qua dissolucione completa, pone desuper alembicum vel in principio quando materiam ponis in balneo potes desuper alembicum aptare. Da ignem et ascendet stilla roris madii, aqua vite, lucida lacrima oculi, lac virginis, purissima materia lapidis

---

113 quintam essenciam] *add.* mercurii M | materiam] medicinam P1 **114** et] sed Cu2 | albiorem nive] *add.* in 15<?> vicibus semper reiterando ut prima vice fecisti P1 **115** Tercia] *antea add. rub.* Tercia operacio Cu2; *antea add. rub.* Digestio lapidis et ejus calcinacio Ko1; *antea add. rub.* Modus calcinationis quinte essentie supradicte M; *antea add. rub.* Calcinacio dicti spiritus in stupha sicca P1 est] *om.* Ka2 | prefatum] *om.* Cu2 M P1 | spiritum…separatum] spiritum argenti vivi predictum et preparatum ut supra P1; mercurium P2 **116** seu] *om.* P1 | quintam…ejus] prefatam quintam essenciam Cu2; predictam quintam essentiam M; *om.* P1; quintam essenciam P2 | mundissime] minutissime Ko1 M | et[2]] Postea M **117** pulverem] paulatime P1 | capsea] vase B; tassa aut tassia Ka2; bocia M; tacea P1 P2 | vitrea] vitreo B; *add.* lutata et fortissima taliter quod mercurius exire non possit Ka2 | mitte] pone seu mitte Cu2; pone P1 | infra] intra P1 | furnum…118 sicca] *add.* et calida M **119** carbonibus] tribus carbonibus Ko1 | dies] dies et noctes Ko1 | ultra] plus B **120** describam] scribam P2 | Ibi] ubi Ka2 **121** quoniam] quoniam quando B; quia Cu2 Ko1 M; in qua Ka2 | non] *om.* B | dissolveretur] solveretur Ko2 P2 | opus perficeretur] opus veniret ad perfeccionem Cu2; perficeretur P2 | pro certo] *om.* B Cu2 M **122** Quarta] *antea add. rub.* Sequitur quarta operacio Cu2; *antea add. rub.* Quarta operacio Ka2; *antea add. rub.* Solucio lapidis in aquam per se et in lac virginis Ko1; *antea add. rub.* Modus dissolutionis quinte essentie calcinate et expressio lactis virginis M | est] *om.* Ka2 Ko2 | predictam] dictam Ka2 | completam] *om.* B; *add.* factam P1 | in…123 sicca] *om.* B; in dicta stupha seu furno P1 **123** ponas] pone B | spiritum] *om.* B; spiritum tritum M; dictum spiritum P1 P2; *add.* calcinatum P1 | urinali clauso] cucurbita clausa superius Cu2 | in balneo] *add.* marie M P1 | infra] intra P1 | ollam] *om.* M | aqua…124 plenam] aquam tepidam in olla plena M; *add.* ad putrefaciendum et dissolvendum in suo vitro P1 **124** donec] *om.* Ka2 | totus] tuus B; *om.* Ka2 spiritus…aquam] *om.* Ka2 **125** completa] facta et completa B Ko2; vero completa P1 | vel] *incipit lacuna* P1 | materiam ponis] urinale ponis B; materia ponitur Ka2 **126** aptare] coaptare M; *finitur lacuna* P1; ponere et aptare P2 | Da] deinde da B; da ergo M | ignem] *add.* competentem Ka2 | et ascendet] ut ascendat Ka2 | stilla roris] et distilla rorem Ka2; stella roris P2 **127** madii] madidi Ko1 aqua vite] in modum aque vite B; id est aquam vite Ka2; aque vite M | lucida lacrima] lucidam lacrimam Ka2; lucide lacrima M | lapidis] lapidis occulti B

B Cu2 Ka2 Ko1 Ko2 M P1 P2 **125** vel] *incipit lacuna* P1 **126** aptare] *finitur lacuna* P1

material of the stone prepared in a second step, incorruptible, volatile not fixed, and whiter than snow.

The third operation is this: Take the aforesaid separated spirit of mercury, or its quintessence, and grind it very cleanly upon a stone, and put the powder in a glass capsule and place it inside the furnace of the philosophers in the dry oven in order to be digested (as some wish to call it, although others say in order to be calcined) with a gentle charcoal fire for eight days or more. I will describe the furnace below. For there it is cooked so that it can be dissolved into a water, because otherwise it will not be dissolved, and then surely the work will not be brought to completion.

The fourth operation is that after the aforesaid digestion is completed in the dry oven you put the spirit in a urinal flask enclosed in a bath within a pot full of water until all the spirit is dissolved into a water. After this dissolution has been completed, put an alembic on top, or you can attach the alembic on top in the beginning when you put the material into the bath. Apply the fire and there will ascend the drop of May dew,[25] the water of life, the clear tear of the eye, the virgin's milk,[26] the purest material of the stone completed by its final preparation,

[25] These first two names given here for this precious distillate appear twinned also in pseudo-Arnald, *Questiones tam essenciales quam accidentales*, where however, they are applied as Decknamen for quicksilver. The printed versions of this text (*BCC*, 1:698-702, on 698 and *Artis auriferae*, 3:151-65) give *madidi* (moist) and *medii* (medium) for *madii* (May), but most early manuscripts (e.g. Vatican, MS Palatinus lat. 1330, fols. 115v-19v, on 116r) read *madii*. A treatise entitled *Practice of May Dew* (*Practica roris madii*) is attributed to pseudo-Arnald; see Calvet, *Oeuvres alchimiques*, 30 and Thorndike, *HMES*, 3:665-66.

[26] On the "virgin's milk" see above Section 2.5.

ultima preparatione perfecta, gustu amara, ad literam acetosa, volatilis non fixa, incorruptibilis fere, per digestionem in pulverem reducibilis, a Gebero nunquam cognita, ab Avicenna ignorata. Hanc Hermes palpavit, Alphidius aptavit, Rosarius intellexit, magister Arnaldus dictavit. Nam de hac quarta operacione, que dicitur expressio lactis virginis, dixit magister Arnaldus quod oportet a terra exaltari filium hominis in aerem in cruce, quia ad literam de materia in tertia operacione digesta que trita ponitur ad dissolvendum in fundo urinalis, ascendit quod ibi purum est et spirituale, et in aerem fumosum convertitur et exaltatur in cruce capitis alembici quasi Christus, ut Arnaldus dicit elevatus in cruce. Et ista distillacione completa, habes tuam materiam ultima preparacione perfectam. Si unam guttam ex hoc lacte virginis super laminam ignitam projeceris, dealbabit ipsam intus et extra. Et ita vidisti quod materia lapidis est una res, non simplex omnino sed composita ex spiritu mercurii et ex spiritu vitrioli romani qui dicitur sulphur invisibile non vulgi sed philosophorum, qui est tinctura intrinseca rubedinis. Licet mercurius in se habeat tincturam ad rubeum intrinsecus absconsam sed per purgaciones, quia nimis est terrea, diminutam et ob hoc indiget sulphure purissimo vitrioli epotari ad plenum. Hac autem distillacione completa, invenies in fundo urinalis tria elementa simul mixta, quia remanet materia grossa combustibilis, et extraxisti ab ea elementum

---

128 ad literam] *om.* Ko1; ad linguam M P2 | acetosa] parum acetosa Ka2 **129** incorruptibilis] incorporabilis P1 | fere] *om.* Cu2 | reducibilis] *om.* B; irreductibilis P1 **131** aptavit] optavit Cu2 magister] *om.* Cu2 M | Arnaldus] *add.* de villa nova vir magne reputationis Cu2 M | dictavit] *add.* Oportet a terra exaltari Cu2; *add.* Oportet inquid a terra exaltari filium hominis M | hac] *om.* P1 quarta] quinta P1 **132** dixit] dicit B P2 | magister] *om.* Cu2 M; idem magister P1 **133** quod] quia B M | a terra] *om.* B; a turre P2 | hominis] *om.* Ko1 | in cruce] id est in cruce P2 | quia] *add.* intelligit P1 **134** trita] trita vel que terra Ka2; terra Ko1 P1 **135** dissolvendum] distillandum Ka2 | urinalis] cucurbite Cu2 | purum] purissimum B **136** fumosum] sursum B M; firmosum Ko2 | capitis] *incipit lacuna* Cu2 Ko2 M | alembici…137 Christus] *add.* jhesus Ko1 **137** ut Arnaldus] ut magister Arnaldus B P2 | dicit…cruce] | cruce] *finitur lacuna* Cu2 Ko2 M | Et] *incipit lacuna* Ko1 | ista] ita Ka2 Ko2 M | distillacione] dissolucione B P1 | completa] perfecta P2 **138** materiam…preparacione] medicinam ultimo preparatam P1 | perfectam] preparatam et perfectam B; *finitur lacuna* Ko1 **139** hoc] isto Ka2 | dealbabit] dealbat B Ka2; dealbabitur P1 | ipsam] eam B; *om.* P1 | intus] ab intus Cu2 **140** lapidis] *om.* P1 | una res] una B | omnino] *om.* Cu2 **141** sed] est B | spiritu[1]…ex[2]] mercurio et P1 | vitrioli romani] leonis M | qui dicitur] quod est Cu2 **142** qui] quia B; quod P2 intrinseca…143 rubedinis] mercurii et rubedinis B **143** Licet] *add.* et Ka2 | mercurius] materia mercurii M | in…habeat] in se non habet B | ad rubeum] rubedinis Cu2 M **144** sed] et quia materia Cu2 | per] propter B Ko1 | purgaciones] purgacionem P1 | quia] *om.* Cu2 P2; que Ka2 | nimis] minus Ka2 | terrea] *om.* Cu2; terra P1 P2 | diminutam] dimittit B; diminuta Cu2 Ka2 M P1 P2 | et] scilicet mercurius Cu2 | ob] ex P1 **145** purissimo] purissimi B | vitrioli] *om.* Cu2; leonis M | epotari] inbibi B; *om.* Ka2; sibi P1 | ad plenum] *add.* ut supra B | autem…146 distillacione] igitur dissolucione Ko1 **146** invenies] tu invenies P1 | urinalis] cucurbite Cu2 | tria elementa] elementa aeris et ignis cum terra Ko1 **147** materia] *om.* B; ibi materia Cu2

B Cu2 Ka2 Ko1 Ko2 M P1 P2 **136** capitis] *incipit lacuna* Cu2 Ko2 M **137** cruce] *finitur lacuna* Cu2 Ko2 M | Et] *incipit lacuna* Ko1 **138** perfectam] *finitur lacuna* Ko1

bitter in taste, literally acrid, volatile not fixed, almost incorruptible, reducible by digestion into a powder, never recognized by Geber, unknown to Avicenna. Hermes[27] touched upon it, Alphidius[28] prepared it, Rosarius[29] knew it, master Arnald taught it.[30] For in regard to this fourth operation, which is called the expression of the virgin's milk, master Arnald said that it is necessary that the son of man be raised up from the earth into the air on the cross, because from the material digested in the third operation which, after it is ground and is put to be dissolved in the bottom of a urinal, there literally ascends whatever is pure and spiritual, and it is converted into a smoky air and raised up upon the cross of the head of the alembic like, as Arnald says, Christ raised up on the cross.[31] And when this distillation has been completed you have your material perfected by the final preparation. If you should cast one drop of this virgin's milk upon a glowing sheet of metal, it will whiten it inside and out.[32] And so you have seen that the material of the stone is a single thing, not entirely simple but compounded out of the spirit of mercury and the spirit of Roman vitriol that is called the invisible sulphur—not the common sulphur but that of the philosophers—which is the inner tincture of redness. Although mercury has in itself a tincture for the red hidden within, yet because that tincture is too earthy it is diminished during the purifications, and on this account the mercury needs to drink up fully the most pure sulphur of the vitriol.[33] When this distillation is completed you will find at the bottom of the urinal flask three elements mixed together, for a gross combustible material remains behind. You have already drawn out from that

---

[27] The term "virgin's milk" is used by "Hermes" in the *Septem tractatus*, a medieval Latin alchemical text believed to be translated from an Arabic original; *Ars chemica* (Strasbourg, 1566), 7-33 on 22.

[28] Little is known of Alphidius, but he has recently been shown to be an Arabic author thanks to the discovery of the Arabic original of his *Liber ad filium suum*; see Sébastien Moureau, "*Min al-Kīmīyā' ad Alchimiam*: The Transmission of Alchemy from the Arab-Muslim World to the Latin West in the Middle Ages," 87-141 in *The Diffusion of the Islamic Sciences in the Western World* (Florence, SISMEL, 2020), on 105-6.

[29] "Rosarius" is out of place in this list of names since it is instead a title for several Medieval alchemical treatises, notably pseudo-Arnald's *Rosarius philosophorum*. If John here refers to that book, then citing both *Rosarius* and Arnald is redundant; moreover, *lac virginis* does not appear in that text. "Rosarius" may instead be a corruption of the author known in the Latin Middle Ages as Rosinus, a corruption via Arabic of the Graeco-Egyptian authority Zosimus. Torino MS E.II.23 (which contains **T**), however, cites Rosarius as an *author* rather than a text in its *Dialogus Rosarii et Alchidius* (fols. 42v-46v), the copyist even provided a small drawing of this alchemical authority teaching Alchidius. This duo is cited again in *Aureum vellus*, 3 vols., (Rorschach, 1599) 3:142. Alchidius might be a Latin rendering of al-Kindī or a corruption of Alphidius.

[30] *Lac virginis* does not appear in any pseudo-Arnald text in Calvet, *Oeuvres alchimiques*. It does, however, appear in the Darmstadt compendium of pseudo-Arnald's metaphorical instructions (Section 2.3.2), where its production is called its "expression" just as John names it here.

[31] Pseudo-Arnald's *Exempla* mentions that "we operate upon mercury according to the example of Christ," but is not as detailed as John; see Calvet, *Oeuvres alchimiques*, 542 and Section 2.3.2.

[32] A similar claim for the power of *lac virginis* to whiten copper into silver appears in pseudo-Aristotle's *Liber perfecti magisterii*, a thirteenth-century Latin composition, but there it is clear that the term refers to mercury; see *Verae alchemiae*, 2 vols. (Basel, 1561), 2:188-225 on 214.

[33] The notion that mercury contains its own tincture of redness is stated by Geber, see Newman, *Summa*, 335. Since Geber cites this redness "of shining brightness" alongside his mention of the fixation of mercury, he may refer to the production of mercuric oxide, a brilliantly red solid made by heating mercury in contact with air for a long time. On reading this line, see Section 4.3.

aque animatum anima spirituali vitrioli romani, et est per se sine aliis tribus sufficiens materia lapidis, maxime ad album et ad rubeum consequenter. Et si vis separare alia tria elementa ab invicem operaberis in hunc modum:

Recipe illas feces terre et pone ad digerendum in stupha sicca ut prius. Post pone in balneo ad dissolvendum et distillandum ut prius cum fortissimo igne et quod ascendet per ignem balnei est elementum aeris de colore olei. Recipe ipsum et serva. Et cum nichil amplius poterit ascendere, transfer vas de balneo ad ignem sublimacionis in cinere, et da ignem bonum et ascendet per alembicum ignis rubicundus in liquore olei rubei. Recipe ipsum et serva ad partem. Et ita habes a mercurio sublimato quatuor elementa ab invicem separata: primo, elementum aque, scilicet lac virginis; secundo, elementum aeris, scilicet oleum resplendens in balneo distillatum; tercio, elementum ignis, scilicet oleum rubeum distillatum ad ignem; et quarto, habes in fundo urinalis elementum terre nigerrimum ut carbo quod *ad nichilum valet ultra nisi ut mittatur foras et conculcetur* ab hominibus. *Intellexistis hec omnia?* Nescio quid clarius potest dici.

---

148 vitrioli romani] simplici vitrioli Ka2; leonis M; vitrioli P1 | et…tribus] *add.* elementis B Ka2 P2 **149** consequenter] *om.* Cu2 M; *add.* ut dicetur B **150** ab…operaberis] operare P1 | in…modum] *add.* qui sequitur. Separacio aliorum trium elementorum. P1 **151** illas] alias Cu2 Ko1 Ko2 | terre] tere eas Cu2; *om.* Ko1; et tere M | et pone] *add.* ad ignem Cu2 Ko2 M | digerendum] *add.* seu calcinandum ut supra P1 | stupha sicca] *add.* prius cum lento igne per viii dies post pone in balneo ad dissolvendum in urinali P1 | prius] supra dictum est et P1 **152** in balneo] *om.* P1; *add.* marie M dissolvendum…distillandum] destillandum P1 | ut prius] *om.* Cu2 **153** ascendet] ascendit Ka2 Ko1 M P1 | ignem] illud P1 | balnei] et balneum Cu2 M; balneum Ko2 P1 | de] in Cu2 P1 **154** olei] aurey P2 | ipsum] ipsum aerem seu oleum P1 | nichil] non P1 | ascendere] distillari M **155** vas] vas tuum P1 | de] a M | cinere] igne cinerum P1 | bonum] lentum et bonum Cu2; lentum bonum M **156** ascendet] ascendit Ka2 Ko1 Ko2 | per alembicum] *add.* elementum tercium P1 | rubicundus] rubicundissimus M; rubicondissimum P1 | liquore] colore Ko2 **157** ipsum] spiritum P1 | ad partem] *om.* P1 | sublimato] *om.* B **158** primo] primum B Cu2 Ko1 P2; videlicet primum P1 | scilicet] quod dicitur B P1 | lac virginis] *add.* in balneo marie M **159** secundo] secundum B Ko1 P2 | resplendens in] splendens per P1 | balneo] balneum igneum P1 | distillatum] distillationis cum igne fortissimo M **160** tercio] tercium B Ko1 P2 | scilicet] videlicet P1 | oleum rubeum] oleum Cu2 P2 | distillatum] resplendens distillatum Ko1 | ad ignem] ad ignem cinerum M; per cineres P1 | et…161 quarto] et quartum elementum P2 **161** habes] *om.* Cu2; habes scilicet M | in fundo] *om.* Cu2 | urinalis] vasis B; *om.* Cu2 | elementum…nigerrimum] scilicet terram nigerrimam P2 | carbo] corvus M; carbones P2 quod] jacens in fundo cucurbite quod Cu2; que P2 **162** ad nichilum] nichil B Cu2 Ko1 P1 | valet ultra] valet P1 | ut] quod Ko2 | conculcetur] concalcandus B | ab hominibus] a pretereuntibus B; ab omnibus Ka2 P1 **163** Intellexistis] Intellexisti B Cu2 Ka2 Ko2 P1 | Nescio] *add.* in mercurio B | quid] quis Ka2 | potest] posset Cu2 Ka2 M; possim P2 | dici] dicere Ka2 P2; *add.* seu potest dici Cu2

B Cu2 Ka2 Ko1 Ko2 M P1 P2

material the element of water animated by the spiritual soul of Roman vitriol, and that element is by itself without the other three a sufficient material for the stone, especially for the white and thereafter for the red. And if you wish to separate the other three elements from one another you will work in this way:

Take those dregs of earth and put them to be digested in the dry oven as before. Then put them in the bath to be dissolved and distilled as before with a very strong fire. Whatever will ascend in the heat of a bath is the element of air in the color of oil. Take it and keep it. And when nothing further is able to ascend, transfer the flask from the bath to a fire of sublimation in ashes and give a good fire and the red fire will ascend in a liquid red oil. Take it and keep it separately. And thus you have from the sublimed mercury the four elements separated one from another: first, the element of water, that is, the virgin's milk; second, the element of air, namely the shining oil distilled in the bath; third, the element of fire, namely the red oil distilled in the fire; and fourth, you have at the bottom of the urinal the exceedingly black element of earth, like a coal, which *is good for nothing further but to be thrown out and trodden upon by men.*[34] *Have you understood all of these things?*[35] I do not know what could be said more clearly.

[34] Mt 5:13.
[35] Mt 13:51.

Materia lapidis preparata perfecte, in hac quinta operacione docebo quomodo materia preparetur ad album, id est quomodo lac virginis digeratur ut redigatur virtute ignis in pulverem tam subtilem ut non senciatur per tactum. Recipe igitur lac cum benediccione Jhesu Christi, et dixerunt ipsum quidam philosophorum animal cum sanguine suo quia crescit sublimando, sanguis quia habet animam sanguinis, id est tincturam rubedinis spiritus vitrioli supradicti, et pone ipsum infra amphoram vitri rotundam in fundo et igneas ei collum et strangula ipsam stringendo cum forpicibus caute ita quod nichil possit quoquomodo exire. Et ponas laminam tegulatam perforatam in collo inter ignem et amphoram ne lac virginis destruatur ab igne. Et pone vas in furno philosophorum inferius describendo ad digerendum in igne ubi vas habeat per ingenium ignem ab omni parte equalem. Nam hec est intencio nostra, ut ab omni parte vas equaliter ignem recipiat ut ad nullam partem lac virginis possit quomodolibet evolare, quod fieret nisi ignis ab omni parte donaretur equalis. Et est ars ad dandum ignem eciam temperatum, non excessivum et

164 Materia] *antea add. rub.* Quinta operacio Ka2; *antea add. rub.* Fixacio lapidis et lactis virginis pulverisacio Ko1; *antea add. rub.* Modus congelationis lactis virginis ut fiat elixir completum. Rubrica. Quinta operacio est ista. M; *incipit lacuna* M | perfecte] perfecta Ka2 Ko1 | in…operacione] *add.* que sequitur P1 | docebo] docetur Ko1 **165** materia] *om.* P1; materia ipsa P2 | id est] Item Ko1 P1 **166** ut[1]] vel B; et Ko2 P1 | redigatur] reducatur B Ka2 | virtute ignis] *om.* Ka2 | tam subtilem] subtilem per virtutem ignis Ka2; tam subtiliter P1 **167** tactum] *finitur lacuna* M; *add.* manus B; *add. rub.* Conjunctio elementorum. Quinta operacio est ut post dictas separaciones quatuor elementorum incipiamus ipsa conjungere. P1 | igitur] enim Ko1; *add.* istud M; *add.* dictum primum elementum scilicet dictam aquam seu dictum P1 | lac] lac virginis B P1; lac virgineum M | Jhesu Christi] domini nostri jhesu christi B; christi P2 | et] quod quidam lac B; quod P1 **168** dixerunt] dixerit P1 P2 ipsum] *om.* B P1 | philosophorum] philosophi B Ka2; *om.* P1 | animal] animam Ka2; alius Ko1; anima P1 | cum…suo] *om.* Cu2 Ko1 Ko2 M P1; *add.* animal B | quia…169 animam] quia crescit in sublimando et habet animam B; *om.* Cu2 Ka2 Ko1 Ko2 M P1 **169** sanguinis] *om.* Ka2; sanguis P2 id est] P2 | tincturam] tinctura Cu2 Ko1 Ko2 P1; *om.* P2 **170** rubedinis] *om.* P2 | vitrioli] leonis M ipsum] *om.* B | infra] in M P1 | amphoram] ampola P1 | vitri] intus Ko1; vitream M **171** rotundam] rotunda P1 | in fundo] *add.* scilicet ovum philosophorum Ko1 | et[1]] quo facto P1 | strangula] claude M **172** forpicibus] tenella M | caute] *om.* Cu2 Ko2; candenti M | ita] tamen ita B; scilicet Ko1; *om.* P1 | nichil] non Ka2 M | quoquomodo] aliquo modo Ko1 M; quomodo P2 | Et] post P1 **173** laminam] *om.* B Cu2 | tegulatam] tegulam Cu2 P2; ferream vel tegulam P1 | perforatam] per foramen Ko2 M | in collo] *add.* furni B; *add.* amphore Cu2; *add.* vitri M; *add.* ampolle P1 | inter] super quam facies Cu2 | et amphoram] *om.* Cu2; et ampolam P1 | ne] ut B **174** destruatur] non destruatur B; defervatur P1 | ab igne] per ignem Ka2; *add.* et cum collum amphore bene fuerit strangulatum cum forpicibus Cu2; *add.* cujus quidam ampole forma talis est P1 | Et] tunc Cu2 inferius…175 describendo] *om.* Cu2; inferius designato P1 **175** ubi] ut Ko1; *om.* P1 | vas] *add.* autem P1 | habeat] habebit B Ko2 P2 | ignem] ignitionem Ko1 | ab…176 parte[1]] cum parte Cu2; *om.* P1 **176** equalem] equaliter vel equale M | hec est] *om.* P1 | nostra] mea B Ka2 P1 | vas] *om.* M **177** equaliter] *om.* P1 **178** evolare] revelare B; revolare P2 | quod] *add.* non M | ab…parte] *om.* Cu2 | donaretur] daretur Ko1; fieret M **179** equalis] equaliter B Ka2 | non] *add.* ita Ko1 M | et] quia aliter esset Cu2; quod esset M; quem esset P1

B Cu2 Ka2 Ko1 Ko2 M P1 P2 **164** Materia] *finitur lacuna* M **167** tactum] *finitur lacuna* M

The material of the stone having been perfectly prepared, in this fifth operation I will teach how the material is prepared for the white, that is, how the virgin's milk is digested so that it is turned by the power of the fire into a powder so subtle that it cannot be sensed by touch. Therefore take the milk, with the blessing of Jesus Christ, (certain of the philosophers have called it an animal with its blood, "animal" because it grows in subliming and "blood" because it has a soul of blood, that is, the tincture of redness of the aforesaid spirit of vitriol) and put it into a glass flask with a round bottom.[36] Heat up its neck and choke it off by pressing it together carefully with tongs such that nothing can escape by any means. And you should put a thin perforated tile around the neck between the fire and the [belly of the] flask so that the virgin's milk is not destroyed by the fire.[37] Put the flask in the furnace of the philosophers to be described below in order to be digested in a fire where by a clever technique the flask may have an equal fire from all sides. For this is our intent: that the flask may receive fire equally from all sides so that the virgin's milk can in no way fly to any part, which would happen if an equal fire were not given from all sides. And there is an art to giving a fire that is also temperate, not excessive or destructive to the

[36] Cf. pseudo-Arnald's *De secretis naturae*, where a master explains to a student why the stone (meaning in this case the material for making the stone) is called "animal" and "blood," in Calvet, *Oeuvres alchimiques*, 490 and 500-2. John's metaphorical explanation of these terms is an implicit rejection of pseudo-Avicenna (see Sébastien Moureau, *Le* De anima *alchimique du pseudo-Avicenne*, 2 vols. [Florence: SISMEL, 2016] 1: 206-15) and other authors such as Roger Bacon who proposed animal substances like blood, hair, etc. as starting materials for the stone--a claim rejected by the more thoroughly metallic chrysopoeia of the fourteenth century. On this line, see the explanation of my editorial intervention presented in Section 4.3.

[37] The idea here is to fit an earthen tile around the neck of the flask temporarily as a heat shield so that the virgin's milk is protected from the heat radiating from the very strong fire needed to melt the neck of the flask shut.

operis destructivum, ut subtus tot accensi carbones ardeant ita tamen quod manus hominis teneri valeat in aere inter latera furni et vas in quo est materia lactis virginis fulgentis. Et continuabis hunc ignem donec videas materiam denigrari quod si nimis tardaret adauge ei ignem. Et cum videbis materiam denigrari, gaude, quia principium digestionis tunc habes. Et continua ignem donec omnes colores transierunt et videris materiam aliquantulum dealbari, tunc augmenta ei ignem paulatim et insensibiliter prout videbis magis ac magis materiam dealbari. Et cum videris materiam dealbatam perfecte, habes lapidem perfectum ad album.

Sexta operacio lapidis docet lapidem jam completum ad album perfici ad rubeum, nichil penitus addito nichil omnino remoto sic: Extrahe de vase predicto quod vocatur ovum philosophorum et magister Arnaldus dicit quod lapis est clausus in eo ut Christus in sepulcro. Recipe partem unam quam volueris pulveris sine tactu ut per ipsum convertas omnia metalla imperfecta in verius argentum quam si esset de minera. Reliquam partem conclude in vase ut prius, et fortifica ei ignem notabiliter magistro Arnaldo

---

180 destructivum] defectivum Ka2 | ut] sed M; et P1 | tot] sint Ko1 | accensi carbones] carbones incensi Ka2; carbones P1 | ita] *om.* P1 | tamen quod] ut B P1; quod M **181** manus hominis] *add.* sine lexione M | teneri] *om.* P2 | valeat] possit B | vas] vasis M **182** virginis] virginalis Ka2 Ko2 M P2 fulgentis] fluentis P2 | hunc] ei Cu2 Ko1 Ko2; *om.* P1 **183** denigrari] denigratam Ko2 | quod] *incipit lacuna* Ka2 Ko1 | tardaret] tardeat Cu2; tardaverit M | adauge] adaugeas M; auge P1 P2 | Et cum] donec P1 | videbis] videris M P2; videas P1 **184** denigrari] *finitur lacuna* Ka2 Ko1; denigratam P2; *add.* tunc M **185** omnes] ad P2 | colores] carbones B | videris] videns B; videbis Ka2 P1; videas Ko1; vidis<?> Ko2 **186** aliquantulum] aliqualiter Ko1 | dealbari] dealbatam P2 | ei] *om.* Ka2 insensibiliter] sensibiliter B **187** prout] donec B; *om.* Ko1; quousque M | videbis] videris M materiam[1]] *om.* Ko1 | dealbari] dealbatam et dealbari M | videris] videbis B Ko2 P2 **188** perfecte] perficis Ka2; *add.* tunc M | habes] *om.* Ka2 | lapidem] *add.* magnum philosophorum Cu2; *add.* philosophorum M | perfectum] perfecte B Ka2 Ko1; perfectum scilicet P2 **189** Sexta] *antea add. rub.* Sequitur sexta operacio Cu2; *antea add. rub.* Sexta operacio Ka2; *antea add. rub.* Qualiter elixir album fit elyxir rubeum M; *antea add. rub.* Rubificacio partis dicti lapidis albi ad rubeum P1 | lapidis] *om.* B; *add.* jam Ko1; *add.* hic P1 | docet] *incipit lacuna* M | lapidem] dictum lapidem P1 | jam] *om.* Ko1 Ko2 P1 | completum] *om.* Ka2 Ko2 P1 | ad album] *om.* B; album Ka2 P1; *add.* perfectius Ka2; *add.* perfectum P1 | perfici] perficere P1 **190** ad rubeum] *finitur lacuna* M; *add.* et Ka2; *add.* est M; *add.* hoc modo P1 | nichil[2]] vel Ka2 Ko1 M; *add.* penitus P1 | omnino] *om.* B Ka2 P1 | sic] sic videlicet B; *om.* M P1 **191** quod vocatur] que dicitur P1 | ovum philosophorum] *add.* dictum lapidem in pulverem redactum sine tactu P1 | et... Arnaldus] *add.* de villa nova Cu2 M | dicit] dixit B Ko2; *add.* super hoc passu P1 **192** lapis] iste lapis P1 | ut] sicut fuit P1 | Recipe] accipe B; *om.* Ko2 P2; *add.* de eo P1 | unam] *om.* Ko1 **193** quam] *om.* Ko2 | volueris] *add.* dicti lapidis seu P1 | ut] *om.* B convertas] convertens B **194** verius argentum] melius argentum Cu2 M; verum argentum melius Ka2 quam] quam illud quod M | si esset] esset B Ko2; est Ko1 M; sit P1 | de] in Ko1 | Reliquam] Recipe reliquam Ka2 **195** conclude] claude Ko2 | notabiliter] vocabitur B; bene P1 | magistro Arnaldo] *add.* de villa nova M

B Cu2 Ka2 Ko1 Ko2 M P1 P2 **183** quod] *incipit lacuna* Ka2 Ko1 **184** denigrari] *finitur lacuna* Ka2 Ko1 **189** docet] *incipit lacuna* M **190** ad rubeum] *finitur lacuna* M

work such that all the lit coals may burn underneath yet such that a man's hand can be held in the air between the walls of the furnace and the vessel containing the material of the shining virgin's milk.[38] And you will continue this fire until you see the material blacken, which if it takes too long, increase the fire. And rejoice when you see the material blacken because you have then the beginning of the digestion. Continue the fire until all the colors have passed and you see the material whiten a little. Then increase the fire gradually and insensibly as you see the material whitening more and more. And when you see the material perfectly whitened you have the stone completed for the white.[39]

The sixth operation of the stone teaches how the stone, now completed for the white, is perfected for the red, with nothing at all added and nothing removed, in this way: Take it out of the aforesaid flask, which is called the egg of the philosophers, and master Arnald says that the stone is shut up in it like Christ in the tomb.[40] Take one part, as much as you wish, of the impalpable powder so that using it you may convert all imperfect metals into silver more true than if it were from the mine.[41] Seal up the remaining part in the flask as before and increase the fire notably, as master Arnald orders, who says that

[38] Cf. the parallel section from *De famulatu* in Appendix 1 in regard to monitoring the temperature with the hand.

[39] Here "for the white" (*ad album*) means "for the white *metal*," that is, for making silver from the base metals. In short, the stone brought to this white stage is *argyropoetic* ("silver-making"). Similarly, in the next line, "for the red" means "for the red metal," i.e., gold.

[40] A comparison of the digesting flask to Christ's tomb occurs in pseudo-Arnald's *Exempla*; see Calvet, *Oeuvres alchimiques*, 544; similiar language appears also in the compendium of the Darmstadt manuscript (see Section 2.3.2).

[41] In asserting that the silver produced by the white stone is "more true" than natural silver, John is adopting the view of Latin alchemy that alchemical metals are superior in quality to the naturally-occurring ones, that is, art can outdo nature. Roger Bacon, for example, was a key proponent of this view of the power of alchemy. On this point, see William R. Newman, "Technology and Alchemical Debate in the Late Middle Ages," *Isis* 80 (1989):423-45.

jubente, qui dicit quod postquam lapis pervenit ad perfectam albedinem non timet penas crescentis ignis, nec est periculum aliquod si fortificetur notabiliter, quia sic ei expedit et est omnino necesse. Cujus causa in veritate est quia per ignem lentum qui fit quando digeritur ad album non potest decoqui sulphur invisibile absconsum vitrioli, scilicet spiritus latens sub albedine lapidis, sed digeritur spiritus mercurii illo igne ad album. Quapropter necesse est ignem augeri ut attingat profundius et digerat tincturam rubedinis sulphuris vitrioli latentis et faciat ipsam exterius apparere et ita, nichil addito nichilque remoto, idem lapis albus per majorem ignem efficitur rubeus et ascendit de sepulcro rex excelsus, resuscitatus a mortuis cum dyademate rutilante, fulgens et gloriosus ut magister Arnaldus excellenter testatur. Sed nota quod in duobus vasis simul operari debes, in uno ad album et in alio ad rubeum in eodem igne et in eodem furno simul et pariter. Nec differt nisi quod vas ad album extrahes albedine inducta, et vas ad rubeum relinques igne notabiliter aucto ut dixi. Et nota quod aliqui operantur per omnia ut dixi de lacte virginis et de elemento aeris et ignis non curant, et est opus eorum per solum lac virginis nobile et perfectum pro certo, et maxime ad album. Alii miscent tria elementa simul, scilicet unum pondus ignis, quatuor pondera aeris, et octo pondera lactis, et hic modus forcior est ad rubeum, debilior ad album. Tres hos modos potes facere simul: in uno vase de solo lacte ad album, in alio vase de solo lacte ad rubeum, et in tercio vase de tribus elementis ad aurum,

---

196 qui] *om.* P1 | quod] *om.* B | perfectam albedinem] perfeccionem albedinis B; perfeccionem Ko1 **197** timet] senciat B | penas] penam Cu2 Ko2; penam seu pugnam M | est] *om.* B **198** notabiliter] *om.* P1 | quia…est] *add.* ei P1 **200** decoqui] coqui P1 | vitrioli] *om.* Cu2 Ka2 M | scilicet] sed B; videlicet P1 | spiritus] *add.* vitrioli Cu2; *add.* leonis viridis M **201** digeritur] *add.* quia P1 **202** augeri] augere Ko1 P1 | profundius] profunde P1 | digerat…203 tincturam] digeratur tinctura P1 **203** vitrioli] *add.* romani Cu2; leonis viridis M | latentis] *om.* P1 | exterius] exteriorem P1 **204** ita] *om.* B P1 | nichilque] vel Ka2 | idem] quia B **205** majorem ignem] ampliorem ignem Ko1; vigorem ignis P1 | efficitur] efficietur Cu2 Ko1 Ko2 | rubeus et] *incipit lacuna* B P2; *add.* sic Cu2; *add.* ita Ka2 P1 | ascendit] ascendet M; *add.* christus P1 | excelsus] *finitur lacuna* B P2 **206** resuscitatus] resuscitans Cu2 Ka2 | rutilante] refulgente B | fulgens…gloriosus] fulgida et gloriosa P1 **207** excellenter] *om.* Cu2 M P2; excellens Ko2 | testatur] attestatur P1; dicit P2 | Sed] *om.* Cu2; Item M **208** debes] debent P1 | uno] uno vase Cu2 | ad[1]] *om.* B | alio] altero B P2 | in[3]] *incipit lacuna* B; cum M **209** nisi] illud Ko1 | extrahes] extrahas Ka2 **210** inducta] induta P1 | rubeum] *finitur lacuna* B | relinques] relinquens B; relinquas Ka2 Ko1 | igne] in igne Cu2 Ka2 | aucto] augmentando Ka2; acuto Ko1; continuando P1 **212** per] propter P2 **213** nobile] notabile Ko1 | et[2]…album] *om.* P2; *add.* Et Ka2; *add.* Item M | Alii] alii vero B **214** simul] insimul Cu2 Ka2 Ko1 Ko2 **215** debilior] *add.* vero B | Tres] Omnes P2 **217** aurum] rubeum Ka2 M

B Cu2 Ka2 Ko1 Ko2 M P1 P2 **205** rubeus et] *incipit lacuna* B P2 | excelsus] *finitur lacuna* B P2 **208** in[3]] *incipit lacuna* B **210** rubeum] *finitur lacuna* B

after the stone has come to perfect whiteness it does not fear the pains of an increasing fire, nor is there any danger if the fire is increased significantly, because that expedites it and is completely necessary.[42] In truth, the reason for this is that the hidden invisible sulphur of the vitriol, namely the spirit hiding beneath the whiteness of the stone, cannot be fully cooked by the gentle fire that is made during the digestion for the white, even though the spirit of mercury is digested for the white by means of that fire. Wherefore it is necessary that the fire be increased so that it may reach more deeply and may digest the tincture of redness of the hidden sulphur of vitriol and make it appear externally. And thus, with nothing being added or removed, the same white stone is made the red stone through a greater fire, and the king rises exalted from the tomb, revived from the dead with a gleaming diadem, resplendent and glorious, as master Arnald excellently declares.[43] But note that you ought to work in two vessels at once: in one for the white and in the other for the red, in the same fire and in the same furnace together and equally. Nor is there any difference except that you take out the flask for the white once the whiteness has been brought forth, and you leave the flask for the red in place with the fire notably increased, as I have said. And note that some operate with the virgin's milk for everything, as I have said, and they do not bother with the element of air and the element of fire, and their work using the virgin's milk alone is noble and perfect to be sure, and especially for the white. Others mix the three elements together, namely, one weight of fire, four weights of air, and eight weights of milk, and this way is more powerful for the red, weaker for the white. You can do all these three methods at one time: in one flask with only milk for the white, in a second flask with only milk for the red, and in a third flask the three elements for gold, and all in

[42] The short pseudo-Arnaldian text in the Darmstadt manuscript (see Section 2.3.2) uses wording that is nearly identical to John's citation of Arnald here and clearly in reference to this same part of the procedure: "thereafter [i.e., having passed the white stage] it withstands the pains of the increasing fire without fear of danger" (*ex tunc sine metu periculi penas sustinet ignis crescentis*).

[43] A similar reference to the stone as an "emperor...crowned with a red diadem" (*imperator pacificus...dyademate rubeo...coronatum*) appears in pseudo-Arnald's *De secretis naturae*, in a series of metaphorical descriptions which Calvet suggests is drawn from Joachite sources; see *Oeuvres alchimiques*, 518-21. A more closely similar expression appears in the compendium of the Darmstadt manuscript, see Section 2.3.2.

et in eodem furno et igne ut probes subtilitates diversorum modorum et videas miracula maxima a deo glorioso concreata in rebus.

Credas vir pauper evangelice quod ante me nullus hominum in aperto hanc veritatem scripsit. Scito autem et animadverte quod philosophi magni quibus revelatum fuit hujus artis magisterium revelantibus imprecati fuerunt malediciones horrendas timentes ne archanum hoc deveniret ad manus indignorum. Ego autem has malediciones non incurro pro certo quia non revelo archanum hominibus nequam neque filiis hominum sceleratis, sed mistico sacrosancto corpori Jhesu Christi, ecclesie romane *que non habet maculam neque rugam* mortalis criminis nec peregrinum colorem heresis nec erroris. Hoc enim solum revelavi pro sanctis in proxima antichristi tribulacione premendis. *Bestia autem carnalis que hunc librum veri luminis tetigerit absque dubio lapidabitur* et iram infernalem incurret, nam sicut hic liber est in remedium sublevacionis calamitatis sacrosancte ecclesie et electis, ita est *infernalis laqueus* reprobandis. Tot ergo mortibus dignus erit qui scienter reprobis revelabit, hoc est communicabit, hunc

---

218 et[1]] *add.* potes facere P1 | et igne] *om.* B Ka2 P2 | subtilitates] sublimitates B P2; subtilitatem Ka2 P1; *add.* sublimacionis Ka2 **219** videas] videbis Ka2 P1 | miracula] mirabilia B Ko1 P1 | maxima... rebus] *add.* naturalibus. M **220** Credas] Credat Cu2; Et credas mihi M | vir pauper] vir B evangelice] euvangeliste P2 | ante me] *om.* P1 | nullus hominum] nullum hominem P1 | in aperto] ita aperte M **221** hanc] *om.* Ka2 | veritatem] scienciam et veritatem B | scripsit] protulit Ko1 | Scito] Dico Ka2 | autem] *om.* Ko1 | animadverte] animadvertite Ko2; adverte P1 | quod] *add.* multi M magni] magistri P1 **222** magisterium] misterium B | revelantibus] revelacionibus veritatis B P2; revelatores veritatis Ka2 P1; *om.* Ko1; revelatoribus Ko2 | imprecati] impetrati B P1; imputati Cu2; intricati Ka2; deprecati Ko1; inplicati P2 **223** malediciones horrendas] deum Ko1 | timentes] *om.* Ko1; et tremendas P1 | archanum...deveniret] veniret archanum B Ka2 Ko2 P2 | ad] in Ka2 **224** indignorum] virorum indignorum et quod maledictiones incurrent horrendas quicumque revelarent alicui magisterium hujus artis et veritatem Ko1 | has] *om.* Cu2 P1 **225** archanum] hoc archanum Cu2 | nequam] *om.* Ko2 P1 P2 | neque] *om.* B Cu2 Ko1 | hominum] *om.* B P2; hominis P1 **226** sceleratis] sceleritatis B; *om.* Cu2 | sacrosancto] sacro P1 | corpori Jhesu] corpori Ka2 | Christi] *om.* P1 | ecclesie] *add.* jhesu christi Ka2 Ko2 P1; *add.* scilicet M; *add.* christi P2 | romane] *om.* B; catholice romane Cu2 **227** maculam] in se materiam Cu2; materiam M | neque rugam] *om.* Cu2 M criminis] cineris P2 | nec] *om.* Cu2 | peregrinum] perequinum P2 **228** heresis] nec eciam heresim Cu2; heresum M | nec] *add.* habet M | erroris] errorem B Cu2 M; errorum Ko2 | Hoc] *om.* P1 solum] *om.* B | pro sanctis] personis Ko1 | in...229 antichristi] *add.* tempora M **229** premendis] serviendis Cu2 M; proveniendis Ka2; puniendis Ko1 | carnalis] rationalis carnalis B | que] qui Ko2 P1 | hunc librum] habet mentem B; non habet mentes Ka2; hunc P1; hunc montem P2 **230** veri luminis] verum luminis P1 | tetigerit] reticescat alioquin B; si tetigerit Ka2 | lapidabitur] lapidatur Ka2 **231** sicut] si P1 | in] *om.* Ko1 | sublevacionis] levationis P2 **232** electis] clericis B | infernalis laqueus] in laqueum infernalem Ka2; *add.* reprobis et M | Tot] Talis Ka2 P1 | ergo] *om.* B | mortibus] morte Ka2; mortis P1 **233** erit] est B | reprobis] *om.* B | revelabit] revelat Ka2 | hoc est] vel B communicabit] qui communicat Ka2 | hunc...234 librum] *om.* P1; librum P2

B Cu2 Ka2 Ko1 Ko2 M P1 P2

one and the same furnace and fire so that you may explore the subtleties of the various methods and see the greatest miracles created within things by the glorious God.

You may believe, poor man of the Gospel, that before me no man wrote this truth openly.[44] Know likewise and observe that the great philosophers to whom the magistery of this art was revealed have called down horrible curses upon those who reveal it, fearing lest this arcanum should come into the hands of the unworthy. Yet I indeed do not incur these curses, because I do not reveal this arcanum to wicked men, nor to the criminal sons of men, but rather to the holy mystical body of Jesus Christ, the Roman Church, that *has neither spot nor wrinkle* of mortal crime, nor the alien hue of heresy or of error.[45] For I have revealed this knowledge only for the holy ones who will be oppressed in coming tribulation of the antichrist. *The beast or carnal man who touches this book of true light shall without doubt be stoned*[46] and shall incur hellish anger, for just as this book is to be a remedy of relief against the distress of the holy Church and for the elect, so it is *the infernal snare* for reprobates.[47] Therefore he who will knowingly reveal this book, that is, communicate it to the wicked, will deserve as

[44] On this section, which I call the "Credas" in reference to its opening word, and which is largely or wholly omitted from both printed editions and significantly shortened or omitted in many of the later manuscript witnesses, see Section 2.8.

[45] 2 Cor 12 and Eph 5:27; John uses the latter verse identically extended in *LO*, 331 and 443.

[46] Heb 12:20, in reference to Ex 19:12-13. The original Biblical text refers to a *mountain* (specifically Mt. Sinai) instead of a *book*. Witnesses from family π^3 and its descendant families ε^2 and ι^2 do read *montem* rather than *librum*, which is probably a correction by the copyist of π^3 who recognized the Biblical text, while many descendants of γ and all of γ^3 corrupt *montem* into *mentem*. The reading *librum* is however found throughout several other families (π^2, φ, ε^1, γ^2, and ι^1), and was certainly John's original since a later "correction" to *montem* based on the original Bibical verse is easily explicable while the substitution of *librum* for *montem* is not. Moreover, the reading "book of true light" (*librum veri luminis*) would seem to allude directly to the title of John's treatise.

[47] 1 Tim 6:9.

librum quot peccatis et culpis ministrabit fomentum. Sed nephas esset in me si ecclesie indigenti hanc veritatem celarem quia librum ad manus malorum devenire timerem. Nam et Christus propter abutentes hereticos nequaquam evangelium revelare dimisit, sicut Paulus dicit *quod enim est illis,* scilicet incredulis, *causa dampnacionis,* scilicet odium quod habent contra evangelium, *nobis autem salutis.* Ita et hic mors et vita in manibus hujus libri, sed non ex libro sed ex libero arbitrio sic utente. Commendo igitur hoc revelatum archanum in manu crucifixi sanctissimi domini nostri Jhesu Christi a quo recepi hoc bonum, et custodie angelorum sanctorum et evangelicis pauperibus qui sollicitudine divitiarum seduci, deo servante, non possunt. Sed caveant ne cum aliis evangelicam paupertatem commendant verbo, ipsi cupiditate harum diviciarum seducti dormiant in inferno. Et recordentur Pauli dicentis *radix omnium malorum est cupiditas quamquidam appetentes exciderunt a fide.* Nec est omnibus simplicibus communicandus hic liber, ne forte seducti amore temporalium *ab aratro retrahant manum suam et aspiciant retro,* sed perfectis timentibus deum qui solum tempore tribulacionis aut famis generalis juvent ecclesiam affligendam et sanctos.

234 quot] *om.* Cu2 P1; qui Ka2; quot et quantis M | peccatis] pro peccatis P1 | ministrabit] ministrat Ka2 **235** hanc...celarem] *add.* ob hoc Ka2 | manus] manum B Ka2 Ko2 P2 | malorum...236 devenire] indignorum venire M **236** abutentes...nequaquam] abeuntes hereticos nequam Ka2 **237** evangelium] evangeliis P1 | revelare dimisit] distulit B | dicit] dixit B | quod] que Ka2 P2; quis P1 | enim] evangelium non Cu2; non Ko1 | illis] in illis P2 | scilicet] videlicet B **238** incredulis] incredibilibus Cu2; incredibilis Ka2 | causa dampnacionis] tam denacionis P1 | scilicet] sed Cu2 habent] *om.* P1 | contra] extra B **239** evangelium] *add.* hoc M | autem] *om.* Ko1; est M; *add.* causa Ko1 M | salutis] ad salutem B | in] *om.* B P1 | manibus] manus Cu2 Ko2 | hujus...240 libri] hominum Cu2 **240** sed[1]] *om.* B Cu2 | non] *add.* tamen Cu2 | sed[2]...libero] *om.* Ko2 P1 | sic] *om.* B; scilicet Cu2 M; sit Ka2 P1 | utente] succedente B; venire Ko1; utentis M P1 | igitur] autem B; *om.* Ko1 **241** hoc] *om.* B | revelatum] *om.* Cu2 | manu] manus M; manibus P1 | sanctissimi...nostri] domini nostri B; dei Ko1; sanctissimi domini Ko2 **242** Christi] *om.* B | bonum] donum Ko1; donum bonum et perfectum M **243** evangelicis] evangelistarum Cu2 P1; *add.* ac eciam Cu2; *add.* viris M pauperibus] *add.* hominibus Cu2 | qui] quia Ko1 P1 | sollicitudine] sollicite Cu2 | seduci] seducti B Cu2 Ka2 Ko1 Ko2 M | servante] servire Cu2 Ka2 Ko1 M P1; *om.* Ko2 **244** evangelicam] qui ewangelicis B; *om.* Cu2 | paupertatem] *om.* Cu2 | commendant] *om.* Cu2; comedent Ko1; commendatur P1 **245** verbo] *om.* Cu2; *add.* ne M | cupiditate] cupidus in P1 | in] *om.* B | Et] Cum P1 **246** Pauli] *om.* Cu2 | dicentis] dicente B; dicentes Cu2 | est] *om.* B **247** quamquidam] ex qua quidam B; quam quid Cu2 | appetentes] potentes B; *om.* P1 | exciderunt] cum cederunt exierunt Ka2; ceciderunt P1 | est] *om.* Ko1 | omnibus] *add.* hominibus bonis B; *add.* bonis Ka2 Ko2 **248** communicandus] commendandus B | hic liber] liber iste B; *om.* P1 | temporalium] temporanei Cu2; temporali P1; *add.* bonorum Ko1 **249** retrahant] trahant Cu2 M | aspiciant] aspicientes B P2; respiciant Ka2 Ko1 | perfectis] tamen perfectis Ka2; pro sanctis P2 | deum] eum B; dominum Ka2 M qui] *om.* Cu2 M **250** tempore...aut] propter indigenciam Ko1 | famis generalis] *add.* ut M juvent] *om.* P1; juvet P2 **251** affligendam] effugiendam B; afflictam M; affligi timent P1 | et sanctos] *add.* ut finaliter ab hac calamitate liberetur amen. Ka2; *add.* viros euvangelicos Ko1

B Cu2 Ka2 Ko1 Ko2 M P1 P2

many deaths as the number of sins and crimes for which he furnishes the fuel. But it would be criminal of me to hide this truth from the needy Church just because I should fear the book would come into the hands of evil men. For likewise Christ did not in any way give up revealing the Gospel on account of the heretics who abuse it; as Paul says, *what is for them* (namely, the unbelievers) *a cause of damnation* (namely, the hatred they have against the Gospel) *is likewise a cause of salvation for us.*[48] So likewise here, *life and death are in the hands*[49] of this book, but not from the book itself but rather from free will being used in such a way. I commend therefore this revealed arcanum into the hand of the most holy Crucified One, our Lord Jesus Christ from whom I received this good, and into the custody of the holy angels and to the poor men of the Gospel who, with God protecting them, cannot be seduced by the desire for riches.[50] But let them beware, lest while they commend evangelical poverty to others in their words, they themselves, seduced by the desire for these riches, sleep in hell. And let them be mindful of Paul where he says that the *love of money is the root of all evil, some in their passion for it have fallen away from the faith.*[51] Nor is this book to be communicated to all simple men, lest perhaps having been seduced by the love of temporal things they withdraw *their hand from the plow and look back*,[52] but only to the perfect God-fearing ones who will only help the afflicted Church and the holy ones in the time of the tribulation or of universal want.

[48] Phil 1:28.
[49] Prov 18:21; here, as above (note 46), John has substituted the word "book" for the original word in the Biblical verse, in this case, "tongue."
[50] On the editorial intervention here, see Section 4.3.
[51] 1 Tim 6:10.
[52] Lk 9:62.

Septima pars hujus operis est praxis virtutis elixir, hoc est prefate medicine perfecte. Virtus autem medicine reservate et perfecte ad album est ut convertat argentum vivum, plumbum, stagnum, cuprum, latonem, ferrum, et calibem in argentum verissimum. Et medicine perfecte ad rubeum virtus est ut predicta corpora et etiam ipsum perfectum argentum convertat in aurum. Et hoc fit ut ponas in crucibulo mercurium et calefac donec quasi buliat et notabiliter fumet, post projice medicinam albam vel rubeam ut voles et auge ignem fortiter usque ad ignem fusionis argenti. Et post competentem decoccionem, extrahe et si nimis est de colore, aut si esset frangibile, est ad aliud medicina perfecta donec tibi placeat color et malleacio. Et super quot pondera cadat experiencia te docebit, quia aliquando cadit super majori, aliquando super minori, prout unus modus operandi excedit alium modum, et prout magister perfectius fuerit operatus. Super plumbum et stagnum fusis projicitur medicina, et augetur ignis ut supra. Super cuprum, latonem, ferrum, calibem, et argentum projicitur dupliciter: primo, super laminas ignitas super quas projicitur pulvis et

252 Septima] *antea add. rub.* Sequitur septima pars hujus operis Cu2; *antea add. rub.* Septima operacio Ka2; *antea add. rub.* Item de virtute elysirum predictorum et modo projectionis M; *antea add. rub.* De projectione dicti lapidis seu elixir P1 | pars] operacio Ka2 M | hujus] *om.* M | operis] *om.* M; libri et operis P1 | praxis] *om.* Ka2 P2; projectio P1 | virtutis] iteratis virtutibus Ka2; *om.* P1 | elixir] *om.* M; *add.* predicti P1 | hoc est] *om.* M; *add.* de medicinis Ka2 **253** perfecte[1]] *om.* B | Virtus autem] virtutis autem Ka2; *om.* M | medicine[2]] *om.* B M | reservate] *om.* M | et perfecte] imperfecte Cu2 Ko2; *om.* M P2 | ad album] *om.* P1 | est] *om.* M **254** argentum vivum] mercurium B | cuprum] *om.* P1 **255** ferrum] *om.* B Cu2 M | et calibem] *om.* Cu2 M | argentum verissimum] argentum verum P1 | medicine] *om.* P1 **256** virtus est] est Ko1; est ejus virtus M | ipsum…argentum] ipsum argentum vivum P1 **257** hoc fit] fit sic B Ka2; hoc sic Cu2; sic ergo Ko1; *om.* P1 | ut] *om.* B; Recipe igitur ipsum P1 | ponas] pone B; *om.* P1 | in[2]…mercurium] argentum vivum P1 | quasi] *om.* M **258** et notabiliter] totaliter P1 | fumet] buliat Ko1 | projice] pone Ko2; *add.* intus P1 | medicinam] medicinam tuam P1 | rubeam] imbibe Cu2 | ut] quod B **259** voles] vis Ka2; volueris M P1 | usque] ut B P2 **260** extrahe] *add.* Ecce formam crucibuli prout melius depingi potest P1 | nimis] nimium Cu2; nimis vel minus Ka2; minus Ko1 Ko2 | est] habeat Cu2; esset P1 | si[2]] *om.* M P1 | esset] est B Ka2; *om.* M **261** frangibile] nimis frangibilis P1 | est…aliud] hoc tunc ad aliud est Cu2; est addenda Ko1; est super alium mercurium M; adde ibi aliud argentum vivum P1; *add.* Et si minus fuerit de colore adde de P1 | medicina perfecta] medicina Ka2; *add.* perficienda Ka2; *add.* projicienda M | tibi placeat] tibi placuerit M **262** Et] donec videris P1 | cadat] cadet Cu2; cadit Ka2; *add.* medicina super corpora et sic P1 | quia] si sis discretus quia projectio P1 **263** aliquando[1]] aliqui P2 | cadit] *om.* Ka2 super[1]] pars super Ko1 | majori] magis B; plus Ka2; ii Ko1 | minori] minus B Ka2; quatuor Ko1; minori quantitate M **264** alium] aliquando Ka2 | modum] *om.* M | prout] *add.* unus M | magister] magis P1; magistri P2 | perfectius] perfecte B; *om.* P1 | fuerit] fuit B Ko1; *om.* P1; fuerint P2 operatus] operati P2; *add.* quam alius M **265** et[1]] vel P1 | fusis] fusum Cu2 Ka2 Ko1 M **266** Super] supra Ko1 P1 | ferrum] *om.* Ko1 | argentum projicitur] *add.* medicina Cu2 **267** dupliciter] *incipit lacuna* B P1 P2 | super[2]…projicitur] super projicitur Cu2; projicitur M; *finitur lacuna* P1 P2

B Cu2 Ka2 Ko1 Ko2 M P1 P2 **267** dupliciter] *incipit lacuna* B P1 P2 | super[2]…projicitur] *finitur lacuna* P1 P2

The seventh part of this work is the practical use of the power of the elixir, that is, of the aforesaid completed medicine. The power of the reserved medicine completed for the white is that it converts quicksilver, lead, tin, copper, brass, iron, and steel into the truest silver. And the power of the medicine completed for the red is that it converts the aforesaid bodies, and also perfect silver itself, into gold. And this is done thus: Put mercury into a crucible and heat it until it fumes notably and almost boils, then throw the medicine, white or red as you wish, upon it and increase the fire powerfully up to the melting point of silver. After sufficient cooking take it out, and if it is too deeply colored or if it is brittle, it is a perfect medicine for other quicksilver until the color and hammering please you.[53] Experience will teach you upon how many times its weight it will fall, because sometimes it works upon a greater and sometimes upon a lesser amount according to how one method of working exceeds another method, and as one master has worked more or less perfectly.[54] The medicine is projected upon lead and tin in fusion, and the fire is increased as above.[55] It is projected in two ways upon copper, brass, iron, steel, and silver: first, upon red-hot sheets, upon which the powder is projected and then the sheets are put to be cooked further;

[53] The idea here (as clarified in the following sentence) is that the practitioner does not initially know the exact transmuting power of the completed material, hence, an excessive amount of the philosophers' stone might unknowingly be added to the mercury, resulting in a brittle or incorrectly colored metal. The excess of elixir thus renders that transmuted metal a "perfect medicine" itself, and thus capable of effecting further transmutation. Thus the brittle metal is to be projected upon more (*aliud*) quicksilver, thereby using up the excess transmuting agent present in it, until the proper color and malleability are achieved.

[54] John uses the tag "experience teaches" (*experiencia docet*) regularly in his prophetic writings as well, e.g., *VM* (Tealdi, 222 and 224; Kaup, 148 and 155)

[55] The verb "to project" and the noun "projection" that are generally used describe the act of bringing about transmutation using a prepared agent of transmutation like the philosophers' stone derive from the Latin *projicere*, meaning "to throw upon." This verb describes the practice in which some of the stone is supposed to be literally *thrown upon* a mass of molten metal, heated mercury, or as in the following sentence (somewhat unusually) on red-hot plates of the more refractory metals like copper and iron.

ponuntur ad recoquendum, aut illa corpora funduntur ut prius et medicina projicitur in ipsis corporibus ita fusis. Et corpora hec purgacione precedente non egent, quia medicina est quasi ignis inferni virtualiter et consumit imperfecciones omnium corporum et convertit in sui naturam quod est bonum in eis. Et aliqui quando volunt operari ad argentum projiciunt primo medicinam super verum argentum et illud argentum medicinatum projiciunt super cuprum vel ferrum. Ad aurum primo projiciunt medicinam super aurum purissimum et illud aurum medicinatum est medicina ad convertendum cuprum vel argentum in aurum et hic modus projiciendi ab utilitate non vacat. Potest autem medicina dissolvi in balneo et congelari ad ignem iteratis vicibus et augetur in infinitum virtus ejus ad placitum magistri opera iterantis.

Octava pars continet magisterium furni, fit autem sic: Recipe de luto magisterii ex argilla et stercore equi et papiro carminato et capillis incisis cum aqua salsa et aceto compasta, et construe furnum magnum sicut vis vel parvum cum porta competenti et sit altitudo circularium parietum in

268 ponuntur] ponitur P1 | recoquendum] coquendum M | aut] et post Cu2; alia Ko1; aliud P1 | illa corpora] corpora Ko1; *add.* scilicet cuprum vel laton et sic de aliis M | medicina] *om.* P1 **269** projicitur] *finitur lacuna* B | ita] *om.* B Ko1 | fusis] *add.* aut non sic P2 | Et] Ita et Ko1 | hec] hac preparatione vel M **270** egent] indigent B Ko1; *add.* ut aliqui purgant P1 | quasi] *om.* B | inferni] infernalis Cu2 P1 | virtualiter] virtualis Cu2 Ka2 M **271** imperfecciones] omnes imperfecciones B Cu2; imperfectionem P1 | corporum] aliorum corporum Ka2; metallorum sive corporum Ko1 | sui] suam B Ka2 **272** quando] qui B Ko1 **273** primo] primam B; *om.* Cu2 M | verum argentum] mercurium B; vivum argentum Cu2 | illud argentum] illum mercurium B | medicinatum] medicinaliter Ko1 **274** super] supra B | vel ferrum] *om.* Cu2 M; et ferrum Ka2 P1 | Ad] Item ad P1 | aurum] *add.* autem B Ka2; *add.* vero M P2 | primo] *om.* P1 | medicinam] *om.* Ko1 **275** super] supra Ka2 M P1 purissimum] verissimum B P2; verum P1 | est…276 convertendum] proiciunt supra P1 **276** vel argentum] *om.* M; *add.* et alia corpora M; *add.* etc. P1 | in aurum] *om.* P1 | ab…277 utilitate] a veritate Ko1 **277** Potest] potes B; Et est notandum quod medicina potest M; *om.* P1 | autem medicina] *om.* M | dissolvi] iterum dissolvi Cu2; dissolve P1 | congelari] coagulare B **278** vicibus] insimul P2 | et augetur] ut augeatur B; et augebitur Ko1; et sic augetur M | ejus] *om.* B Ko2 P1 P2; lapidis Cu2 | placitum] voluntatem Ko1; beneplacitum P1 **279** opera iterantis] opus iterantis B; opera reiterantis Ko1; operantis vel opus iterantis M; operantis P1; *add.* si experte operatus sit Cu2 **280** Octava] *antea add. rub.* Pars octava de furnis Ka2; *antea add. rub.* De magisterio furni qui dicitur athanor philosophorum M; *antea add. rub.* De furnis philosophorum et vasis neccessariis huic operi P1 | continet…furni] hujus libri est de furnis seu de furno faciendo pro dicto opere P1 | fit autem] *om.* M | sic] furnus sic Ko1; *om.* M | Recipe] *om.* Ka2; Fiat igitur M P2 | de…281 magisterii] lutum M **281** ex] et de P1 | argilla] argilla bona B | capillis] pilis P1; *add.* humanis B **282** salsa] salsata Cu2; *add.* vel incisuris vel cimaturis pannorum vel borra bastorum P1 | et[1]] cum P1 | compasta] inpasta B; compacta Ko2; cum pasta P1 | construe] construes P2 | magnum] *om.* Ko1 | sicut vis] ad libitum Ko1; *om.* Ko2; si vis M; sicut volueris P1; sicut voles P2 **283** porta competenti] sua porta convenienti P1; *add.* juxta terram ubi immittantur carbones Ko1 | circularium] circularis P1

Cu2 Ka2 Ko1 Ko2 M P1 P2 **269** projicitur] *finitur lacuna* B

or those bodies are melted as before and the medicine projected upon them thus in fusion. And these bodies do not need any foregoing purification because the medicine is like a fire of hell in power and it consumes the imperfections of all bodies and converts what is good in them into its own nature. And some, when they want to work to make silver, project the medicine first upon true silver, and then project that silver made into a medicine upon copper or iron. For gold, they first project the medicine upon the purest gold and that gold, thus made into a medicine, is then the medicine for converting copper or silver into gold, and this method of projecting is not without utility.[56] The medicine can also be dissolved in a bath and congealed on the fire multiple times, and its power is increased to infinity as it pleases the master repeating the work.

The eighth part contains the magistery of the furnace, and it is made thus: Take the lute of the magistery, made of clay and horse dung and carded paper and cut hair compounded with salt water and vinegar, and build a furnace, large or small as you wish, with an adequate opening. And let the height of its circular

[56] This secondary method reprises what was advised a few lines earlier in regard to transmuting mercury before the exact potency of the stone is known. In that case, an excess of transmuting agent produced a metal of faulty color or malleability, and so that metal, imbued with an excess of the agent, was projected upon more mercury in order to use up the excess. In this case, a "medicated" silver or gold is produced intentionally, and then that material is projected upon a base metal in order to transmute it into silver or gold. While it is not immediately clear what source John drew upon for this method, that is, the identity of the "some" (*aliqui*) who advise this method, this alternative method would satisfy those who, like pseudo-Arnald, claimed that a "ferment" of natural gold or silver is a necessary ingredient for a transmuting agent. Further, this same method, employed upon mercury, useful especially when the potency of the stone is very high (and hence an enormous quantity of mercury needs to be used), is the recommendation of the section of pseudo-Arnald's *Rosarius* which became inserted into John's text in several manuscripts (and also the printed editions) either here or at the end of the book, see above, Section 2.6.

altitudine palmi vel minus cum auribus. Et dicitur basis furni, et ecce forma ejus prout depingi potest. Aptentur enim ei aures hinc inde ut fumus egrediatur exinde et non ascendat sursum. Consequenter autem fac fieri laminam fortem ferream habentem quatuor brachia ad modum crucis sub hac forma, et apta eam super caput prefati furni quod est in rotundo apertum sic ut non totaliter orificium furni claudatur, sed per circuitum predicte lamine inter laminam et furni latera sursum ascendat calor. Super laminam autem hanc situabis gobeletos seu ciatos de vitro quot poteris vel volueris in hanc formam, et super quemlibet gobeletum potes unum ovum philosophicum sive unam amphoram collocare ad lapidem digerendum. Deinde vides quod quatuor brachia lamine ideo sunt ut super furni parietes laminam teneant elevatam circulariter. Parietes furni eleva per circuitum per

284 palmi] duarum palmarum Cu2 | vel minus] *om.* Cu2 P2 | auribus] auriculis M | Et] que M; et hoc P1 | dicitur] *incipit lacuna* Cu2; dicuntur M | basis] vas B Ko1 P2; bases M; furnus bassus P1; *add.* id est fundamentum Ka2 | furni] *om.* P1 | et ecce] *om.* M; prout ecclesia P1 | forma…285 prout] *om.* M P1 **285** depingi potest] dependi potest B; *finitur lacuna* Cu2; in fine est depicta Ka2; *om.* M P1 | Aptentur] hic aptantur Ko1 | enim] nam B; *om.* P1 P2 | ei] *om.* B; furno predicto Ko1 | aures] auricule M | fumus] furnus Ko2 **286** exinde] *om.* Cu2 | ascendat] exeat B | sursum] *add.* ad caput furni Ka2; *add.* ad caput furni hoc est quod fiant aliqua foramina in furno per que fumus possit exire cujus forma partis basse seu inferioris talis est que est prima pars furni P1 | fac fieri] facies fieri Cu2; fac M **287** fortem] *om.* B | ferream] rotundam ferream B Ka2 P1 | habentem…brachia] *om.* B Cu2 Ko1 Ko2 M P2; *add.* seu maniculas P1 | ad…crucis] *om.* B Cu2 Ko1 Ko2 M P2 | sub…288 forma] *om.* Cu2 Ko1 M; sub forma in fine depicta Ka2; ut hic sequitur ipsius forma P1 **288** et] *add.* hanc laminam P1 | super] supra Ka2 P1 | caput prefati] dictam primam partem P1 | quod] *om.* Cu2; *add.* capud B | est] *om.* Cu2 Ko1 **289** apertum] *om.* Cu2; aptum Ka2 Ko1 M | sic] *add.* fac P1 | ut] *om.* B; quod Ko2 P1 | non] *om.* B | totaliter] notabiliter B; taliter P2 | furni] dicti furni P1 | sed…circuitum] et quia circuitus B **290** predicte] dicte B P1 P2 | et] *incipit lacuna* B | calor] *add.* ignis usque ad partem superiorem coperculi dicti furni ut hic sequitur P1 | Super…291 laminam] *om.* P1 **291** autem hanc] *finitur lacuna* B; super quam Cu2; *om.* P1 | gobeletos] vasa tua Cu2 | seu] *incipit lacuna* Cu2 | ciatos] cyphos B; siacos Ka2; violas Ko1; fiacos Ko2 | de vitro] *om.* P1 | quot] quos B; *om.* M P1 | poteris…292 volueris] *om.* M; vel ampolas supra dictam laminam ferri prout volueris duas vel tres vel unam P1 **292** in…formam] *om.* B M P1; in forma quam vide in fine Ka2 | super] supra Ka2 P1 | quemlibet] *om.* P1 | gobeletum] *om.* Ka2; dictos gobelletos P1; ciatum P2 | potes] poteris Ka2; ponas Ko1; pone M | unum] *om.* P1 P2 | ovum…293 amphoram] tua vasa vitrea cum medicina P1 **293** collocare] *finitur lacuna* Cu2; *add.* ad capud sive ollam fornelli B | lapidem] ipsam P1 | digerendum] *add.* quorum gobelletorum et ciatorum vel ampolarum forma hic infra designatur supra dictam laminam super quam debent teneri. P1 **294** Deinde] *om.* Cu2; *add.* quia Ka2 Ko2 M P2 | vides] *om.* Cu2; scias P1 | quod] *om.* Cu2; quia Ko1 | quatuor] Et habeat predicta lamina quatuor Cu2 | lamine] que laminam Cu2; dicte lamine P1 | ideo] que ideo B; *om.* Cu2 Ko1; vero Ko2 sunt ut] *om.* Cu2 | super] supra B Ka2 M; *om.* P1 | furni parietes] *om.* P1 **295** laminam] *om.* Cu2 elevatam] levatam P1 | circulariter] circulenter Cu2 | Parietes] tunc parietes Cu2; super pariete P1 furni] dicti furni ut P1 | eleva] *om.* Cu2 P1 | per circuitum] *om.* Cu2 | per[2]] dicte lamine ad latitudinem P1

B Cu2 Ka2 Ko1 Ko2 M P1 P2 **284** dicitur] *incipit lacuna* Cu2 **285** depingi potest] *finitur lacuna* Cu2 **290** et] *incipit lacuna* B **291** autem hanc] *finitur lacuna* B | seu] *incipit lacuna* Cu2 **293** collocare] *finitur lacuna* Cu2

walls be that of a palm or less, with ears, and this is called the base of the furnace, and here is its shape as it can be depicted.[57]

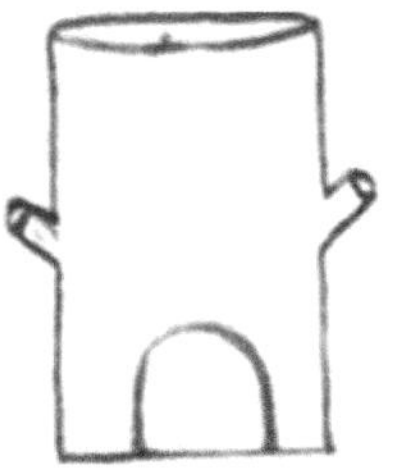

Let the ears be fitted to it here and there so that the smoke exits from them and does not ascend upwards. Then also have a strong plate of iron made, with four arms in the form of a cross according to this shape,[58] and fit it upon the top of the aforesaid furnace (which is open and round) so

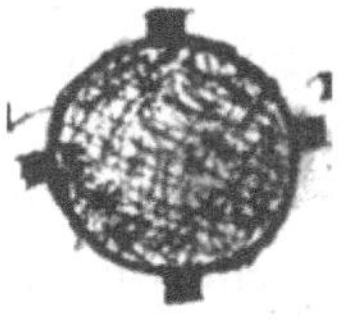

that it does not completely close up the opening of the furnace, but such that the heat may ascend upwards around the edge of the aforesaid plate between the plate and the walls of the furnace. Upon this plate you will place glass goblets or cups in this form,[59]

as many as you can or as you wish, and upon each goblet you can position one philosophical egg or one flask for digesting the stone. Then you will see that the four arms of the plate are thus for the purpose of holding the plate raised up upon the walls of the furnace all the way around its circumference. Raise up the walls of the furnace all around for five fingers at least and according to

---

[57] A Medieval palm is a little less than 10 inches, or about 25 cm. "Ears" (*aures* or *auricula*) was a standard term for furnace vents; see for example, Newman, *Summa perfectionis*, 375-77 and 687-88. The image given here is taken from **P²**; on the drawings, see above, Section 2.7.

[58] The image inserted here is taken from **P¹**.

[59] Despite the appearance in many manuscripts of the first two drawings, the promised third drawing appears rarely; the image inserted here is taken from **P¹**. On this issue, see above Section 2.7.

quinque digitos ad minus et ad quantitatem furni. Fac fieri de luto aut cupro vel ferro ollam rotundam quasi cooperculum pyxidis ita ut possit poni amoveri et reponi quotiens opus fuerit ad materiam videndam. Debes autem intus et extra ollam illam lutare tenuiter ut magis teneat calorem. Et facies in ea fenestram competentem et portam aptissimam per quam immittas manum ad probandum mensuram ignis an tepidus sit vel forte nimis fortis. Dixi enim tibi quod in principio lapidis sufficit ignis ut inter latera furni et vasis in aere possit teneri manus. Cave tamen a tactu vasis quia nimis est calidum et valde venenosum. Debes autem quando materiam intus posuisti lutare juncturam olle et furni ac porte cum luto sapiencie et panno lineo ne calor respiret et ne fumus ascendat quia ideo non potest fumus ad furnum ascendere quia non potest a loco aliquo respirare. Intellige hoc totum. Et sic habes ergo stupham siccam in qua potes ponere omnes spiritus et medicinas ad digerendum ut in balneo possint dissolvi in

---

296 quinque] decem Cu2; trium P1 | digitos] digitorum P1 | ad minus] aut minus B Ka2 Ko2 P2; *om.* P1 | et] *add.* per intus dictum furnum ascendat dictus calor ignis ad superiorem coperculum. Deinde fac fieri coperculum dicto furno super quo jam est posita laminam ferri cum suis vasis supra ordinatis et fac fieri P1 | ad[2]] juxta P2 | quantitatem furni] latitudinem dicti furni P1 | Fac] fac insuper B; post fac Cu2; *om.* P1 | fieri] *om.* P1 | luto aut] *om.* P1 **297** cupro] luto P2 | ollam...cooperculum] videlicet ad modum unius olle ad modum P1 | pyxidis] *add.* cum suis manicis a latere ipsius coperculi P1 **298** poni] deponi B | amoveri] *om.* B | reponi] recipi Ko1; *om.* P1 | quotiens] quando P1 | ad] pro P1 | videndam] videndo P1; *add.* vasorum tuorum in dicto furno super laminam positorum cum tua medicina vel materia P1 **299** autem] enim P2 | ollam illam] ollam istam B; predicta olla Cu2; ollam Ko1 M; dicta olla cupri vel ferri P1 | tenuiter] tenaciter Ko1; subtiliter P1 | magis] *om.* B Ka2 P1 P2; melius Cu2 **300** facies] *om.* P1 | ea] predicta olla Cu2; dicto coperculo fac fieri unam P1 competentem] *om.* P1 | et...aptissimam] *om.* Cu2; et portam apertissimam Ka2; apertam M; cum sua clausura P1 **301** immittas manum] possis ponere manus intus coperculum P1 | probandum mensuram] senciendum calorem P1 | tepidus] *om.* P1 | sit] *om.* P2 | vel] an Ko1 M; *om.* P1 | forte... 302 fortis] nimis fortis B Ko1; fortis Ka2; fortis an moderatus M; nimis fortis vel nimis tepidus P1; *add.* Cujus coperculi seu olle forma debet esse talis ut sequitur P1 **302** Dixi enim] Quia dixi Cu2 quod in] a B | lapidis] *om.* Ko1; digestionis lapidis M; operacionis lapidis P1 | sufficit] *om.* B | ignis] *om.* B P1 | ut] ut tunc Ka2; quod P1 **303** possit teneri] possis tenere P1 | manus] manus intus P1 **304** quia] qui si P1 | nimis] *om.* P2 | est] *om.* Ko1 | Debes] *om.* B | autem quando] aliquando postquam M | materiam] *add.* tuam Cu2 **305** intus posuisti] imposuisti Cu2 M | lutare] omnes lateres B | juncturam] tunc juncturam Cu2; ollam sive juncturas M | olle] *add.* bene et apte Cu2; *add.* seu coperculi P1 | et furni] *om.* Cu2 | ac porte] *om.* Cu2 P2; et fenestre P1 **306** ne[2]...ascendat] *add.* quod fieret nisi firmissime et optime lutaretur junctura olle et furni Cu2 | quia] *incipit lacuna* Cu2; *om.* Ka2 M | ideo...307 fumus] *om.* M **307** ascendere] *om.* M | a...aliquo] *om.* P1 | respirare] habere respiraculum P1 **308** totum] *finitur lacuna* Cu2 | Et sic] *om.* B Ko1 Ko2 P2; Et Ka2 | habes] *incipit lacuna* Cu2; habeas Ko2 | ergo] *om.* Ka2 M | potes] potest B Ka2 | ponere] inponere B **309** omnes] omnis B; *om.* P1 | medicinas] medicinam B | ut in] et B; ut P1 | balneo] balneum B; *om.* P1 | possint] in quo possunt B; possit M P2; *add.* postea P1

B Cu2 Ka2 Ko1 Ko2 M P1 P2 **306** quia] *incipit lacuna* Cu2 **308** totum] *finitur lacuna* Cu2 | habes] *incipit lacuna* Cu2

the size of the furnace. Have a round bowl like the cover of a pyx made out of clay or copper or iron so that it can be set in place, removed, and replaced as often as is necessary in order to view the material. You should also cover this bowl thinly with lute inside and out so that it holds the heat better. And you will make in it an adequate window and well-fitted door through which you may stick in a hand for the purpose of checking the degree of fire, whether it is tepid or perhaps too strong. For I have told you that in the beginning of the stone the fire is sufficient such that a hand can be held in the air between the walls of the furnace and the vessel. Beware however of touching the vessel because it is too hot and extremely poisonous. When you have put the material inside you ought also seal up the juncture of the cover and the furnace and the door with the lute of wisdom and linen cloth so that the heat does not escape and so that the smoke does not ascend, because in this way the smoke cannot rise into the furnace because it is not able to get out anywhere. Understand all of this! And so you have therefore the dry oven in which you can put all spirits and medicines to be digested so that they can be dissolved into water in a bath.

aquam. Habes furnum philosophorum a paucissimis scitum et cognitum in quo omnia que ponuntur habent ignem equaliter ab omni parte quia lamina prohibet ne dyametraliter calor ignis ascendat ad vasa sed per latera lamine ascendat ad caput olle et inde refunditur ad vasa que stant in medio aeris super ciatos aut tripedem elevata. Considera hoc totum quia ignem habent undecumque equalem. Sed non vacat a ministerio maximo quia per unum solum foramen parvulum aut duo, aut tria, aut quatuor, sicut tibi placebit, circa latera lamine ignem poteris minuere vel augere. Clarum enim esse puto cuilibet quod major calor ascendit per duo equalia foramina quam per unum ex illis. Sed nota quod super ciatum aut tripodem in lamina potes situare scutellam terream et infra scutellam facere tripodem et ibi ponere ovum philosophicum et desuper aliam collocare scutellam et ita vides duas scutellas conjunctas erectas in aere et in medio earum ovum in aere elevatum non tangens latera scutellarum.

---

310 aquam] *finitur lacuna* Cu2; aqua P1; *add.* Nunc M; *add.* Item P1 | Habes furnum] Habes igitur nunc furnum B; habes hic furnum P1 | paucissimis] paucis Cu2 P1 | scitum et] *om.* P1 | cognitum] incognitum B | in...311 que] *add.* ibi P1 **311** ponuntur] imponuntur B | habent ignem] hunc ignem B; *om.* Cu2 | equaliter] equalem Ko1 P1 P2 **312** ne] ut Ka2; ire M | calor...ascendat] calorem ignis M | vasa] *incipit lacuna* Ka2; fundum vasorum P1 | per] ad B **313** ascendat] ascendit Cu2 M; ascendit calor ignis P2 | ad[1]] *finitur lacuna* Ka2 | caput olle] *add.* seu coperculi P1 | stant] sunt Cu2 P1; *add.* super laminam P1 **314** super] dicti furni supra P1 | ciatos] *om.* Ko1; fiacos Ko2; gobeletos P1; *add.* vel cyphos vitri B | aut] *om.* Ko1; *add.* super M P1 | tripedem] trepides Cu2; capedem Ko2 elevata] elevatam B; *add.* in medio dicti furni super laminam P1 | ignem] *om.* Cu2 | habent] habeas B; *om.* Cu2; habet Ka2 M; hunc Ko2 **315** undecumque] undique B Ka2 P1; *om.* Cu2; unumquodque M equalem] *om.* Cu2 | Sed] *om.* Cu2 | ministerio] misterio Cu2 Ko2 M; misterio alius a magisterio Ka2; magisterio Ko1 | maximo] maxime P2; *add.* ignem componere ex omni parte equalem Cu2 | quia] quod B Cu2 M | per] *incipit lacuna* Cu2 | unum] ipsum M **316** solum] *add.* unicum M | foramen] carbonem B | parvulum] parvum P1; *om.* P2 | duo] duos B | tria] tres B | quatuor] decem B | sicut] si M | placebit] placuerit B M; placet P1 **317** circa] intra Ko1 | latera...ignem] *add.* seu calorem ignis P1 | poteris minuere] potes diminuere M | enim] *om.* B M P2 | esse] est enim B **318** puto cuilibet] cuilibet B; puto M P1 | major] amplior Ko1 | ascendit] ascendet B | per[1]...foramina] *add.* lamine predicte ferri P1 | quam] quasi Ko1 | per[2]] *om.* P2 **319** illis] *finitur lacuna* Cu2 | Sed] Et Cu2; Item M | super] semper Ko2 | ciatum] cyphum vitri B; unum siatum Ka2; fiacum Ko2 | aut] *add.* gobelletum seu P1 | in lamina] positos supra laminam P1 | potes] poteris Cu2 **320** et[1]...scutellam[2]] *add.* potes M **321** philosophicum] philosophorum vel philosophicum B P2; *add.* id est vas vitreum supra nominatum in quo est medicina nostra perficienda P1 | et desuper] *add.* potes P1 | collocare] *om.* Cu2; ponere P1 | vides] videbis Cu2 P1; videas Ka2 **322** erectas...aere[1]] erectas et in aere elevatas Cu2 M | et...medio] *om.* Cu2 M | earum] ita quod Cu2; *om.* M; dictarum scutellarum videbis P1 | ovum] *om.* M; ovum philosophorum P1 | in aere[2]] *om.* Cu2 M **323** elevatum... scutellarum] non tangat latera scutellarum Cu2; non tangentes latera furni M

B Ka2 Ko1 Ko2 M P1 P2 **310** aquam] *finitur lacuna* Cu2 **312** vasa] *incipit lacuna* Ka2 **313** ad[1]] *finitur lacuna* Ka2 **315** per] *incipit lacuna* Cu2 **319** illis] *finitur lacuna* Cu2

You now have the furnace of the philosophers, known and recognized by very few, in which everything that is placed inside has fire equally from all sides because the plate prevents the heat of the fire from ascending directly to the vessels; it must instead ascend around the edge of the plate to the top of the cover and from there it is poured back upon the vessels which stand raised up in the middle of the air upon the cups or a tripod. Reflect upon all of this, because they have fire equal from all sides. But it is not lacking in the best regulation, because by means of one small opening alone around the edge of the plate, or two, or three, or four, howsoever it pleases you, you can decrease or increase the heat. I think that it is clear to anyone that more heat ascends through two equal openings than through just one of them. But note that upon the cup or tripod on the plate you can position an earthen dish, and within the dish make a tripod and place there the philosophical egg, and then position another dish upon the first, and so you will see two dishes joined together standing upright in the air and with the egg in the middle of them raised up in the air and not touching the walls of the dishes.

Vide mira ingenia patule revelata in tempore isto tribulacionis instantis *quod aliis generacionibus non est agnitum filiis hominum sicut nunc datum est agnosci* pauperibus fidelibus filiis ecclesie sacrosancte romane et prelatis ejusdem ecclesie deprimendis ad tempus et depauperandis a tyrannis crudelibus a futuris puniendis a Jhesu Christo domino nostro cui est honor, decus, fortitudo, gloria, et imperium per omnia secula seculorum. Amen.

Explicit liber lucis magisterii magni editus per fratrem Johannem de Rupescissa anno domini m ccc liiii die xiiii octobris.

---

324 Vide] Enim B; Unde sic vides Cu2; Unde Ko2 M P1 | ingenia] *om.* P1 | patule] spiritualiter B; patule et manifeste Ka2; patent M; possunt P1 | revelata] *om.* M; revellari P1 | isto] *om.* Cu2 **325** instantis] proxime instantis Cu2; presentis P1 | generacionibus] *om.* P1; gravationibus P2 agnitum] traditum Cu2; cognitum Ka2 P1 | filiis hominum] *om.* Cu2 **326** est] *om.* Cu2 | agnosci] noscere P1 | filiis] *om.* B M **327** ejusdem] eidem P2 | ad…327 tempus] *om.* Cu2 P1 | et depauperandis] *om.* P1 P2 **328** a futuris] futuris B Ko1; afflicturis P1 | puniendis] solvendis B; perveniendis perimendis id est puniendis Ka2; perimendis Ko2 M; prohimendis P2 | Jhesu…328 Christo] christo Cu2 P1 **329** nostro] *om.* B | cui…honor] *add.* virtus Cu2 | decus] *om.* B P1 P2 fortitudo] *om.* P1 P2 | gloria] *om.* Ka2 | et imperium] *om.* P2; *add.* hic et ubique Cu2 | per] in P2 **330** omnia] infinita Cu2 M; *om.* P2 | secula seculorum] secula Ko1; *om.* P2 | Amen] *om.* Ko1 M **331** liber…magni] liber qui vocatur magisterium lucis Cu2; liber lucis M; liber lucis et tribulacionis P1 | editus] *om.* Cu2 Ko1 | per…Johannem] ab egregio philosopho divinitus illustrato fratre Johanne B; *om.* Cu2 P2; a fratre Johanne Ka2 Ko2; fratris Johannis Ko1 | de…331 Rupescissa] *om.* Cu2 P2; de Petrescisa Ka2; *add.* ordinis minorum B Ka2; *add.* cordigeri Ko1 **332** anno…xiiii] Anno Domini m^{o} ccc^{o} lxxx et die xiiii B; *om.* Cu2 Ka2 Ko1 M P1; *add.* mensis B | octobris] *om.* Cu2 Ka2 Ko1 M P1; septembris Ko2; *add.* qui latuit multis annis qui utinam lateat in perpetuum malos homines avaros et tyrannos precipue sed sub costodia crucifixi servetur viris ewangelicis in tempore necessitatis B; *add.* anno domini m ccc xci in bona sexta feria Ka2; *add.* Deo gratias. Ko1 P1; *add.* compilatus per me Walterum Vallrevere Anno domini m^{o} cccc^{o} xxviii^{o} protunc parisiis studentem in artibus Ko2; *add.* Deo gratias. Amen. M

B Cu2 Ka2 Ko1 Ko2 M P1 P2

Behold the marvelous techniques openly revealed in this time of present tribulation *which in other generations were not known to the sons of men, as are now given to be known*[60] to the faithful poor ones, the sons of the holy Roman Church and the prelates of the same Church, who will be oppressed for a time and impoverished by cruel tyrants who will thereafter be punished by Jesus Christ our Lord, to whom is honor, majesty, strength, glory, and power for all ages of ages. Amen.

Here ends the *Book of the Light of the Great Magistery*, written by friar John of Rupescissa, in the year of Our Lord 1354, on the fourteenth day of October.[61]

[60] Eph 3:5; on the significance of this line see above, Section 2.8.
[61] On the dating in this colophon, see above, Section 1.3.

Part 3

Manuscripts

6 WITNESS FAMILIES AND MANUSCRIPT DESCRIPTIONS

This section provides descriptions of each of the 106 currently known witnesses of the *Liber lucis magisterii magni* grouped by families, beginning with the two largest and earliest. I have also included here, in augmented format, the individual branches of the full stemma for each family. In some cases, more information (especially about contaminations) can be displayed here than was possible in the more crowded full stemma.

6.1 Provençal-North Italian Family π

Be, **Br**, **D**, **G**, **Ka**^{1}, **M**, **P**^{1}, **P**^{2}, **Po**, **Rc**, **T**, **W**^{1}, **W**^{2}, and **Y**^{1}; abridged, **Mp**$^{1}{}_{a}$; in Italian translation, **Fn**$^{3}{}_{i}$; in German translation, **Go**$^{3}{}_{d}$, **Ka**$^{3}{}_{d}$, **Ka**$^{4}{}_{d}$, and *Thesaurus mundi*; three descendant families: φ (**Ko**^{2}), ι^{2} (**Fn**^{2}, **L**^{3}, and **Mp**^{1}), and ε^{1} (**Cu**^{2}, **L**^{2}, **X**^{4}, and, with significant contamination, **L**^{6}); and, with significant contamination, the published Cologne version and its forerunner **Ua**.

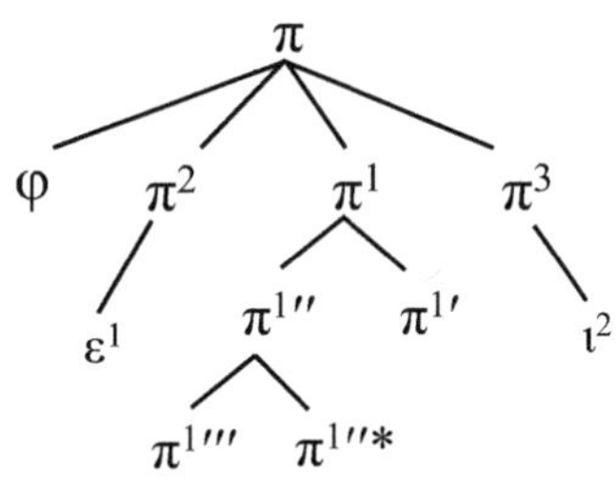

The family descending from π is large, complex, and important. It contains twenty-eight known manuscript witnesses, including some of the earliest and most complete, two of which are of Provençal origin, thus close to the locale of John's authorship in Avignon. It engendered subsequent lines of descent into three other geographical regions that represent some of the earliest diffusions of the work: φ in Northern France (Paris), ι^{2} in Italy, ε^{1} in England, and also ε^{2} (via ι^{2}) in England. It also eventually spawned two of the printed versions of the *Liber lucis*, one Latin and one German. Family π split into several distinct subfamilies early, certainly before 1400. Given the importance of this extended family, its position close to the original source, and the completeness and reliability of several of its witnesses, five of its members have been used in the critical edition.

Multiple translations emerged from this family. The German version of the *Liber lucis* embedded in the composite work *Thesaurus mundi oder Schatz der Welt* (Halle, 1621)—whose full text appears in three earlier manuscripts—descends from **Y**^{1}. An Italian translation, **Fn**$^{3}{}_{i}$, likewise descends from π, although with contamination from family ι'. A third vernacular translation, τ, of which no extant witnesses are known, is postulated as the origin of **Mp**^{1}, whose variant wordings and rephrasings identify it as a back-translation into Latin. This Southern Italian witness, interleaved with the abridged **Mp**$^{1}{}_{a}$ (from a different line of

DOI: 10.1201/9781003666547-9

descent from π), provided the amalgamated text of **Ua**, which was the basis for the 1579 Cologne edition.

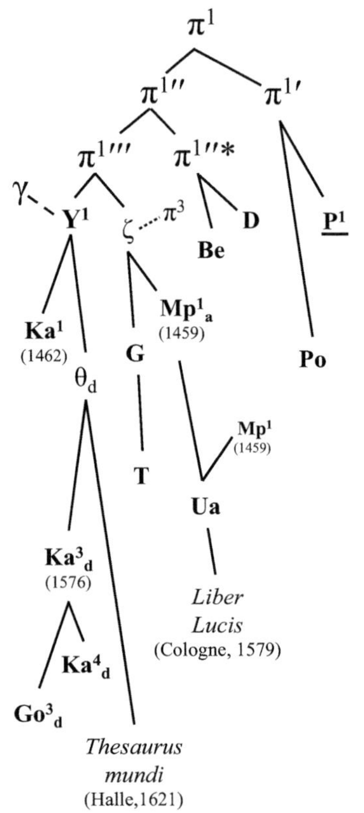

6.1.1 Subfamily π^1

Be, D, G, Ka^{I}, P^{I}, Po, T, and **Y^{I}**; abridged, **Mp^{I}$_a$**; with significant contamination, **Ua**; in German translation, **Go^3$_d$**, **Ka^3$_d$**, **Ka^4$_d$**, and *Thesaurus mundi.*

Subfamily π^{I} has the most complex line of descent. All witnesses share distinctive characteristics that indicate their descent from a common ancestor and fall into four subgroupings, each of which displays further degrees of commonality. Hence, I propose six lost exemplars:

π^{I} from which all fourteen witnesses descend;
$\pi^{I'}$ as the parent of Provençal **P^{I}** and Italian **Po**;
$\pi^{I''}$ as an intermediate with no known immediate descendants, but which seems to have carried the text into Germanic regions;
$\pi^{I''*}$ as the parent of Germanic **Be** and **D**;
$\pi^{I'''}$ as the parent of Germanic **Y^{I}**, **Ka^{I}**, and the German translation θ_d;
ζ (produced by contamination from π^3) as the parent of Italian witnesses **G, T, Mp^{I}$_a$**, and **Ua**.

The major characteristics of family π^{I} include:

- ll. 85–91: a section detailing how to distill aquafort. (Abbreviated in **Be**, present in **M** and **S** through contamination.)
- l. 117: *paulatim* for *pulverem*.
- l. 137–38: *dissolucione completa habes tuam medicinam ultimo preparatam* instead of *distillacione completa habes tuam materiam ultima preparacione.*
- l. 205: *vigorem ignis* for *majorem ignem.*
- l. 247: *ceciderunt* for *exciderunt.*
- l. 250: *timent* for *juvent.*

6.1.1.1 *Subfamily* $\pi^{1'}$

P^I^ and Po

The two known witnesses of this subfamily contain additional details about the practical processes, presumably the result of glosses inserted in $\pi^{I'}$ by an active practitioner. In addition to the textual characteristics common to all descendants of π^{I}, the major distinguishing features of subfamily $\pi^{I'}$ include:

- l. 34: after *in ea*, a further direction "to put the said material, once cooled, into a stone mortar and incorporate them with a wooden pestle."[1]
- ll. 73–74: *mercurium marcidum effectum* replaced by *materiam que tinctura rubedinis appellatur.*
- l. 87: *ymo suavissimum per unum diem et noctem scilicet per viginti quatuor horas*, added after *da ignem suavem.*
- l. 96: *cum parvo igne carbonum* added after *distilla.*
- l. 97: *cum nil plus destillabit amove cappam et pone antonarium/anthenerem* added after *distillacionem.*[2]
- l. 256: *argentum vivum* in place of *argentum.*

P^{I} Paris, BNF, MS NAF 4141, fols. 29r–38v (1st q. 15th c.)[3]

P^{I} has been cited as the earliest witness of the *Liber lucis* and thought to date from the late fourteenth century.[4] While that dating might be correct, it is based upon an extension of the date of the first text in the collection to the rest of its contents. The first text, an Occitan version of pseudo-Arnald's *Rosarius philosophorum*, does indeed date from the late fourteenth century.[5] But **P**^{I} is in a different hand on a different paper. It contains the prologue and all eight operations and is written in *gothica rotunda cursiva* and entitled *Liber lucis et tribulacionis.* Like the document that precedes it, **P**^{I} was produced in Provence. Although written in a French hand, the text contains multiple Italianisms, such as *fondere*, *plombum*, *mixtico*, *rexuscitatus*, *combuxit*, *lottone*, and *rotondus*. Its exemplar was apparently difficult to read, especially toward the end of Operation 6, and the copyist left several blank spaces that were filled in later. A signature at the end, J. Guode, may be that of the copyist or of an early owner. **P**^{I} contains more illustrations than

1 P^{I}: "Et postea pone dictam materiam cum infrigidata fuerit in mortario de petra et incorpora cum pistello ligni"; Po: "postea pone cum refrigeratum fuerit in mortario de petra et cum pistello ligni bene incorporetur."

2 The unique term *antonarium* appears to be a Latinized form of the French *entonnoir* (funnel) whose shape is similar to that of a subliming head.

3 Corbett, *Catalogue*, 1:277–78; Jean-Baptiste Camps, *Les manuscrits occitans à la Bibliothèque nationale de France* (Paris: Enssib, 2010), 36; Clovis Brunel, *Bibliographie des manuscrits littéraires en ancien provençal* (Liège, 1935), 226.

4 E.g., Halleux, "Rupescissa," 262.

5 Antoine Calvet, *Le rosier alchimique de Montpellier: Lo rosari (XIVe siècle)* (Paris: Presses de l'Université de Paris-Sorbonne, 1997) and "Étude d'un texte alchimique latin du XIVe siècle: Le "Rosarius philosophorum" attribué au médecin Arnaud de Villeneuve (ob. 1311)," *Early Science and Medicine* 11 (2006): 162–206, on 181–82.

any other witness: ten depictions of the furnace and its parts and three smaller figures of apparatus drawn within the lines of text.

Po Poppi, Biblioteca Communale Rilliana, MS 90, fols. 224v–32v (1470s–80s)[6]

Bearing the anomalous title *Opus lune*, **Po** is part of a finely produced collection of alchemical texts written in a single Italian hand on paper with large initials in gold with red and blue foliage and first lines in gold. The headings are in *gothica rotunda* while the main text is *humanistica*, although retaining some features of *gothica rotunda*. A rubric in its copy of the *Vade mecum* attributed to Frater Elias (fol. 6v) provides a *terminus a quo*. This rubric probably originated as a pair of marginal notes by two successive owners of the manuscript from which this tract was copied. The first reads, "I think that the projection of the powder ought to be done upon luna fixa," and the second, "I, Borso, Duke of Ferrara, tried it; it came out a beautiful golden color."[7] Borso d'Este (1413–1471) did not acquire the title of Duke of Ferrara until just before his death in 1471.

Po contains the prologue and all eight operations. Like **P**^{I}, it provides added practical directions, several of which are unique to this witness and reflect practical laboratory experience, such as the advice to dry the salts "by stirring with a wooden paddle in a dish until [the mixture] is dissolved and becomes like a sauce, then continue mixing and grinding with a piece of wood made in the shape of horse's hoof, grinding constantly until it becomes an earth and the earth begins to whiten and the wood to become dry."[8] **Po** opens with a prayer not found in any other witness.[9] It does not contain the illustrations of **P**^{I}, although space was left in Operation 7 for a depiction of a crucible. It has a single crude drawing of the furnace, produced by a later reader not the copyist.

6.1.1.2 Subfamily $\pi^{1''*}$

Be and **D**

The parent of these two witnesses, $\pi^{I''*}$, descends from the lost exemplar $\pi^{I''}$ that brought the lineage of the Provençal-North Italian family into Germanic lands. The two known witnesses show all the characteristics of π^{I}, none of $\pi^{I'}$, some of $\pi^{I'''}$, and their own variants that include:

6 Emanuele Casamassima, *I manoscritti della biblioteca communale di Poppi* (Milan: Giunta Regionale Toscana, 1993), 72–74.

7 Fol. 6v: "Puto quod projectio pulveris debet fieri super lunam fixam" and "Ad idem. Ego Borsius dux ferrariae probavi. Quod venit color aureus pulcher."

8 Fol. 225v: "...movendo...cum baculo ligneo et <*sic*; an error for *in*> una patella donec dissolvatur et fit ut salsa deinde continua miscendo et terendo cum uno ligno facto ad modum pedis equi semper terendo donec efficiatur terra et incipiet terra dealbari et desiccari lignum."

9 Fol. 224v, incipit: "O innumeranda divinitas, O inseparabilis Trinitas, O indivisa charitas. Ego licet indignus..."; explicit: "...philosophus alchimista volens formare lapidem pretiosum ut supra scripsi Exemplarium."

- l. 67: *distillatum* added after *acetum.*
- ll. 73–74: *materiam mercuriam/mercurii* in place of *mercurium marcidum effectum.*
- ll. 129–30: *a Gebero nunquam cognita, ab Avicenna ignorata* omitted.
- ll. 93, 108, and 109: the times for processes differ; 2 for 12 hours and 20 for 12.
- ll. 172–74: lacuna between *et ponas* and *ab igne.*
- l. 184: *punctum* for *principium.*
- l. 193: *quantum* for *quam.*
- Operation 7 is highly edited and concludes with a line beginning with *Multiplicacio ipsius.*
- Operation 8 omitted.

D Darmstadt, Universitäts- und Landesbibliothek Darmstadt, Hs 2226, Faszikel III, fols. 45r–50r (mid-15th c.)

This witness, written in Germanic *gothica cursiva* on paper, transmits the prologue and Operations 1–7. It is part of a miscellany of alchemical and astrological works mostly in Latin but with some German. At ll. 33–41, the text is entirely rewritten, but fully preserving the sense. Of special interest is the short text of pseudo-Arnald's metaphorical instructions for making the stone that is attached to the end of the text and that contains most of the phrases the *Liber lucis* attributes to *magister Arnaldus* (see Section 2.3.2).

Be Berlin, Staatsbibliothek zu Berlin, MS lat. oct. 472, fols. 96r–101r (2nd half 15th c.)[10]

Part of an alchemical miscellany of Silesian or Bohemian origin written in *gothica cursiva* by a single hand, **Be** is more edited than **D.**[11] It omits the prologue, most of the details for distilling aquafort, the "Credas," and Operation 8; Operation 7 is similarly pared down. Unique features include its title, *De arte benedicta et lapide philosophorum*, the use of *parchanum* (i.e., *barchanum*, a type of linen cloth) for *panno lineo* (ll. 31 and 33); and the ending to Operation 4: "I do not know what I could have said more clearly to you since I have placed everything explicitly and lucidly in the light for you."[12]

6.1.1.3 Subfamily $\pi^{1'''}$

G, **Ka^{I}**, **T**, and **Y^{I}**; abridged, **Mp^{I}$_{a}$**; with significant contamination, **Ua**; in German translation: **Go^{3}$_{d}$**, **Ka^{3}$_{d}$**, **Ka^{4}$_{d}$**, and *Thesaurus mundi*

10 This manuscript is discussed in Otto Štemberka and Pavel Zahradník, *Cesty Vlachů: Mysterium Antonia z Florencie* (Vrchlabí: Green Mango, 2022); I thank Sabina Kwiecień for bringing this witness and this publication to my attention.

11 For the distillation protocols, **Be** preserves the opening words followed by "as is fitting" [*ut decet*] and retains the last words.

12 "Nescio enim quod clarius dixerim vobis cum omnia expresse et lucide posui vobis in diem."

The parent $\pi^{1'''}$ of this subfamily is a sibling to $\pi^{1''*}$. Its descendants survived through a long and complex transmission in two geographical regions and contributed to two of the published versions of the *Liber lucis*. Its dissemination began in Germany and, after contamination with a descendant of π^3, continued in Italy. Four of its Latin witnesses (**G**, **Ka^1**, **Mp^1_a**, and **Y^1**) date from the fifteenth century, while **T** is mid-sixteenth; the three German witnesses are late sixteenth and early seventeenth. All members, both Latin and German, show substantial lacunae in Operation 1, suggesting that $\pi^{1'''}$ was damaged. It shares the anomalous times for processes present in **Be** and **D** and a few other readings.[13] The major textual characteristics of this subfamily include the following:

- l. 4: *graviter flagellanda* rather than *plurimum affligenda.*
- ll. 13–15: garbled into *Et quia intentionis mee est probative recitatione verborum.*
- ll. 18–27 lacuna.
- l. 31: Mercury added through *panno lineo duplicato vel corio leporis.* These are the only witnesses to cite hare's leather that are not part of the German family γ.
- l. 41, following *nix*: *in quo mercurio sublimato dicitur contineri spiritus vitrioli rubeum*<sic> *qui sulfur invisibile appellatur. Et hic est de prima operacione. Et istud quod sequitur est addicio prime operacioni(s) ut sequitur.* (**T** omits the last sentence.)
- ll. 42–72: lacuna up to *in fundo.*
- l. 284: *nasus furni* instead of *basis furni.*
- text ends at l. 319, following *unum ex illis.*

Y^1 New Haven, Yale University Beinecke Library, Mellon MS 5, fols. 115v–19r (1st q. 15th c.)[14]

This important witness is part of an alchemical miscellany written by a single hand in badly formed *gothica semiquadrata* with rubrics and blue and red initials on reused parchment. A later *gothica cursiva* hand supplied alternate marginal readings that resemble those of Germanic family γ. **Y^1** is headed "Secuntur dicta Johannis de Cappa scissa de lapide philosophorum." It contains the prologue and Operations 1–8, although the text breaks off suddenly at l. 319, as noted above, and is followed by a brief, unrelated text on sublimation. Two rubrics highlight the "Credas": "Author's warning that no one share this book with the wicked" (before *Bestia*, l. 229) and "The entrusting of this book" (before *Commendo*, l. 240).[15]

[13] E.g., at l. 110 "dissolve illum *mercurium id est* spiritum mercurii."

[14] Laurence C. Witten and Richard Pachella, *Alchemy and the Occult: A Catalogue of Books and Manuscripts from the Collection of Paul and Mary Mellon Given to the Yale University Library*, 4 vols (New Haven: Yale University Library, 1968–77), 3:26–41. https://brbl-dl.library.yale.edu/vufind/Record/3441262

[15] "Monicio auctoris quod nullus cum reprobis communicet hunc librum" and "Commendacio hujus libri."

Due to its many texts by Johannes von Teschen, it has been suggested that this alchemical collection may be his own production or that of his immediate circle.[16]

Ka^1 Kassel, Universitätsbibliothek Kassel-Landesbibliothek und Murhardsche Bibliothek der Stadt Kassel, 2° Ms. chem. 10, fols. 16r–19r (1462)[17]

This witness, part of an alchemical miscellany written on paper in a Germanic *gothica cursiva*, is copied—possibly directly—from **Y**^{1}. It displays all the same lacunae and all the variants unique to **Y**^{1}, such as the two rubrics in Operation 6, John's name as "de Cappacissa," and the same short text on sublimation that is appended to the end of **Y**^{1}. It does not reproduce the later corrections found in **Y**^{1}, and was thus presumably copied before those additions were made.

Ka^3_d Kassel, Universitätsbibliothek Kassel-Landesbibliothek und Murhardsche Bibliothek der Stadt Kassel, 4° Ms. chem. 33[6, fols. 1r–12r (1576)[18]

This is the earliest known witness of *Thesaurus mundi, oder Schatz der Welt* (see Section 3.4). **Ka**^3_d forms part of an alchemical collection compiled by Wilhelm Vernuken (1542–1607), master workman and sculptor at Kassel from 1577, and written in his Netherlandish dialect.[19] Given the evidence of transcription difficulties, this witness was copied from an earlier exemplar; hence, the translation and *Thesaurus mundi* must predate **Ka**^3_d; I have positioned this presumed predecessor as θ_d in the stemma. **Ka**^3_d, titled *Dey Rede und unterweisung von Johannes Rupecisse zue magen den stein der Philosophen*, includes the prologue and all eight operations. The text is truncated exactly as in **Y**^{1}/**Ka**^{1}. All the characteristic readings of subfamily $\pi^{1'''}$ are present, and wherever the Latin witnesses of this subfamily differ, **Ka**^3_d resembles **Y**^{1}/**Ka**^{1}. The corrections in **Y**^{1} are not present in **Ka**^3_d, indicating that the German version was produced from **Y**^{1} either prior to the corrections or from a descendant that predated them. **Ka**^3_d contains two carefully-rendered (if imaginative) drawings of the furnace, one full page (Figure 2.7), plus smaller drawings of vessels.[20]

[16] On Johannes von Teschen, see Joachim Telle, "Johannes von Teschen," in *Die deutsche Literatur des Mittelalters: Verfasserlexicon*, 4:774–76, and Thorndike, *HMES*, 3:642–43.

[17] Hartmut Broszinski, *Katalog der Alchemica in Folio* (typescript at Kassel library); https://orka.bibliothek.uni-kassel.de/viewer/image/1673529107238/39/LOG_0007/

[18] Broszinski, *Handschriften*, 108–10. https://orka.bibliothek.uni-kassel.de/viewer/image/1343975290453/304; Wolfram Schmitt, "Eine handschriftliche Sammlung alchemistischer Traktate aus Böhmen, *Stifter-Jahresbericht* 7 (1962): 177–95; and Karl Sudhoff, *Versuch einer Kritik der Echtheit der Paracelsischen Schriften, Theil 2: Paracelsus-Handschriften* (Berlin, 1899), 736–39.

[19] Vernuken is responsible for over a dozen alchemical manuscripts preserved at Kassel; see Broszinski, *Handschriften*, xvii–xviv and references on 212. On Vernuken, see Carl Scherer, "Der niederländische Bildhauer Wilhelm Vernuken in hessischen Diensten," *Repertorium für Kunstwissenschaft* 31 (1908): 218–26 and Wolfgang Heuer, *Wilhelm Vernuken: Ein Bildhauer und Baumeister des Manierismus in Deutschland* (Ph.D. diss., Universität Düsseldorf, 2000).

[20] The extravagant furnace drawings of fols. 10r–v appear also in a textually unrelated c. 1760 manuscript, having been cut out of a copy of **Ka**^3_d and pasted into New Haven, Yale University Beinecke Library, Mellon MS 110, fols. 121r–22r.

Ka⁴d Kassel, Universitätsbibliothek Kassel-Landesbibliothek und Murhardsche Bibliothek der Stadt Kassel, 4° Ms. chem. 32, fols. 1r–14v (end of 16th c.)[21]

$Ka^4{}_d$ gives a slightly shorter version of *Thesaurus mundi* than **$Ka^3{}_d$**; it omits Operation 8 and the illustrations. It was produced in Kassel and possibly copied from **$Ka^3{}_d$** since, in both witnesses, a German prose version of Johannes von Teschen's *Lumen secretorum* immediately follows.

$Go^3{}_d$ Gotha, Forschungsbibliothek Gotha der Universität Erfurt, Chart. B 1233, fols. 75v–79r (early 17th c.)

Titled *Alius Tractatus Joannis Rupecisse: Thesaurus Mundi, id est, Schatz der Welt*, this witness was copied directly from **$Ka^3{}_d$**, as demonstrated by the cruder renderings in **$Go^3{}_d$** of the unique furnace drawings in **$Ka^3{}_d$**. As in the collections containing **$Ka^3{}_d$** and **$Ka^4{}_d$**, in this collection the *Lumen secretorum* of Johannes von Teschen immediately follows **$Go^3{}_d$**. The copyist of **$Go^3{}_d$** extensively edited and paraphrased *Thesaurus*: the introduction is summarized as three numbered points, the prologue is omitted, Operation 1 is reduced to one sentence, and the other operations run together. The text is condensed, references to authorities and theoretical principles removed, and the materials appended to the *Liber lucis* translation omitted. The copyist, however, gave special attention to the furnace and apparatus, producing further explanations and drawings, an emphasis that corresponds with his editing down to practical elements, nearly to a recipe.

TM *Thesaurus mundi*, 294–320 in *Heptas alchymica* (Halle, 1621); see Section 3.4.

6.1.1.4 Descendants of ζ

G and **T**; abridged, **Mp_{1a}**; with significant contamination, **Ua** and printed *Liber lucis* (Cologne, 1579).

The final branch of subfamily π^1 descends from an exemplar labeled ζ that combined a descendant of $\pi^{1'''}$ with one of π^3. The latter contributed spurious Operation 9 and the missing conclusion of Operation 8; there is no contamination elsewhere in the text. All the other textual characteristics of $\pi^{1'''}$, including the large lacunae in Operation 1, remain.

G Glasgow, Glasgow University Library, MS Ferguson 49, fols. 134r–42r (3rd q. 15th c.)

Written in a rapid *humanistica cursiva* on heavily ruled paper, this witnesses's Italian origin is clear from paleography and orthography. It contains the prologue and Operations 1–9. Its most unusual characteristic is "earthworm" (*lubricus*) used as a Deckname for mercury. Space was left for an illustration in Operation 5.

[21] Broszinski, *Handschriften*, 103–4. https://orka.bibliothek.uni-kassel.de/viewer/image/1533292518200/3/

T Torino, Biblioteca Nazionale Universitaria, MS E.II.23, fols. 17r–21r (mid-16th c.)[22]

Written in *humanistica cursiva* on paper as part of a collection of alchemical treatises by a single hand, this witness is titled *The most secret preparation of both stones for gold and silver by the most celebrated philosopher, Brother John of Rupescissa, of the observant order of friars minor.*[23] **T** is clearly derivative from **G**, given its identical use of *lubricus* as a Deckname for mercury. It likewise contains the prologue and Operations 1–9. Nevertheless, the text shows substantial alteration throughout. Some rewordings suggest that it might be a back-translation from Italian (e.g. the replacement of *accipe* with *tolle*, cognate with *togli* as used in Italian recipes), but this is not clear.

Mp$^{I}_{a}$. See Section 6.9.

Ua Amsterdam, Universiteit van Amsterdam, Allard Pierson Depot, MS PH338, fols. 49r–55v (ca. 1550)[24]

This crucial witness, written on paper by an Italian hand, is the only manuscript that transmits the text of the *Liber lucis* as published in the 1579 Cologne edition. As described in Section 3.3, it is a pastiche of **Mp^{I}** and **Mp$^{I}_{a}$** and thus a hybrid of subfamilies $\pi^{I'''}$ and ι^2.

6.1.2 Subfamily π^2

M, **Rc**, and **W^{I}**; in Italian translation, **Fn^3_i**; also family ε^{I} (see Section 6.7.1)

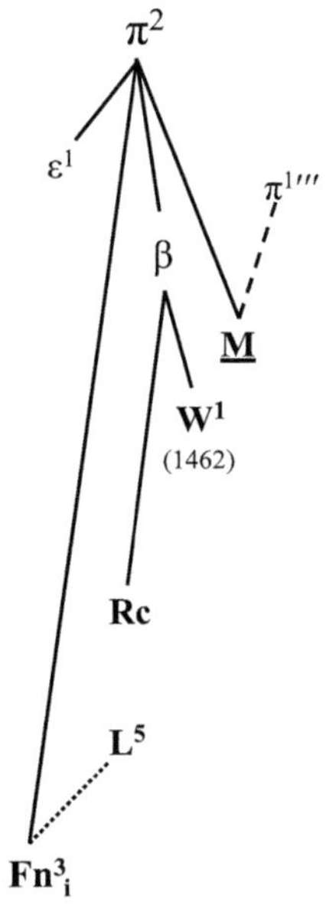

This subfamily shares many readings with the descendants of π^{I} but has its own unique textual characteristics. Two witnesses descend through the lost intermediate β owned by the Florentine goldsmith and empiric Lorenzo da Bisticci. This subfamily is also the source for ε^{I}, one of the transmissions of the *Liber lucis* into England, apparently from Italy, since all known witnesses of subfamily π^2 are Italian.

- ll. passim: The Deckname *leo* or *leo viridis* ("green lion") for vitriol is used frequently, and each operation has a summarizing heading.

[22] Albano Sorbelli, *Inventari dei manoscritti delle biblioteche d'Italia*, Vol. 28: Torino (Florence: Olschki, 1923), 68.

[23] Fol. 17r: "Celeberrimi Philosophi fratris Johannis de Ruppescissa ordini minorum observantie confectio utriusque lapidis ad solem et lunam ecce secretissima."

[24] Lawrence M. Principe, "Bibliotheca Philosophica Hermetica MS 338: The 'Missing Link' in the Transmission of John of Rupescissa's *Liber lucis*," *Quaerendo* 53 (2023): 187–97.

- ll. 21: *cloacis veteribus* rather than *cavernis*, and *aliqui* is replaced by *antiqui philosophi dicentes est omnium rerum vilius quod nos philosophi veneramur.*
- l. 28: *libram i* rather than *libras ii.*
- ll. 164–67: lacuna *Materia...tactum.*
- l. 227: *materiam* instead of *maculam.*
- l. 235: *indignorum* rather than *malorum.*
- l. 251: *afflictam* instead of *affligendam.*

M Manchester, John Rylands Library, MS 65, fols. 164v–68v and 172r (mid-15th c.)[25]

M is the most elegant and formal of all the known witnesses. Written on high-quality vellum in a carefully executed *gothica antiqua* book hand with large multicolored initials, rubrication, and a uniform thirty-eight lines to a page, it is part of a substantial collection of Latin alchemical texts. Its Italian origin is clear from the script, decorative elements, and frequent Italianisms. It is titled *Liber lucis magistri Johannis de rupescissa* and provides the prologue and all eight operations. Frequent insertions and marginalia (in the hand of the copyist of the main text) give alternate readings (often prefaced with *aliter*) that correspond closely with unique variants of subfamily $\pi^{1'''}$, indicating access to a witness in that line of descent, and from which **M** also obtained the instructions for distilling aquafort (uniquely for a descendant of π^2). There are six illustrations: on fol. 166r, the *ovum philosophorum* in its furnace; on 168v, the aludel, alembic, *lamina*, and complete furnace (highly stylized); and on 172r, an almost identical rendering of the furnace, pointed to by the marginal note on fol. 168v: "after three folios you will find this shape better" (*post tres cartas invenies hanc formam meliorem*).

W^{I} London, Wellcome Library, MS 117, pp. 261–74 (1462)[26]

This witness is part of a collection of alchemical treatises and medical receipts, mostly in Latin but with some Italian, written on vellum in a careful *humanistica cursiva* by a single copyist. **W**^{I} contains the prologue and eight operations, but attributes the work to Lull: "The divine work of the most learned and wise Ramon Lull which is called the *Book of Light of the Art of Alchemy* because it revealed the secrets most clearly and faithfully to the Church."[27] The explicit repeats this attribution. The collection also contains *De famulatu* (pp. 110–207), likewise attributed to Lull. Extended colophons report that Bartolomeo Marcello produced the manuscript in September 1462 from documents belonging to Lorenzo da Bisticci (d. 1478), a Florentine gold- and silversmith who attained fame for medical

[25] Montague Rhodes James, *A Descriptive Catalogue of the Latin Manuscripts in the John Rylands Library at Manchester* (Manchester: Manchester University Press, 1921; expanded reprint ed., ed. Frank Taylor, Munich: Kraus Reprint, 1980), 24* and 124–27.

[26] Moorat, *Catalogue*, 1:79–82; https://wellcomelibrary.org/item/b19295108

[27] On 261: "Divinum opus doctissimi et sapientissimi Raymundi lugli qui<*sic*> de arte alchimiae liber lucis dicitur eo quod lucidissime ac fideliter secreta ecclesie apperuit."

cures based partly upon *De famulatu*.[28] Marcello obtained his exemplar furtively: "I, Bartolomeo Marcello, excerpted this copy by stealth from Bisticci's books, which were stained with corrosives and smoke, and for eight days I transcribed with a swiftly running pen."[29] Marcello used a poor-quality ink for the rubrics, headings, and colophons, which faded badly. A later owner traced over the faded words with dark ink, but by his time, whole lines were already indecipherable, and now—except under close scrutiny—they appear to be blank. Consequently, the full colophon cannot be retrieved.[30] The later owner had more success preserving the rubrics heading each operation; they are fuller versions of the rubrics found in **M**, each of them beginning "Scientia ad..." ("The knowledge for...") mimicking the headings in *De famulatu*. **W¹** has one illustration of the furnace (p. 276) due to a later owner. He added "a bottom portion so that the ashes may be more conveniently removed" and, rather amusingly, drew the vents of the furnace—which **M** and **W¹** uniquely call *auriculae*—literally as *little ears* of a pig.

Rc Rome, Biblioteca Casanatense, MS 3567, fols. 274r–79r (late 15th c.)[31]

Like **W¹**, **Rc** is attributed to Lull and copied from β (or one of its descendants); nearly all its textual variants are identical to those in **W¹**, although it shows minor readings absent from or incorrect in **W¹** and therefore was not copied from that witness.[32] It forms part of a collection of alchemical treatises by Christopher of Paris and pseudo-Lull. Some text is nearly illegible due to bleed-through. **Rc** has many lacunae not present in **W¹**: it jumps from line 27 to 167, omitting most of Operation 1 and all of Operations 2–4; Operations 6 and 7 suffer from numerous eye-skips; Operation 8 is summarized as a single sentence ending "as we have known [*ut scivimus*]."

28 On Bisticci, see Lynn Thorndike, "Bisticius (or Bistichius) of Florence," *Romanic Review* 19 (1929): 244–49; *Science and Thought in the Fifteenth Century* (New York: Columbia University Press, 1929), 41–44 and 265–68; and Michela Pereira, "*Medicina* in the Alchemical Writings Attributed to Raimond Lull (14th–17th Centuries)," in Piyo Rattansi and Antonio Clericuzio, eds. *Alchemy and Chemistry in the 16th and 17th Centuries*, ed. Piyo Rattansi and Antonio Clericuzio (Springer, 1994), 1–15, on 1–2 and "Mater Medicinarum: English Physicians and the Alchemical Elixir in the Fifteenth Century," in *Medicine from the Black Death to the French Disease*, ed. Roger French, Jon Arrizabalaga, Andrew Cunningham, and Luis Garcia-Ballester (Burlington,VT; Ashgate, 1998), 26–52.

29 On 207: "Ego Bartolomaeus Mar. Abiat cirra furtim hoc exemplum ex bisticii corrosis ac fumo tinctis codibus excerpsi et diebus octo transscripsi currenti admodum calamo." The words *abia(t) cirra* always follow Marcello's name, but their meaning is unclear.

30 Venice, Biblioteca Marciana Lat. VI. 282 (= 2859) is a copy of part of the Wellcome manuscript, including the colophons, produced in 1521; see Valentinelli, *Catalogo*, 5:111–13. The red ink was apparently more legible at that date, thus the readings differ from the retraced letters in the original. Unfortunately, the scribe did not copy the *Liber lucis* or its colophon.

31 Jose Perarnau Espelt, "Los manuscritos lulianos en las Bibliotecas Casanatense y Angélica," *Anthologica annua* 21 (1974): 185–248, on 225.

32 E.g., ll. 222–23 where **W¹** reads *in peccatis fecerunt*, **Rc** correctly reads *imprecati fuerunt*.

Fn³ᵢ Florence, Biblioteca Nazionale Centrale di Firenze, Magliabechi Cl. XVI, cod. 76, fols. 315v–21v (last q. 16th c.)[33]

This Italian translation, written in *humanistica cursiva* on paper, is part of a collection of alchemical treatises, recipes, and correspondence in Latin and Italian. Considerable material from Agnolo della Casa, alchemist and assistant to Antonio Neri in the employ of the Medici, suggests that the volume may be a Florentine production related to him.[34] The title assigns John to the wrong Order: "Here begins the *Liber lucis* of Brother John of Rupescissa of the Order of Preachers concerning the philosophical stone."[35] **Fn^3_i** contains the prologue and all eight operations but depends upon two different families. Operations 1 and 3–8 descend through π^2, while the prologue and Operation 2 descend through Italian family ι′, specifically **L^5**, whose unique marginal note *In centum diebus totum opus perficitur* reappears in **Fn^3_i** as *In Cento Giorni se compisse l'opera*.[36] **Fn^3_i** bears a few particular readings. The lengthiest applies male/female language to the elements of fire and water in Operation 4, and a marital metaphor to their combination.[37] Curiously, this same addition appears *in German* at the same point in **Le^5_d**. I have not identified the source of this added text, nor the link between the Italian and German translations.

6.1.3 Subfamily π^3

P^2 and **W^2**; also family ι^2 (see Section 6.6)

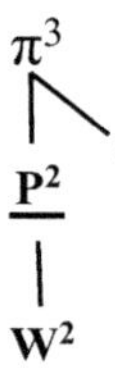

The two witnesses of this subfamily are the most distinct from the other members of family π. Most noticeably, they lack the prologue and the lengthy description of how to distill aquafort and include the spurious Operation 9. As argued previously (Section 2.6), π^3 probably initiated the linkage with the *Liber lucis* of the pseudo-Arnald text that became "Operation 9" in subsequent copies.

33 Giovanni Targoni Tozzetti, *Catalogo generale dei manoscritti Magliabechiana*, 14 vols. (Manuscript: Florence, Biblioteca Nazionale Centrale, Sala Manoscritti e Rari, Cat. 45), 5:239r–41r.

34 On Agnolo della Casa, see Paul Engle, *Conciatore: The Life and Times of the Seventeenth-Century Glassmaker Antonio Neri* (Hubbardston, MA: Heiden & Engle, 2014).

35 Fol. 315v: "Incipit Liber Lucis Fratris Johannis de rupescissa ordinis fratrum praedicatorum de Lapide Philosophico." The explicit repeats the error; incipit and explicit are Latin, the text Italian.

36 Readings unique to family π^2 include: in Operation 1, mercury is found in "cloache antique et vechie"; the Deckname "leone verde" is used for vitriol, and several characteristic headings, e.g., for Operation 1: "Che cosa sia la materia del lapis Et come si debe prepararla." Readings unique to family ι′ include: in the prologue, "Contiene adonche questo libro 8 capitoli" and in Operation 2, "perche la dissolutione del mercurio in lacqua del vitriolo se fa in 12 hore: In quella dissolutione beva quanto il potra bevere."

37 Fol. 318r: "Poi conjungerai matrimonialmente esso focho con la femina per equale peso di uno et di laltro. Perche il focho e maschio et e di calida natura lo latte virgineo e femina et e de natura frigida. Et lo maschio opere in essa femina: Et cosi la femina havera lhomo consi. Adunche quando sia fatta tale conjunctione bisogna che tu consideri che sono cose diverse: Et perho non possono bene stare insieme. Adonche bisogna ponere lo aere sopradditto o sia lo oleo che sia mediatore fra loro: accio che se conjugono pacificamente et possano habitare insieme."

Both **P²** and **W²** were produced in Provence and are part of collections of alchemical texts in Latin and Occitan. **W²** was probably copied directly from **P²**: there are no lacunae in **P²** that are not in **W²**; **P²**'s erroneous and ungrammatical readings recur in **W²**; where **P²** abbreviates *vitrioli* as *vit'*, **W²** reads *vitri*, and where **P²** mistranscribes *sub* as *sbā*, **W²** reads *substantia*. Moreover, the manuscript containing **P²** was later owned by the copyist of **W²**, since a second explicit added to **P²** is clearly in the same hand as **W²**.[38] The two manuscripts remained together long thereafter, as indicated by the marginal notes, insertions, and extravagant manicules made throughout both volumes in the early sixteenth-century French hand of a subsequent owner. Distinguishing textual characteristics include:

- ll. 1–17 (prologue) and 65–78 ("demonstratio") omitted.
- l. 126: *stella roris* in place of *stilla roris*.
- ll. 227–28: *mortalis cineris nec perequinum colorem* in place of *mortalis criminis nec peregrinum colorem*.
- l. 229: *montem* in place of *librum*.
- l. 249: *sed pro sanctis* in place of *sed perfectis*.
- l. 325: *aliis gravationibus* rather than *aliis generacionibus*.
- Operation 9 included (see Section 2.6).

P² Paris, BNF, MS Lat. 11202, fols. 61r–69r (1st q. 15th c.)[39]

One of the earliest known witnesses, **P²**, is part of a compilation of alchemical treatises and recipes in Latin, Catalan, and Occitan, most written in *gothica cursiva* with red initials by a single hand. Brunel localizes the compilation to Avignon, thus near the place of John's authorship. **P²** was originally titled *Compositio lapidis mineralis* (*Composition of the Mineral Stone*), to which a later hand added "made by master John of Rupescissa" (*facta a magistro Johannes de Rupe cissa*) both at the start of the witness and in the contemporaneous list of contents on fol. 1r. The explicit (fol. 68v) gives the more usual title and the date of composition (see Section 1.3): "Here ends the *Liber lucis magisterii magni* written in AD 1354, the 14th day of October" (*Explicit liber lucis magisterii magni editus anno domini m^{o} iii^{mo} liiii^{to} die xiiii^{ta} octobris*). After the explicit and the heading "How the said medicine is projected" (*Quomodo dicta medicina prohiciatur*), Operation 9 follows. This section is then followed (fol. 69r) by "the form of the furnace and the plate" (*Sequitur forma furni et lamine*) with three drawings, although these

38 Fol. 69r: "The end of this little book of the mineral philosophy of master John of Rupescissa" (*Finis huius libelli philosophie mineralis magistri Johannis de rupecisa*). The same hand also (erroneously) reattributed to John the following text from Jacob de Garandia (fols. 69v–70v) and added a title to another alchemical tract on fol. 91r: "Liber de summis pontificibus super Lapide philosophali."

39 Corbett, *Catalogue*, 1:147–58; Jean-Baptiste Camps, *Les manuscrits occitans à la Bibliothèque Nationale de France* (Paris: Enssib, 2010), 39; Clovis Brunel, *Bibliographie des manuscrits littéraires en ancien provençal* (Liège, 1935), 64.

illustrations, slightly differently rendered, already appear on fol. 68r.[40] An extract from *De famulatu* immediately precedes the *Liber lucis*.[41]

W² London, Wellcome Library, MS 707, fols. 1r–10v (1430s–40s)[42]

This witness exists in what was originally the second part of a large manuscript, later divided into three separate documents that are now Wellcome Library MSS 523, 707, and 418.[43] The original collection contained alchemical treatises and recipes in Latin, French, and Occitan written by several hands on paper. All the Latin treatises (a majority of the material) are written in a single hand, a carefully executed and "spiky" *gothica bastarda cursiva*.[44] The opening line of **W²** is written in *gothica quadrata*. Spaces were left for initials but not filled in. Like **P²**, it was produced in Provence; Moorat localizes it to Aix-en-Provence. **W²** has been dated to ca. 1430, but while this dating seems roughly correct based on paleography and watermarks, it was based on the presence of the date 7 May 1430 on fol. 16r of MS 523. But that date is not relevant since it refers to the composition of the treatise on fols. 2r–16r, Jean Saulnier's *De la transmutation des metaux*, not its copying.[45] Nor does this date give a *terminus a quo*, because Saulnier's treatise was *not* part of the original manuscript ensemble; it was added considerably later.[46]

6.1.4 Descendant of π, Indeterminate Subfamily

Br Bern, Burgerbibliothek, MS 480, fols. 24v–30r (1st half 15th c.)[47]

This unusual witness is written entirely in the margins of a slightly earlier witness of *De famulatu*, surrounding the processes that parallel the *Liber lucis* (see

40 Corbett, *Catalogue*, 1:152 is incorrect to attribute this heading to "a slightly later hand," although this is possibly just a typographical slip, since the final line on the folio *is* in "a slightly later hand."

41 Fols. 60r–v: "Nota. Quomodo debemus extrahere 4or elementa ab omni re et separare unum ab alio bene et naturaliter," listed on fol. 2r as "liber modicus qui docet modum et practicam extrahendi quintam essenciam." This is a version of the section "Non pretermittam."

42 Moorat, *Catalogue*, 1:521; https://wellcomelibrary.org/item/b19208364

43 Moorat, *Catalogue*, 1:367–8 (MS 523) and 283–84 (MS 418). https://wellcomelibrary.org/item/b19270483 and https://wellcomelibrary.org/item/b19295121, respectively.

44 The French and Latin medical recipes of MS 707, fols. 10r–20r may be the work of the same copyist but using a *cursiva currens* for materials lower on the scale of textual hierarchy; the spiky *gothica bastarda cursiva* used for the Latin treatises makes a brief appearance on fol. 13v.

45 Its first publication—*Le grand miracle de la nature metallique* (Paris, 1615)—does not give a date of composition, but the second (Paris, 1660) provides the similar date 7 May 1432 on 10 and 57.

46 These folios are in a different hand on different paper; the "repertorium" at MS 523, fol. 98r and written in the same hand as the rest of the manuscript and of **W²**, does not list Saulnier's treatise. (The Wellcome catalog erroneously considers the "repertorium" written in a different hand than the Rupescissa and pseudo-Arnald text.) Instead, it records that the *original* fol. 2r contained "Capitulum primum secreti quinte essencie aque vite," that is, the first chapter of *De famulatu*. But the *De famulatu* now present is acephalous, beginning on the *original* fol. 13r (now fol. 18r); thus, at some time *after* the completion of the original manuscript, the first twelve folios (the first quire) were lost and Saulnier's treatise put in their place. A feeble attempt was made to "back-foliate" the added leaves, crudely imitating the hand of the original copyist's foliation, but this enterprise was abandoned, possibly when the foliator realized that he would run out of numbers before reaching the first folio. On pseudo-Arnald's *Liber de aquis*, which follows *De famulatu* here, see Calvet, *Oeuvres alchimiques*, 82–91, esp. 82 n.3 in regard to this witness.

47 Hermann Hauge, *Catalogus codicum Bernensium* (Bern, 1875), 413.

Figure 3.1). It contains the prologue, all eight operations, four small illustrations of the furnace and its parts, and a colophon giving the date of composition as 14 October 1380. The text is augmented with a suggested alternate mixture for the sublimation and advice about distilling the aquafort that reprises some of the directions found in otherwise omitted text. It appears to belong to family π but cannot be situated in the stemma due to the lack of textual variants indicating any one subfamily; it shows some variants of π^1, but only a minority. It also contains the ending to l. 251 found otherwise only in family γ^3, but no other features of that family. It might be contaminated or the only known witness of an additional subfamily of π.

6.1.5 Family φ

Ko² Copenhagen, Kongelige Bibliotek, GKS 236 folio, fols. 52r–54v (1428)[48]

The readings of **Ko²** most resemble those found in descendants of π^1 and π^2 (but not of π^3), yet without displaying any of the variants characteristic of either subfamily. **Ko²** does not show, for example, the additional practical advice of **P¹/Po**, the lacunae of **Y¹/G**, or the headings and Decknamen of **M/W¹**. I therefore suggest that it is the only known descendant of a hyparchetype φ, dependent upon π but independent of π^1, π^2, and π^3.

Ko² is written in a carefully formed *gothica cursiva* on paper as part of a large collection of alchemical treatises, mostly Latin but with some German. **Ko²** provides the prologue and all eight operations. The first four texts of the collection (fols. 1r–80v)—part of Geber's *Summa perfectionis*, an *Exposicio omnium rerum alkamye*, the *Liber lucis*, and pseudo-Lull's *Vade mecum*—are all by one copyist, who identifies himself in two colophons. The *Liber lucis* colophon reads: "Here ends the *Liber lucis magisterii magni*, written by Brother John of Rupescissa on 14 September 1354, copied by me, Walter Vallrevere, in 1428, presently a student in the arts in Paris."[49] Vallrevere's colophon to pseudo-Lull's *Vade mecum* reports that the text "was collected at Paris" (*collectus parisiis*) by him, implying that the arts student was seeking out and copying alchemical materials in Paris, and thus that φ was present in Paris in 1428.[50]

6.2 Germanic Family γ

B, Bo, Ge, Hb, Km, Le¹, Le², P³, Pr, U, V, W⁷, Z, and the printed Basel version *De confectione*; with significant contamination or significantly edited, **Fn¹, Ma, Mp²**,

[48] Ellen Jørgensen, *Catalogus codicum latinorum medii aevi bibliothecae regiae Hafniensis* (Copenhagen: Gyldendals, 1926), 447–48.

[49] Fol. 54v: "Explicit liber lucis magisterii magni editus a fratre johanne de ruppescissa Anno domini mille trecentesimo quinquagesimo quarto decima quarta die septembris compilatus per me Walterum Vallrevere Anno domini m°cccc°xxviii° protunc parisiis studentem in artibus."

[50] Fol. 80v: "Explicit liber magistri raymundi lalii qui vade mecum sive clausula testamenti nuncupatur. Scriptus et collectus parisiis per manus Walteri Vallrevere Anno domini m°cccc° vicesimo nono mensis marcii die visesima quarta hora quasi vesperarum."

Mu, and **S**; in German translation, **Bg**$_{d}$, **Ha**$_{d}$, **Hb**$^{2}_{d}$, **Mu**$^{2}_{d}$, **Wb**$^{2}_{d}$, and **Wb**$^{3}_{d}$; in Czech translation, **Le**$^{3}_{cz}$; in French translation, **Or**$_{f}$. Abridgments **Fr**$_{a}$, **Hb**$_{a}$, **Le**$^{I}_{a}$, and **Z**$_{a}$ also belong to this family (see Section 6.9 for their descriptions), as do the five witnesses in the collections of Frater Johannes Baptista, **W**^{3}, **W**^{4}, **Y**^{3}, **Y**^{4}, and **Y**^{5} (Section 6.2.5).

This family contains the largest number of extant manuscript witnesses (thirty-five), most of them (except **Bo**, **Fn**^{I}, **Or**$_{f}$, and the collections of the Northern Italian Frater Johannes Baptista) written in Germanic hands. The core text for the 1561 Basel edition emerged from this family; the alternate title under which it was published originates in the incipit unique to this family (Section 1.4). The earliest precisely dateable vernacular translation (1442 or 1443) also descends from γ. The numerous descendants of γ are easily recognizable thanks to an extremely large number of unique textual variants that render German family γ the most distinct grouping of witnesses. The most significant of these readings include

- A unique incipit (omitted in some later witnesses and translations): *Incipit liber magisterii confeccionis veri lapidis philosophorum clare et absque quacumque palliacione editus per fratrem Johannem de Rupescissa ordinis minorum virum utique a deo illustratum quem libellum composuit ad sublevandas inopias pape et cleri tempore future tribulacionis ecclesie eciam ad hoc divina premonitus revelacione.*[51]
- l. 1: *ecclesia* for *evangelio* (except **V**).
- l. 2: A clause added to the prologue suggesting that God might delay the appearance of the antichrist(s): *...tam mistici quam realis et veri satis cito instare et quasi in januis nisi deus pro consolacione bonorum et correctione malorum velit tempus cum possit misericorditer prorogare....*[52]
- l. 18: *materia seu natura* in place of *materia.*
- l. 23: *Hec autem natura operatur* in place of *Hec autem materia preparatur.*
- l. 31: mercury is passed through “hare’s leather” (*corio leporis*) rather than linen cloth.
- l. 94: *septem partes aque* in place of *septimam aque partem.*[53]
- ll. 142–43: *mercurii et rubedinis* for *intrinseca rubedinis.*
- l. 162: *concalandum ab pretereuntibus* instead of *conculcetur ab hominibus.*

[51] “Here begins the book of the magistery of making the true philosophers’ stone, written clearly and without any concealment by brother John of Rupescissa of the Order of Friars Minor, a man surely illuminated by God, who composed this little book for the purpose of alleviating the want of the pope and clergy in the time of the coming tribulation of the Church, being forewarned of it by divine revelation.”

[52] “...[the tribulations] of both the mystical and the real and true antichrist, are sufficiently close at hand and stand nearly at the gates, unless God—for the consolation of the good and the correction of the wicked—should wish, since he is able, mercifully to defer the time....” This qualification was presumably added to γ in order to explain the failure of the antichrist to appear when John had predicted.

[53] This error yields the operationally impossible direction to dissolve seven parts of sal ammoniac in one part of aquafort. Recognition of the implausibility of this direction or contamination guided corrections in a few

- l. 185: *donec omnes carbones transierint* instead of *donec omnes colores transierint*.[54]
- ll. 208–10: eye-skip between the two instances of *ad rubeum*.
- ll. 229–30: *Bestia autem rationalis carnalis habet mentem veri luminis recitescat alioquin* (on *mentem* here, see 139 n. 46).
- l. 273: *super mercurium et illum mercurium* instead of *super verum argentum et illud argentum*.[55]
- l. 281: *capillis humanis* in place of simply *capillis*.
- l. 289: *notabiliter* in place of *non totaliter*.
- l. 316: *carbonem* instead of *foramen*.
- l. 324: *spiritualiter*, *spiritualia*, or *specialiter revelata* instead of *patule*.

On the basis of groups of additional shared variants, this populous family can be subdivided into three subfamilies: γ', γ'', and γ'''.

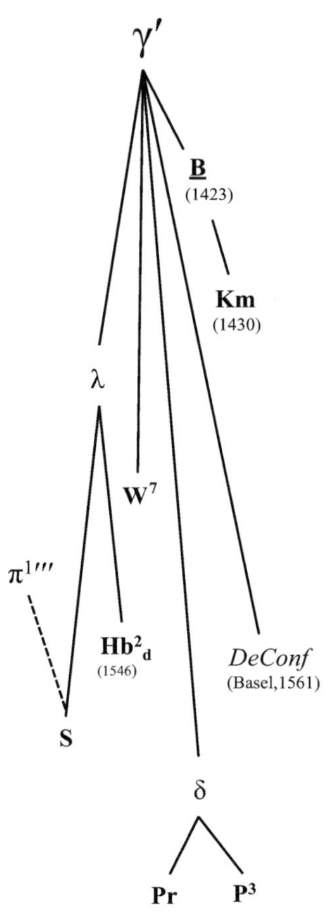

6.2.1 Subfamily γ'

B, **Km**, **P**^{3}, **Pr**, **S**, **W**^{7}, and *DeConf*; probably also **Ma** and **Mu**; in German translation, **Hb**^2_d; in French translation, **Or**$_f$.

The descendants of γ' are identifiable by several variants and especially the inclusion of Operation 9. The witnesses fall into three groups. The first contains **B**, **Km**, **P**^{3}, **Pr**, the significantly altered **W**^{7}, and *DeConf*. The second contains **Hb**^2_d and **S**; this pair descends from a common intermediate ancestor λ, more distantly related to γ' than the first group. **Ma**, **Mu**, and **Or**$_f$ can also be linked to this subfamily, although with less confidence due to their significant editing and/or contamination. They display most or all of the features of family γ, but they show only some of those of subfamily γ'. Yet they show no features of subfamilies γ'' or γ''', and so belong either to subfamily γ' or to a now-unidentified subfamily; they have not been situated in the stemma. Three witnesses (**B**, **Hb**^2_d, and **S**) share a lengthy colophon dating the *Liber lucis*'s composition (see Section 1.3).

witnesses (**Bo**, **Ma**, **Mp**^{2}, and **S**).

[54] *De confectione* (1561) attempts to correct the error by changing the verb to *consumentur*.

[55] The error is probably due to misreading *verum argentum* as *vivum argentum* (this error exists as such in **X**^{4}/**Cu**^{2}) then replacing *vivum argentum* with its cognate *mercurius*.

Characteristics of subfamily γ′, beyond the variants common to the larger family γ, include:

- Operation 9 is present, with the reading *rosas utiles et odoriferas...extractum* near the end (cf. family ι^2).
- l. 21: *ubilibet* in place of *vilibus.*
- l. 121: *non* omitted.
- l. 127: *lapidis occulti* instead of *lapidis.*
- l. 167: *tactum manus* instead of *tactum.*

B Klagenfurt, Bischöfliche Bibliothek, MS XXIX d 24, fols. 202v–14v (1423)[56]

This Klagenfurt manuscript offers the earliest and fullest witness not only of subfamily γ′ but also of the larger family γ. It is part of a quarto collection of alchemical treatises, written on paper by a single hand in *gothica cursiva* and with extremely large margins and unfilled spaces for initials. It contains nineteen alchemical treatises in Latin (with some German on fol. 336r), and has been used as a source for editions of other alchemical texts.[57] It contains a version of *De famulatu* (fols. 292r–324v) not attributed to John. According to a colophon (fol. 21v), the collection was begun in May 1421 at Castro, on the shores of Lake Iseo. Later colophons (fols. 95r, 122r, 188v, 190v, 197v) indicate that most of the manuscript was compiled in 1423 at Vallerendena or simply Rendena, in the South Tyrol west of Trento. An endorsement on fol. 1r shows that it came into the possession of "Dux Sigissmundus," the Habsburg ruler of the Tyrol (1427–1496), sometime between 1439 (when he became duke) and 1477 (when he became archduke). As the earliest member of this family, and one that originated on the Italian frontier, **B** may stand early in the transmission of the *Liber lucis* into Germany, although family γ had already diverged significantly from π by this time. Based on its early date and completeness, it is used in the critical edition to represent family γ.

B opens with the invocation "Jhesus, Maria, Franciscus" (suggesting a Franciscan copyist) and continues with the incipit, prologue, and Operations 1–9. It ends with the colophon

> Here ends the *Liber lucis magisterii magni*, written by the excellent, divinely illuminated philosopher John of Rupescissa of the Order of Friars Minor, on the fourteenth day of the month of October AD 1380; it lay hid for many years, and would that it may remain hidden forever from evil greedy men and especially tyrants, but be preserved under the custody of the Crucified One for men of the Gospel in the time of their need.[58]

[56] Hermann Menhardt, *Handschriftenverzeichnisse Österreichischer Bibliotheken, Kärnten*: vol. 1, *Klagenfurt, Maria Saal, Friesach* (Vienna: Österreichischen Staatsdruckerei, 1927), 14–16, and Lynn Thorndike, "An Alchemical Manuscript: Klagenfurt, Bischöfl. Bibl. XXIX.d.24," *Ambix* 11 (1963): 95–96.

[57] See, for example, Antoine Calvet, "Le *De arte alchemica* (inc: Dixit Arturus explicator huius operis) est-il une oeuvre authentique de Richard de Fournival?," in *Richard de Fournival et les sciences au XIIIe siècle*, ed. Joelle Ducos and Christopher Lucken, Micrologus Library 88 (Florence: SISMEL, 2018), 243–82.

[58] Fol. 214r: "Explicit liber lucis magisterii magni editus ab Egregio philosopho divinitus illustrato fratre Johanne de Rupescissa ordinis minorum. Anno domini m° ccc° lxxx et die xiiii mensis Octobris qui latuit multis annis

B adds a unique line in the margin to Operation 4, warning the reader, "Therefore, if you do not understand the sayings from that which is written above, do not blame me but vilify your own ignorance because you are unfortunate like the one to whom God never gave anything good."[59] This line comes from the *Rosarius philosophorum* of pseudo-Arnald.[60] **B** also contains one illustration (fol. 214v) of the furnace, mostly unlabeled and crudely drawn.

Km Kremsmünster, Benediktinerstift, Codex Cremifanensis 72, fols. 51v–52r, 54r–v, and 227v (extr., 1430)[61]

These extracts are written mostly in the margins of a partial copy of *De famulatu* containing only the metallic and mineral sections (cf. **Br**). A fuller copy of *De famulatu* appears much later in the collection (fols. 206r–26r), a large alchemical miscellany on paper written in a small, crabbed hand. Although the sections that would contain the characteristics of γ' are not present, **Km**'s readings are very nearly identical to those of **B**, including the same unique eye-skip at ll. 267–69. The *De famulatu* excerpt also follows the text of that work present in the Klagenfurt manuscript that contains **B**; therefore, **Km** may derive directly from **B**.

P^3 Paris, BNF, MS Lat. 7173, fols. 76r–91v (early 17th century); and **Pr** Prague, Library of Metropolitan Chapter of the Cathedral of St. Vitus, MS O. LXXIX, fols. 210r–25r (early 17th century)[62]

These two witnesses, written on paper, are treated together because they were produced by the same copyist, identifiable by his elegant italic cursive bookhand. In almost all cases, both leave the same blanks for indecipherable words and display the same unique variants, indicating their origin from the same exemplar, shown in the stemma as δ.[63] Chemical substances are cited by initial letters only: V.r. for *vitriolum romanum*, A.v. for *argentum vivum*, etc. **P**^{3} includes the incipit, prologue, and Operations 1–9, while **Pr** omits the incipit and prologue. **P**^{3} is part of a collection of eleven alchemical treatises in Latin and German; **Pr** is bound into a collection of Latin and German alchemical tracts.

W^7 London, Wellcome Library, MS 526, fols. 388r–90r (1525–27)[64]

This item falls between a witness and a reworking of the *Liber lucis*. It is part of a large volume of alchemical tracts and recipes in Latin, German, and some Italian collected 1515–30, mostly by a Benedictine and doctor of canon law from

qui utinam lateat in perpetuum malos homines avaros et tyrannos precipue sed sub costodie crucifixi servetur viris ewangilicis in tempore necessitatis."

59 Fol. 207v: "Idcirco si suprascriptis dicta non intellexeris, non me reprehendas sed tue ignorancie calumpnieris quoniam infortunatus es sicud ille cui deus nunquam aliqui boni tribuit."

60 Calvet, *Oeuvres alchimiques*, 352; or *BCC* 1:676.

61 Hauke Fill, *Katalog der Handschriften des Benediktinerstiftes Kremsmünster*, 2 vols. (Vienna: Österreichischen Akademie der Wissenschaften, 2000), 2:360–67.

62 Antonín Podlaha, *Soupis rukopisů knihovny metropolitní kapitoly Pražské, F-P* (Prague: Czech Academy, 1922), 557–60.

63 For example, *anxietate* for *antichristi* (l. 229) and *mihi satabit* for *ministrabit* (l. 234).

64 Moorat, *Catalogue*, 1:374–79.

"Wiszbern" (probably Weissenborn) who calls himself "Vulcanus." Written on paper in *gothica cursiva*, **W7** contains most of John's text but in a partly rearranged order with added text and a new title (*Via universalis*). The introduction invokes Saints Anne and Joachim, then borrows several fragmented and rearranged phrases about secrecy from the "Credas" and mentions mercury and sulphur as the "natural principles" of minerals. Then follows "De vase et fornace"; its first section describes the generation of metals underground and cites the *Lumen luminis*, while the latter part ("Furni preparacio") is Operation 8. The next heading, "Dicit Rupicissa," introduces Operations 1–7 and 9; the incipit and prologue are absent.

6.2.1.1 Descendants of λ: Hb^2_d and S

Unique characteristics shared by **Hb^2_d** and **S** indicate their descent through a common ancestor, denoted here as λ. While λ must predate 1546, a further estimate of when it originated cannot be made without additional data. In addition to all the characteristics of both family γ and subfamily γ′, these two witnesses display the following further features:

- Operation 9 is merged into Operation 7; Operation 8 is omitted.
- ll. passim: unique Decknamen are used erratically: *lapis sonans* for saltpeter, *ros maii* for common mercury, and *serpens romanus*, *leo*, and *serpens albus* for vitriol.
- l. 1: *tribulationes clericorum / die trubsal der gelehrten Priesterschafft* rather than *tribulationes electorum*.
- l. 128: *gutta / tropffen* instead of *gustu*.
- l. 139: *laminam veneris* rather than simply *laminam*.
- l. 142: *incombustibile / unverbrinlicher* rather than *invisibile*.

Hb^2_d Hamburg, Staats- und Universitätsbibliothek, Cod. Alch. 656, pp. 1–36 (1546)[65]

This German translation is part of a collection of German alchemical materials and is dated 17 February 1546 and titled *Liber quintae essentiae*. It contains the incipit, prologue, and Operations 1–7. The incipit, chapter headings, and colophon remain in Latin, as do many short phrases; the translator was not entirely comfortable with Latin and occasionally misunderstood it. The colophon dates the composition to 1370 rather than 1380 but otherwise varies only slightly from the colophon in **B**.[66] The translator rewrote part of the prologue, removing references to the antichrist and his defeat, and instead implying that the predicted

[65] Julian Paulus, *Katalog der alchemischen Handschriften der Staats- und Universitätsbibliothek Hamburg* (unpublished, 1995), 66–68.

[66] On 36: "Explicit lucis magisterii magni editus ab egregio philosopho divinitus illustrato fratre Johanne de Rupescissa Ordinis fratrum minorum Anno domini millesimo trecentesimo septuagesimo die quarta decima Mensis Octobris, qui latuit multis annis utinam lateret in perpetuum ab Gentibus malis avaris et tyrannis praecipue sud[*sic*] sub custodia crucifixi domini nostri Jesu Christi servet viris Evangelicis et reveletur in tempore necessitatis. Amen."

persecution and despoiling of the church has already happened at the hands of unbelievers (*Unglaubigen*), possibly a reference to Protestantism.

S Schlierbach, Zisterzienkloster, MS 85, fols. 37r–48r (1560s)[67]

The first part of this manuscript (fols. 1–76), which contains **S**, is comprised mostly of excerpts and dicta from alchemical authorities in Latin, German, and Italian.[68] **S** is written in a florid and professional *gothica cursiva* bookhand using forms from *Kurrentschrift* and bears the title *Rosarius Joannis de rupe scissa*. It contains the incipit, prologue, and seven operations. The colophon resembles that of $\mathbf{Hb^2_d}$ with some flourishes added.[69] The copyist often substituted synonyms and rewordings, seemingly for stylistic purposes, and glossed the Decknamen from λ with common names; he was not especially careful except as a calligrapher. **S** contains some text found in no other member of family γ—most significantly, the lengthy description of how to distill aquafort (ll. 85–91), a text otherwise found only in descendants of π^{I}. Judging from the wording, a descendant of $\pi^{I'''}$ was probably used along with λ (or a close descendant thereof) to produce **S**.

6.2.1.2 Probable Members of Subfamily γ′: Ma, Mu, and Or$_f$

Ma Marburg, Universitätsbibliothek, MS 23, fols. 136v–39r (1496)[70]

Ma forms part of a collection of alchemical treatises, most of them heavily edited, from pseudo-Geber, pseudo-Roger Bacon, pseudo-Arnald, and others—including extracts from *De famulatu* (fols. 232r–47v)—as well as a miscellany of alchemical recipes, predominantly in Latin but with some German. The text is written in a *gothica rotunda cursiva* on paper with rubrication. The scribe identifies himself as Peter Holzruter, "master of liberal arts of Paris and rector of the school at Konstanz." **Ma** is headed simply *Johannes de rupecissa ordinis minoris*. **Ma** is significantly edited, reorganized, and reworded. The incipit, prologue, and first part of Operation 1 are missing. A highly condensed version of Operation 9

[67] Benedict Hofinger, "Verzeichniss der Handschriften der Bibliothek des Stiftes Schlierbach," in *Xenia Bernardina 2, Handschriften-Verzeichnisse der Cistercienser-Stifte der Österreichisch-ungarischen Ordensprovinz*, 2 vols. (Vienna, 1891), 2:481–511, on 501–502; Donald Yates, *Descriptive Inventories of Manuscripts Microfilmed for the Hill Monastic Manuscript Library: Austrian Libraries*, 3 vols. (Collegeville, MN: Hill Monastic Library, 1981), 1:199–201.

[68] The second part (fols. 77–119) is written in a different hand (*humanistica cursiva*) and contains but a single alchemical treatise in Italian whose colophon reports that it was copied in Vienna in 1538.

[69] Fol. 48r: "Et sic est finis magisteri magni libri de lapide philosophorum benedicta editus ab egregio philosopho divinitus illustrato fratre Joanne de rupe scissa ordinis minorum. Anno domini Millesimo Tricentesimo septuagesimo, die quartadecima mensis Octobris, qui latuit multis annis. O utinam lateret in perpetuum ab omnibus malis et avaris hominibus, et precipue a Tyrannis, sed desidero ut sub custodia crucifixi Jesu Christi servetur viris evangelicis et deum timentibus ut habeant in tempore tribulationis et necessitatis."

[70] Sirka Heyne, *Die mittelalterlichen Handschriften der Universitätsbibliothek Marburg* (Wiesbaden: Harrassowitz, 2002), 62–70. The catalogue misidentifies the work as *De consideratione quintae essentiae* and Operation 8 as an unrelated text by a different author. Note also that Halleux, "Rupescissa," 284, gives the folios incorrectly as 122v–25r.

intervenes *between* Operations 7 and 8, following the order found also in **Mu**, **Hb²d**, and S. Overall, **Ma** contains less than 40% of the *Liber lucis*. Three small illustrations of furnace parts are embedded in Operation 8.

Mu Munich, Bayerische Staatsbibliothek, Clm 27438, fols. 10r–18r (early 17th c.)[71]

This late witness exists in a collection of Latin and German alchemical texts and recipes. **Mu** lacks the prologue but contains the incipit and Operations 1–9. Operation 9 follows 7 as "Operation 8" and the furnace chapter is renumbered Operation 9. It contains substantial marginalia dealing with specific practical instructions, indicating the continuing practical use of John's work into the seventeenth century. Several alternate readings are interlined or placed in the margin, sometimes prefaced with *alias*. While descending predominantly from γ′, it incorporates readings from other lines. It contains two small marginal illustrations of the subliming vessel (with and without the *foramen capitis*) next to Operation 1.

Or$_f$ Orléans, Bibliothèque municipale, MS 291, fols. 20r–25r (1516)[72]

The only known French translation of the *Liber lucis* is written in *gothica bastarda cursiva* on paper as part of a collection of alchemical materials in Latin and French. **Or$_f$** omits the incipit, prologue, and first part of Operation 1 but contains all eight operations plus the spurious ninth. The last line of Operation 9 provides the title for **Or$_f$**: *Le tresvray Rosier des Rosiers compose par frere Jehan de Rupe scissa de lordre des cordeliers touchant lart de philosophie au blanc et au rouge.* A colophon gives the date 1516, and other colophons in the collection (in the same hand) identify the copyist as Pierre Bureteau (ca. 1485–ca. 1537), Celestine monk, poet, and humanist at the abbey of Sens.[73] The descent of **Or$_f$** through German family γ is rather surprising, but the evidence is clear: nearly all of the textual characteristics of the family appear in French translation.[74] Its Operation 9 bears the specific variants found only in subfamily γ′. But it also displays a few readings found otherwise only in Mp²; it has not been possible to work out satisfactorily a line of transmission that would account for these latter similarities.[75] What may

71 Hermann Hauke, *Katalog der lateinischen Handschriften der Bayerischen Staatsbibliothek München*, Volume 4, Part 5: *Clm 27270–27499* (Wiesbaden: Harrassowitz, 1975), 229–30.

72 *Catalogue générale des manuscrits des bibliothèques publiques de France*, vol 12: *Orléans* (Paris: Plon, 1889), 143–46; Corbett, *Catalogue*, 2:147–58.

73 Maximilien Quantin, "Notice sur P. Bureteau religieux celestin, chroniqueur et poète," *Bulletin de la société des sciences historiques et naturelles de l'Yonne* 29 (1875): 407–41 and François Secret, "Littérature et alchimie," *Bibliothèque d'humanisme et Renaissance* 35 (1973): 499–531, on 499–500.

74 For example, l. 31, *cuyr dun lievre*; l. 185, *jusques ad ce que les charbons soient passez et degastez*; l. 273, *getter sur mercure pour le blanc*; l. 281, *cheveulx de personne humaine*; and l. 324, *espirituellement revelée.*

75 For example, in Op. 8, where **Mp²** uniquely reads "habeat quattuor brachia...quibus super parietes furni sustentari debeat et sed tanta distantia inter circumferentiam dicte lamine et latera orificii dicti furni quod calor libere per tale interstitium ascendere circundare possit," **Or$_f$** has "ayt quattre bras pour la soustenir sur ladite paroy du four... mais quil y ait ung peu de distance entre les paroys et la lamine de fer dessusdite et soit faicte per compas pour monter la challeur par compas sudessus."

be an illustration of the furnace appears, separated from the text, on fol. 26v; however, it is neither drawn nor labeled clearly enough to be connected unambiguously with the *Liber lucis*.

6.2.2 Subfamily γ″

Z and **Le**^{2}; in German translation, **Bg**$_{d}$, **Wb**$^{2}{}_{d}$, and **Mu**$^{2}{}_{d}$; and in Czech translation, **Le**$^{3}{}_{cz}$

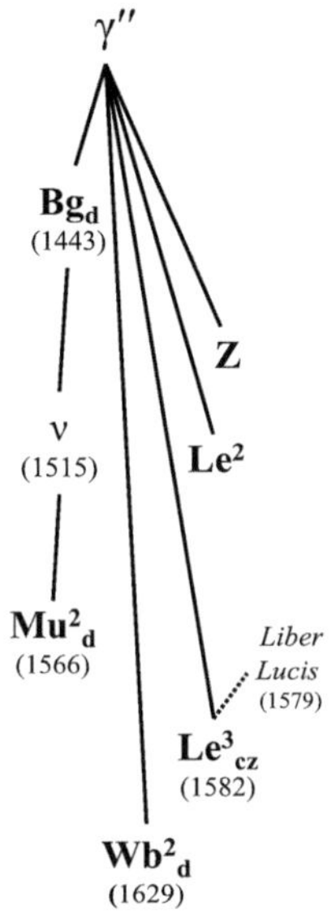

The descendants of γ″ show variants beyond those characteristic of the larger family γ; most reliably by the presence of explanatory glosses incorporated into the text. **Z** contains five such glosses, **Wb**$^{2}{}_{d}$ four, **Bg**$_{d}$ two, and **Le**^{2} one. (Later marginal additions to **Ge** provide parts of three.) The fact that **Z**, **Le**^{2}, **Bg**$_{d}$, **Wb**$^{2}{}_{d}$, and the marginalia of **Ge** all share these glosses indicates that they were present in γ″. These glosses were added *after* the text entered German-speaking lands, since one of them uses the Germanic measure of weight *Loth* (half an ounce). These glosses are presented in Section 6.2.6. Subfamily γ″ contains a high proportion of German translations. **Bg**$_{d}$ is both the earliest known translation of the *Liber lucis* and the oldest member of the subfamily; it dates from 1442 or 1443, thirty to forty years prior to **Z**, the earliest Latin witness of this subfamily. **Wb**$^{2}{}_{d}$, although late, preserves the most text from the characteristic glosses. All six witnesses originated in Bohemia or Bavaria.

Z Venice, Biblioteca Marciana, Lat. fondo antico 323 (=1579), fols. 114v–19r (1470s–80s)[76]

Z is part of a collection of Latin and German alchemical materials. It is written on paper in a rough and highly abbreviated *gothica cursiva*. The copyist omitted sections he considered less relevant and condensed the text, sometimes rendering the text rather telegraphic and sometimes ungrammatical. The manuscript collection originated in the region of Bohemia; it refers to the town of Elbogen (Loket, Czechia) and Schweidnitz (Świdnica, Poland). **Z** contains the incipit, prologue, and all eight operations. It also bears a short version of the colophon found scattered throughout family γ, although with a unique alternate title for the work: "Here ends the *Liber lucis magisterii*, otherwise the *Lumen amenitatis magni*, written by the excellent divinely illuminated philosopher John of Rupescissa of the Order

[76] Joseph Valentinelli, *Bibliotheca manuscripta ad S. Marci Venetiarum: Codices mss. latini* (Venice, 1872), 5:161–165, on 162. This catalog dates the manuscript only to the fifteenth century, but internal evidence narrows the dating to the 1470s–80s: fol. 183r cites a recipe given to a margrave in 1466, and on 272v a colophon to a text of pseudo-Albert states that it was copied in 1482.

of Friars Minor on the seventeenth day of October AD 1370."[77] **Z** preserves five glosses that draw upon parallel material from *De famulatu* and practical attempts to follow the original processes. Unfortunately, the copyist's edits render some texts difficult to understand, and he places them out of order; three of the glosses are run together at the end of Operation 3. **Z** also provides a single illustration of the philosophical furnace, densely labeled. The alchemical collection to which **Z** belongs also contains $\mathbf{Z_a}$ (fols. 185v–86v) and a transmutational recipe attributed to John (fols. 186v–87v), both of which reappear in the collections containing **Hb** and $\mathbf{Le^1}$ (see Appendix 2).

$\mathbf{Le^2}$ Leiden, Leiden University Library, Vossianus Chym. Q. 25., fols. 2r–8v (first q. 16th c.)[78]

This manuscript, written on paper in a careful *gothica cursiva* with rubricated chapter headings and initials, is a much more formal production than **Z**. Accordingly, $\mathbf{Le^2}$ shows many fewer lacunae, condensations, and obvious copyist's mistakes. It likewise contains the incipit, prologue, and eight operations, but only one of the glosses characteristic of this subfamily. Like **Z**, it originated in Bohemia; its first owner was the Bohemian nobleman Jan Zajíc of Házemburk (1486–1553).[79]

$\mathbf{Bg_d}$ Bamberg, Staatsbibliothek, Msc. Nat. 6, fols. 6r–21r (1442–43)[80]

This witness is part of a collection that includes alchemical treatises and recipes as well as calendrical materials, in both German and Latin. The presence of German translations of two of the characteristic glosses unambiguously situates $\mathbf{Bg_d}$ in family γ''. After significant redating, $\mathbf{Bg_d}$ becomes the earliest known translation of the *Liber lucis*. Earlier catalogs assign it to either the sixteenth century or "around 1500." The almanacs that the collection contains (in the same hand as the *Liber lucis*) provide precise dating. These documents cite the years "44" to "56," which Leitschuh and Fischer interpret as 1544 to 1556. But the almanac for "48" (fol. 147r) predicts a solar eclipse on 29 August; this event corresponds with the annular eclipse visible in Central Europe on 29 August 1448. The later catalog considers these documents to be "astronomical data," but they are almanacs, which are useful only *in advance* (they are predictive, not documentary), and since the earliest almanac is for 1444, the manuscript must date from just *before*

77 Fol. 119r: "Explicit liber lucis magisterii alius lumen amenitatis magni editus ab egregio philosopho divinitus illustrato Johanne de Ruppa scissa ordinis minorum Anno domini m° ccc° lxx° et die xvii mensis Octobris."

78 P. C. Boeren, *Codices Vossiani Chymici* (Leiden: Leiden University Press, 1975), 159–64.

79 On whom see Ivo Purš and Vladimir Karpenko, "Alchemy at the Aristocratic Courts of the Lands of the Bohemian Crown, 47–92 in *Alchemy and Rudolf II: Exploring the Secrets of Nature in Central Europe in the 16th and 17th Centuries*, eds. Ivo Purš and Vladimir Karpenko (Prague: Artefactum, 2016), on 87–89.

80 Friedrich Leitschuh and Hans Fischer, *Katalog der Handschriften der Königlichen Bibliothek zu Bamberg*, 3 vols. (Bamberg, 1887–1912), 2:412–13 and Hella Frühmorgen-Vos and Norbert H. Ott, with Ulrike Bodemann and Gisela Fischer-Heetfeld, *Katalog der deutschsprachigen illustrierten Handschriften des Mittelalters* (Munich: Beck, 1991), 55–56. See the new entry in the forthcoming Susanne Rischpler, *Die Handschriften und Drucke des 15. und frühen 16. Jahrhunderts der Staatsbibliothek Bamberg* (*Katalog der illuminierten Handschriften der Staatsbibliothek Bamberg*, 4).

that year, that is, 1443 or 1442.[81] The copyist may have been the translator: his German text contains several garbled Latin words, apparently places where the exemplar was unclear, and Operation 8 is left untranslated, probably due to its many technical terms.

Written in *gothica cursiva* on paper, **Bg$_{d}$** includes all eight operations, but omits incipit and prologue. The opening lines of Operation 1 are garbled and fragmented—unlike the rest of the translation—suggesting that the exemplar was damaged at its beginning, which would equally explain the missing incipit and prologue. The text is prefaced simply "Of the first work and sublimation of quicksilver as master Hans Rupecissa taught" (*Von dem Ersten Werkch und erhochung des lemptigen Silber/Als Maister hanns Rupecissa gelernt hat*) without indicating a title. Two folios are missing between the present fols. 15 and 16, truncating Operation 6 in mid-sentence at l. 199: "des ursach furwar ist mit lindem fewr...," but see **Mu$^{2}_{d}$** below. The translator attempted to make the text more secretive by rendering *sal petri* as *Anatron* and *vitriolum* as *weisses Leben*. The collection also contains a German translation of an excerpt from *De famulatu* (fols. 56r–60v) parallel to Operation 8: "Der Maisterlich ofen id est Athanor."[82] There are three simple drawings of the furnace, the plate (*lamina*), and the sublimatory vessel (*aludel*) on fol. 42r, along with drawings relating to other texts in the collection.

Mu$^{2}_{d}$ Munich, Bayerische Staatsbibliothek, Cgm 3741, fols. 200r–5r and 264r (ca. 1550)

Mu$^{2}_{d}$ reproduces **Bg$_{d}$** nearly word for word, including the reference to "Hans" Rupescissa and even preserving its abbreviations. It ends, however, with Operation 6 and contains neither the glosses nor Operations 7 and 8 present in **Bg$_{d}$**. The copyist working from **Bg$_{d}$** may have seen the heading that prefaces the gloss at the top of fol. 16r in the Bamberg manuscript and assumed that it marked the beginning of a different treatise. **Mu$^{2}_{d}$** is part of a voluminous collection of alchemical processes owned by the Bavarian duke Ottheinrich (1502–1559) whose image appears on the binding with the date 1553. **Mu$^{2}_{d}$** begins a section of the collection (fols. 200r–312v) that is identified as having come from Count Philipp of Nassau-Weiberg. A preamble reports that Kuno von Wallbrunn (1458–1522) originally collected that material in 1515–16 during his search for alchemical texts; I have designated Kuno's now-lost copy of the *Liber lucis* as *ν*.[83] This same

[81] A clear *terminus a quo* of 1437 is provided by the references on fols. 144r and 151v to "Benedictus Abbas in Seittersteten," that is, Benedict I, who became abbot of the Benedictine abbey of Seitenstetten in 1437. But since the copyist would have begun his almanac series with the one for the upcoming year, it is certain that he copied the almanacs on fols. 145r–51r in 1443. The only question is how much earlier he recorded the *Liber lucis* on fols. 6r–21r. Given the uniformity of handwriting, ink, and paper, it seems plausible that it was within a narrow timeframe, hence my dating of 1443, or possibly 1442.

[82] This *De famulatu* translation is roughly contemporaneous with the oldest previously known German translation of the work in Salzburg, Universitätsbibliothek, Cod. M II 180, edited in Benzenhöfer.

[83] Fol. 200r: "...durch vil versuchung diser Kunst, und darzu, auß vil alten Lateinischen und teutschen Buchern, auch under weysungen von andern empfangen, hab ich Chuno von Walpron zu Newen eglofsheim, die zeit haumptmann zu Burkhausen, in alchemia diß buch besamlet und teutschen lassen des funfzechen hunderten funfzechenden und sechzehenden jars."

section of Cgm 3741 also preserves extracts from *De famulatu* (fol. 215r–17v) and a brief quotation (headed "Johannes Rubricissa," fol. 264r) from a later section of the *Liber lucis* (ll. 269–72) not present in **Mu²d** and that does not reproduce **Bgd**. This witness also preserves the text from the folios now missing from **Bgd**, indicating that their loss occurred after Kuno's copy ν was made.

Wb²d Wolfenbüttel, Herzog August Bibliothek, Cod. 185 Blankenburg, fols. 124v–30v (1629)[84]

This witness preserves a German translation independent of **Bgd/Mu²d**. It contains all eight operations but omits the incipit and prologue. **Wb²d** contains four of the glosses characteristic of γ″, and in their fullest form. The witness is part of a multivolume collection of alchemical treatises in German made by Paul Zazer in Nuremberg in the 1620s and 30s. This volume also contains *Eine grosse Heimblichkeit,* spuriously attributed to John of Rupescissa (see Appendix 2).

Le³cz Leiden, Leiden University Library, Vossianus Chym. F. 3., fols. 199v–205r (1585)[85]

Le³cz is part of a large collection of alchemical treatises translated into Czech by the alchemist Bavor the Younger Rodovský of Hustiřany (1526–1592) while imprisoned for debt in Prague during the 1580s.[86] In terms of its relation to the Latin *Liber lucis*, **Le³cz** is a Frankenstein. It is divided into ten chapters with descriptive headings, following the 1579 Cologne version. But the text does not come from that version but instead belongs to German family γ, as signaled by the characteristic incipit. Given the adoption of divisions and headings from one of the printed editions, it would seem plausible that the text came from the other printed version, the 1561 Basel *De confectione*, which descends through γ. But the Czech text contains the full prologue, the "demonstratio" (ll. 65–77), the "Credas" (ll. 220–51), the entirety of Operations 7 and 8, and many other less obvious features, none of which exists in the Basel version. Instead, the textual variants are certainly those of subfamily γ″, even though **Le³cz** does not contain the characteristic glosses.[87] This assignment is strengthened by the fact that both Latin members of this subfamily—**Z** and **Le²**—originated in Bohemia, as did **Le³cz**, and hence, a member of subfamily γ″ was possibly the most readily

[84] Hans Butzmann, *Kataloge der Herzog-August-Bibliothek Wolfenbüttel 11: Die Blankenburger Handschriften* (Frankfurt: Klostermann, 1966), 181–82.

[85] Boeren, *Codices*, 7–13 and Purš and Karpenko, "Alchemy at the Aristocratic Courts," 86–87.

[86] See Purš and Karpenko "Alchemy at the Aristocratic Courts," 75, 82–87 and Jennifer M. Rampling, *The Experimental Fire: Inventing English Alchemy, 1300–1700* (Chicago: University of Chicago Press, 2020), 301.

[87] Subfamily γ‴ can easily be ruled out because Le³cz contains the sublimation instructions in Operation 1 which are always missing from descendants of γ‴, and it contains the equivalent of *sine parabolis* (*bez podobenství*) rather than of *sine pallacione* at l. 9. Le³cz is unlikely to descend through γ′ since it does not include Operation 9, contains the crucial negation at l. 121 (dilo *ne*dokonalo = opus *non* perficeretur), and shows none of the readings from λ.

available source. Rodovský left two spaces in Operation 8 for illustrations of the furnace, which were not filled in. The proportions of these spaces suggest that the two illustrations were to be taken, like the headings and divisions, from the Cologne edition.

6.2.3 Subfamily γ'''

V, Bo, Ge, Hb, and **Le**[I]; in German translation, **Ha**$_d$ and **Wb**$^3{}_d$

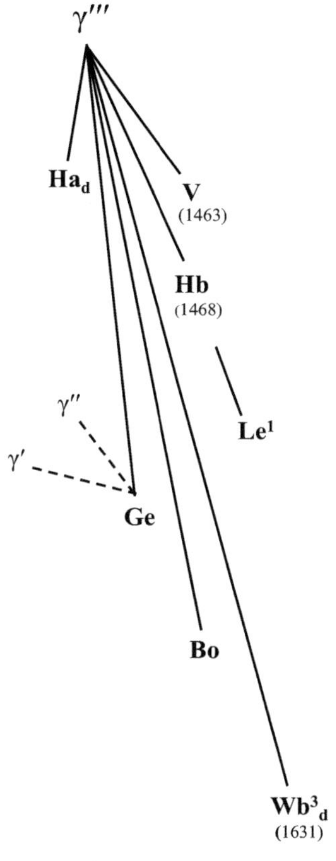

A third subfamily, descending from a hyparchetype designated γ''', contains at least seven witnesses—five in Latin and two in German. Members of this subfamily are easily identified by several distinctive readings and omissions, including:

- l. 9: *sine palliacione* instead of *sine parabolis.*
- l. 13–14: *ut recitem/reticere* instead of *aut recitative.*
- l. 26: *ego nominabo tibi eum nomine suo et certe* omitted.
- l. 31: *comprime subtiliter per coreum* instead of *involve in panno lineo.*
- ll. 35–41 on how to carry out the sublimation omitted and replaced by *ut scis* ("as you know how").
- l. 44: *invisibilem* in place of *miscibilem.*
- l. 82: *volubilis* in place of *volatilis* (corrected in **Bo**).
- l. 134: *in quarta operacione* rather than *in tercia operacione.*

V Vatican City, Biblioteca Apostolica Vaticana, Palatinus latinus 1330, fols. 119v–25r (1463–64)[88]

The earliest witness of this subfamily is part of a collection of more than a dozen alchemical treatises, many connected with pseudo-Albert or pseudo-Arnald, and including excerpts from *De famulatu* (fols. 124r–30r). It is written in a single hand in an angular *gothica cursiva* on paper with large, richly decorated initials. **V** contains the incipit, prologue, and eight operations, plus a condensed version of the colophon given by **B** and **Z**: "Here ends the *Liber lucis magisterii magni*, written by the excellent, divinely illuminated philosopher John of Rupescissa, may it be

[88] Ludwig Schuba, *Die medizinischen Handschriften der Codices Palatini Latini in der Vatikanischen Bibliothek* (Wiesbaden: Reichert, 1981), 444–50; http://digi.ub.uni-heidelberg.de/diglit/bav_pal_lat_1330

preserved under the custody of the Crucified One."[89] The copyist identifies himself as Heinrich Waltpod, a member of the household of Nicholas of Cusa, who worked on this manuscript in 1463–64, possibly for Nicholas himself. Originally a cleric from Cologne, Waltpod lived for a time in Bolzano and Rome.[90] A lengthy text of uncertain origin (fol. 124r–25r) describing a transmutational process divided into nine steps (*gradus*) intervenes between the end of Operation 8 and the colophon. Its style is nothing like John's.[91]

Hb Hamburg, Staats- und Universitätsbibliothek, Cod. Alch. 659, fols. 122v–29r (1468)[92]

This witness, written in *gothica cursiva* on paper, contains the incipit, prologue, and eight operations but breaks off in mid-sentence at l. 304. It is part of a collection of alchemical treatises and recipes in Latin and German. The colophon to the immediately following text provides the date and indicates that it was written in Ehrenfriedersdorf (a mining town in Saxony) by Vitus of Wasungen.[93] **Hb** is accompanied by the same abbreviated version (**Hb**$_{a}$, fol. 122r–v) and transmutational recipe that appear along with **Z** and **Le**^{I}.

Le^{I} Leiden, Leiden University Library, Vossianus Chym. Q. 2., fols. 123v–27v (c. 1500)[94]

This manuscript is a collection of recipes with only a few treatises. It is mostly Latin (but with some German), written on paper in a crabbed and highly abbreviated *gothica cursiva*. **Le**^{I} contains the incipit, prologue, and all eight operations, although Operation 8 is greatly edited. **Le**^{I} and **Hb** form a close pair: **Le**^{I} certainly descends through **Hb**, possibly directly. They share many unique readings and nearly identical lacunae throughout.[95] While Operation 8 is more condensed in

89 Fol. 125r: "Explicit liber lucis magisterii magni editus ab egregio philosopho divinitus illustrato fratre Johanne de rupecessa qui sub custodia crucifixi servetur."

90 On Nicholas of Cusa and alchemy, see Witalij Morosow, *Das Erbe des Nikolaus von Kues im Spiegel der Alchemie* (Münster: Aschendorff, 2018); on Waltpod, see Erich Meuthen, *Die letzten Jahre des Nikolas von Kues* (Cologne: Westdeutscher Verlag, 1958), 313.

91 Incipit: "Hec est ars que ad practicam nobilissimam et secretissimam omnium philosophorum de qua dictum est quod sit levis brevis et modicum causat"; explicit: "Et de rubeo super 60 partes [mercurius?] tinctus et mutabitur in sol, sic partes in partes mutantur usque in infinitum."

92 Paulus, *Katalog Hamburg*, 69–73.

93 Fol. 132v: "anno domini m° cccc° lx octava in villa ernfridistorff per vitum de wasungen sabato ante purificationis marie."

94 Boeren, *Codices*, 115–16. The catalog does not identify the Rupescissa texts correctly; it should be: fols. 122r–v, abbreviated version of the *Liber lucis*; fols. 122v–23v, transmutational recipe attr. to Rupescissa; fols. 123v–27v, *Liber lucis*.

95 For example, in the incipit, they alone give the erroneous duplication "diminutus divina"; in the prologue, they uniquely cite "tyrannos *invidos*," and where others read "domina generalis," these two give "doctrina generalis"; in Operation 1, mercury is generated *in lacrimis*; in Operation 2, both omit the crucial ingredient sal ammoniac, thus reading "pone intus septem partes aque"; in Operation 3, the powder is put into a *concha*; in Operation 4, Geber is omitted from the list of authorities and "Hermes palpavit Alphidius optavit" reads instead "Hermes palliavit apprehendit aptavit"; in Operation 6, the unworthy practitioner incurs "terram infernalem" rather than "iram infernalem"; in Operation 7, tin is omitted from the list of transmutable metals, as well as "color" from the phrase "color et malleacio."

$\mathbf{Le}^{\mathrm{I}}$, it nevertheless breaks off abruptly at the same point as **Hb**. Furthermore, both **Hb** and $\mathbf{Le}^{\mathrm{I}}$ are immediately preceded by the same abbreviated version and the same transmutational recipe with the same introduction ($\mathbf{Le}^{\mathrm{I}}_{\mathbf{a}}$, fol. 122r–v; see Section 6.9 and Appendix 2). The copyist of $\mathbf{Le}^{\mathrm{I}}$ had mediocre Latin—he often writes wrong endings or genders, and uncommon words are garbled or indecipherable.

Ge Geneva, Bibliothèque de Genève, Ms. lat. 82, fols. 98v–104v (1546–60)[96]

The manuscript bearing **Ge** is written in a neat *humanistica cursiva* bookhand on paper and contains twenty alchemical texts, including the first book of *De famulatu* (fols. 68r–98r).[97] The original text of **Ge** bears extensive corrections and marginalia in a later sixteenth-century Germanic hand. The first is probably that of Petrus Raymundus Amadonus, whose name appears on the inside front cover—a physician and, in the first half of the 1550s, professor of Greek at the University of Vienna.[98] The original text transmits all eight operations, omits the incipit and prologue, and displays all the characteristics of subfamily γ'''. A second hand later added the incipit, prologue, and spurious Operation 9, and revised the original text substantially throughout. Most additions come from a descendant of γ'', including parts of three of the characteristic glosses. But the presence of Operation 9 in the form unique to γ' indicates that the later owner drew also upon a descendant of that subfamily. Thus the completed document reintegrates all three subfamilies within family γ.

Bo Bologna, Biblioteca Universitaria, MS 1496, busta 1, fols. 42r–47v (16th c.)[99]

Written on paper in an informal *humanistica cursiva* by an Italian hand, **Bo** omits the incipit and begins instead with the title "Joannis de Rupescisa: Opus lapidis philosophorum." A preamble with no connection to John argues in favor of transmutation based upon the "fourth book of Aristotle's *Meteors*," actually the part of Ibn Sīnā's *Kitāb al-Shifā'* that circulated under the Aristotelian title during the Middle Ages. Here the originally anti-alchemical "Sciant artifices" is turned on its head to provide a means of attaining transmutation. The prologue and eight operations follow. Operation 8 is as highly abbreviated as in $\mathbf{Le}^{\mathrm{I}}$. **Bo** shows considerable scribal intervention through condensation and rephrasing; nearly half of the text is omitted, and thus **Bo** stands apart from the other Latin witnesses of this

96 Isabelle Jager, *Catalogue des manuscrits latins 1–376* (Geneva: Bibliothèque de Genève, 2016), 377–81. My thanks to Didier Kahn for alerting me to this witness.

97 In place of the second book, the manuscript reads on fol. 98v, "The second book of the *Quintessence* is contained in the printed book of the works of Ramon Lull; it may therefore be sought thence." This is probably a reference to *Raimundi Lullii Maioricani de alchimia opuscula* (Nuremberg, 1546), giving a *terminus post quem* for Ge.

98 Joseph Ritter von Aschbach, *Die Wiener Universität und ihre Gelehrten, 1520 bis 1565* (Vienna, 1888), 251–53 and Stephan Weszprémi, *Succincta medicorum Hungariae et Transilvaniae biographia* (Vienna, 1787), 4:6. These sources disagree on whether he was Swiss or Italian.

99 Albano Sorbelli, *Inventari dei manoscritti delle biblioteche d'Italia*, vol. 21: Bologna (Florence: Olschki, 1914), 91–92. My thanks to Matteo Martelli for photographing this witness for me.

subfamily whose readings closely resemble one another. **Bo** also shows minor corrections relative to **V**, **Ge**, **Hb**, and **Le^{1}**, which the copyist could have divined from context and experience rather than from contamination. While contamination remains possible, the substantial omissions do not leave sufficient text to compare with other families. **Bo** has one small and simplified drawing of the furnace.

Ha$_{d}$ Halle, Universitäts- und Landesbibliothek Sachsen-Anhalt, 1/C a/2, fols. 297r–308v (1450s)[100]

Part of a large collection of German translations of Latin alchemical treatises, this early translation, produced in Bohemia or Saxony, is elegantly written in a carefully executed *gothica bastarda* bookhand with polychrome initials on a mixture of paper and parchment. Its incipit and variants clearly identify it as a descendant of γ and it shows most of the unique variants of γ‴. It contains the incipit, prologue, and eight operations but is increasingly edited beginning midway through Operation 5 (Operation 8 is reduced to one sentence).

Wb$^{3}_{d}$ Wolfenbüttel, Herzog August Bibliothek, Cod. 187 Blankenburg, fols. 164r–73v (1631)[101]

This translation, independent of **Ha$_{d}$**, displays most characteristics of family γ (while correcting some of the more obviously erroneous readings) and most variants of subfamily γ‴.[102] It contains the incipit, prologue, and all eight operations. Like **Wb$^{2}_{d}$**, which is an independent translation descending through a different subfamily, **Wb$^{3}_{d}$** was part of the multivolume collection of alchemical treatises made by Paul Zazer in Nuremberg in the 1620s–30s.

6.2.4 Descendants of γ, Indeterminate Subfamily
Fn^{1}, **Mp^{2}**, and **U**

Three further witnesses—not counting the five witnesses connected to Frater Johannes Baptista (Section 6.2.5)—contain enough characteristic readings to show that they descend, at least in part, through γ but cannot be confidently assigned to any of its currently identified subfamilies. All are highly edited and/or contaminated, and the contamination sources cannot be securely identified. Subfamily γ‴ can be ruled out since none of that group's distinctive readings is present in any of these witnesses. Likewise, none contains any of the glosses that identify γ″, but it is possible they were omitted. They might descend through γ′, but they do not

[100] Brigitte Pfeil, *Katalog der deutschen und niederländischen Handschriften des Mittelalters in der Universitäts- und Landesbibliothek Sachsen-Anhalt in Halle (Saale)*, 2 vols. (Halle: ULSAB, 2007), 1:164–86; https://opendata.uni-halle.de/explore?bitstream_id=06f8c7bd-547e-48a1-9ab8-601a0df3a915&handle=1981185920/36385&provider=iiif-image

[101] Butzmann, *Blankenburger Handschriften*, 183–84.

[102] The features linking this witness to γ‴ include: l. 9, "sine pallacione" / "ahne Vertunkelheit"; l. 31, "comprime…per coreum" / "druck durch ein semisch leder"; the replacement of ll. 35-41 with "ut scis"/ "Wie du den woll weiß"; and l. 82, "volubilis" / "beweglich."

contain Operation 9. Some may be the only known witnesses of yet unidentified subfamilies. None has been included in the stemma.

Fn^1 Florence, Biblioteca Nazionale Centrale di Firenze, Magliabechi Cl. XVI, cod. 110, fols. 14r–17v (1533)[103]

This witness is part of a collection of alchemical texts written on paper by a very shaky Italian cursive hand; each page is headed *yhs*. **Fn^1** is titled *Rupescissa secundum Arnoldum de villa nova*. It omits the incipit and prologue, beginning instead at l. 28, then continuing with eight operations. It contains all the readings characteristic of family γ (except where the relevant section is omitted), bears two small and imperfect drawings of the furnace plate within the lines of text, and ends with a reflection on the work:

> Here ends the treatise of the philosophers' stone composed of mercury alone for both gold and silver, brief and open among the other books of alchemy. And likewise here you have treated how the water of mercury is made and how it is sublimed with aquafort, which water of mercury is hidden by all the doctors of this art, and without which nothing is done. For it dissolves and reduces all bodies into water, that is, into their first matter. And without this the complete philosophers' stone cannot be made. Wherefore take note of it and hide it as the greatest treasure, and always praise God for the sake of so great a treasure having been opened to you.[104]

Mp^2 Montpellier, Bibliothèque Universitaire Historique de Médecine, MS H 479, fols. 121v–26v (1558)[105]

This manuscript contains more than two dozen alchemical tracts and was owned in 1558 by Giovanni Battista Zapata (ca. 1520–1586), a Spanish physician and surgeon working in Italy.[106] It is written on paper in a compact *gothica cursiva* displaying some features of *bastarda*. **Mp^2** rearranges the order, beginning with the "Credas" as an introduction and with considerable alteration. The prologue and eight operations follow; the incipit is omitted. **Mp^2** extends Operation 6 by repeating lines from Operation 5 almost verbatim. The place ordinarily occupied by the "Credas" simply quotes Psalm 50:8.[107] There are numerous rewordings throughout, as well as unique and lengthy insertions dealing with practical details.

103 Tozzetti, *Catalogo Magliabechiani*, 5:250r–51r.

104 Fol. 17v: "Explicit tractatus de lapide philosophorum composito ex solo mercurio tam ad solem quam ad lunam brevis et apertus inter alios libros alkimiae. Et similiter[?] in isto habes tractatum quomodo fit aqua mercurii et quomodo sublimetur cum aqua forti quae aqua mercurii est occultata ab omnibus doctoribus hujus artis quam sine ipsa factum est nihil. Nam ipsa solvit et in aquam reducit, id est, ad primam materiam omnia corpora. Et sine ipsa non potest fieri lapis philosophorum perfectus. Quare nota et occulta pro maximo thesauro et semper lauda deum de tanto thesauro tibi aperto."

105 Ada Alessandrini, *Cimeli Lincei a Montpellier* (Rome: Accademia Nazionale dei Lincei, 1978), 283; Corbett, *Catalogue*, 2:99–109. https://ged.scdi-montpellier.fr/florabium45/jsp/nodoc.jsp?NODOC=2019_DOC_MONT_MBUM_6

106 Fol. vi: "Variorum, Tum Prisciorum, Tum Recentium Philosophorum Opuscula. Jo. Bap. Zapata...1558." Zapata is author of *Li maravigliosi secreti di medicine e chirurgia* (Rome, 1577).

107 "David dicente Etenim veritatem dilexisti incerta et occulta sapientie tue manifestasti mihi."

Mp^2 draws upon at least three sources. First, it shows many of the variants of family γ while correcting some standard errors. Some readings correspond closely with **W**^{1}; in fact, unique scribal peculiarities of **W**^{1} are reproduced in **Mp**^{2}. For example, at ll. 129–30 **Mp**^{2} reads *A Gebero cognita ab Avicenna non ignorata* ("known to Geber, not ignored by Avicenna"), reversing the negation found in all other witnesses. Strikingly, **W**^{1} alone transposes two words to read *a gebere cognita Numquam ab avicenna ignorata* where *Numquam* is both capitalized and begins a line, thus linking it incorrectly with Avicenna rather than Geber. Similarly, at *primo* in l. 267, the copyist of **W**^{1} could not read his exemplar and wrote the meaningless "a ut ti<?>"; correspondingly, **Mp**^{2} is the only witness to omit the word *primo*. It thus seems likely that the copyist of **Mp**^{2} referred either directly to **W**^{1} or to one of its close relations. Both **Mp**^{2} and **W**^{1} were produced in Italy. Finally, a few insertions and alternate wordings are borrowed from *De famulatu*.

U Philadelphia, University of Pennsylvania, MS Codex 69, fols. 75v–80r (3rd q. 15th c.)[108]

The volume containing **U** consists of nineteen alchemical texts. It is written in a regular *gothica cursiva* on paper, almost all by a single German hand. Uniquely for a member of family γ, **U** attributes the *Liber lucis* to pseudo-Lull: *Liber lucis Raimundi de Insula Majoritarum*. Accordingly, the text lacks the incipit that signals Rupescissa's authorship, as well as the prologue. A later interlineation—perhaps by the same hand—states "according to others, by Brother John of Rupescissa of the Friars Minor; the prologue is placed on the sheet."[109] This "sheet," a paper of a different size and quality from the rest of the manuscript, is inserted between folios 75 and 76, and contains both the incipit and prologue. The text of **U** is highly edited, including only Operations 1–7, and with about 50% of the usual text missing through paraphrase or omission. A few readings match those in **V**, **Bo**, and **Le**^{1}; therefore, it might be part of subfamily γ''.

6.2.5 The Collections of Frater Joannes Baptista: W^3, W^4, Y^3, Y^4, and Y^5

Five witnesses exist in a multivolume collection of alchemical materials produced by one friar Joannes Baptista in the 1570s–80s.[110] My search for manuscripts of the *Liber lucis* reunited long-separated volumes of this collection. Three volumes are now at the Wellcome Library (MSS 383, 384, and 385, acquired at three auctions, 1934–1939) and three at Yale University's Beinecke Library (Mellon MSS

[108] Norman P. Zacour and Rudolf Hirsch, *Catalogue of Manuscripts in the Libraries of the University of Pennsylvania to 1800* (Philadelphia: University of Pennsylvania Press, 1965), 150–52 (MS Lea 19); Seymour de Ricci, *Census of Medieval and Renaissance Manuscripts in the United States and Canada*, 3 vols. (New York: H. W. Wilson, 1935–40), 2:2106; and W. J. Wilson, "Catalogue of Latin and Vernacular Alchemical Manuscripts in the United States and Canada," *Osiris* 6 (1939): 617–33. https://colenda.library.upenn.edu/catalog/81431-p3h36z

[109] Fol. 75v: "secundum alios per fratrem Johannem de rupescissa ordinis minorum prologus ponitus in cedula."

[110] The date 1576 appears on fol. 7v of Wellcome MS 385.

34, 35, and 36, previously part of the Duveen collection).[111] As indicated by the original enumeration on the flyleaves or pastedowns, the correct sequence of the volumes is as follows.

Volume 1	Mellon MS 35	(contains **Y**^{4})
Volume 2	Wellcome MS 383	
Volume 3	Wellcome MS 385	(contains **W**^{4})
Volume 4	Mellon MS 36	(contains **Y**^{5})
Volume 5	Wellcome MS 384	(contains **W**^{3})
Volume 6	Mellon MS 34	(contains **Y**^{3})

The collection's texts are mostly Latin, with idiosyncratic orthography and frequent vernacular insertions, but with some Italian and Spanish. There is substantial duplication throughout. Based on the languages present, the writer's peculiar orthography, and a table of prices in Genovese currency, Genoa may be the place of composition.[112] The collection deserves closer investigation: Joannes Baptista both collected and wrote alchemical texts. He authored at least seven compositions in these volumes, including an extended text critiquing Giovanni Bracesco's interpretation of Geber.[113]The witnesses of *Liber lucis* are both corrupt and considerably abbreviated, but all contain authentic text absent from the 1561 Basel edition, demonstrating that they were not copied from print. They fall into two groups. **W**^{3}, **W**^{4}, and **Y**^{5} descend through German family γ: they preserve a line from the incipit unique to this family, the characteristic opening "natura seu materia," and various more minor readings, although due to their corrupt character, they do not display all the family's characteristics. They contain shortened forms of the prologue, the eight operations, and Operation 9 (**W**^{3} contains only the first lines of 8 and 9). The presence of Operation 9 indicates their linkage with subfamily γ′. **Y**^{5} is the most complete of the three.[114] **Y**^{3} and **Y**^{4} form a second group; their texts are yet more highly edited.[115] They omit the prologue, ll. 18–27, the "demonstratio" (ll. 65–78), and Operation 8, and have a short version of the "Credas." **Y**^{3} has a shortened Operation 9, which **Y**^{4} does not. Both are followed by highly edited versions of pseudo-Arnald's *Liber de secretis naturae*. These witnesses are so abbreviated that their family remains uncertain, although some readings strongly suggest

[111] Moorat, *Catalogue*, 1:244–55, and Witten and Pachella, *Alchemy and the Occult*, 3:220–48.

[112] Witten and Pachella, *Alchemy and the Occult*, 3:234 and 248.

[113] Wellcome MS 384, fols. 146v–215r: "Apologia Joannis Baptistae contra Joannem Brasesco pro erroribus per ipsum fatis in espositione Geberis." The reference is to Bracesco's *Espositione di Geber* published in Venice in 1544, 1551, and 1562, and in Latin translation in 1548.

[114] The last page of the witness however is missing. Earlier in the volume, at fols. 24r–25r, the friar inserted parts of Operations 2 through 4 into the middle of an unrelated text, written in a different hand, from which its own Chapters 2 and 3 were missing.

[115] **Y**^{3}: "Libelum ad confisienda lapidem Jo. Rupesisa"; **Y**^{4}: "Ad confisiendum lapidem abreviatum clarum."

they also descend ultimately through γ, but distantly and/or with extensive contamination and corruption.[116]

6.2.6 The Glosses of Subfamily γ″

Gloss 1

Z, fol. 116r; following Operation 2:

Secundum quod videris ixes debes sublimare vel donec nullas ponat feces sine calcinacione solvere et post solucionem imponere sal commune et distilla ut scis.

(According to what you have seen, you ought to sublime nine times or until it deposits no feces, to dissolve without calcination, and after dissolution to put in common salt and distill as you know how to do.)

Le², fol. 4r; following Operation 2:

Nota, si videris novies debes sublimare, vel donec nullas ponat feces, sine calcinacione solve, et post solucionem impone sal commune, et distilla ut scis. Sed author quinte essencie dicit, hoc fac tociens, donec non plus feces post se relinquat.

(Note, if you see, you ought to sublime nine times or until it deposits no feces, dissolve without calcination, and after dissolution put in common salt and distill as you know how to do. But the author of the *Quintessence* says to do this many times until it leaves no feces behind itself.)

Ge, fol. 100r; bottom margin, keyed to an insertion at l. 110 in Operation 2:

secundum quod videris novies debes sublimare, vel donec nullas ponet feces, sine calcinacione solveris, et post solucionem impone sal commune, et distilla ut scis.

(According to what you see, you ought to sublime nine times, or until it will deposit no feces, you should dissolve without calcination, and after dissolution put in common salt and distill as you know how to do.)

$Wb^2{}_d$, fol. 126v; inserted at l. 110:

Zum anderen dissolvir den spiritum [mercury] solum bis er keine feces mehr davon lest, und nach der solution due darzu gemeines Salz und distilirs wie du weist

(Then dissolve the spirit of mercury alone until it no longer leaves behind feces, and after the solution add thereto common salt and distill as you know how to do.)

Gloss 2

Z, fol. 116r; following Operation 3:

Quomodo lapis solvatur balneum in balneo marie et debet poni in vase vitreo vel vitreato lato materiam nudam tangat.

(How the stone is dissolved, a bath in a bain-marie, and it ought to be placed in a wide glass or glazed vessel; let it touch the naked material.)

[116] For example, both squeeze the mercury through hare's leather, characteristic of γ.

Ge, fol. 100v; bottom margin, keyed to the end of Operation 3:
Quomodo lapis solvatur in Balneo in aperto philosophica [*sic*]. Nota autor quintae essentiae inquit quod materia est poni in isto loco in vas vitreo aut vitreato lato vel amplo, et non cooperiess eam, sed aer calidus materiam nudam tangat.
(How the stone is dissolved in a bath in the open. Note that the author of the *Quintessence* says that the material is to be put in this place in a large, wide, glass or glazed vessel, and you will not cover it but let the warm air touch the naked material.)

$Wb^2{}_d$, fols. 126v; following Operation 3:
Und hore was der Author sagt, das die Materia der quinta essentia in ein grosses weit glessern geschirn solle gedan werden und dz selbige nicht bedecken womit der blosse warme lüft die Materia beruren könne.
(And hear what the author says, that the material of the quintessence ought to be put in a large, wide, glazed vessel, and the same ought not to be covered, so that the naked warm air can touch the material.)

Gloss 3
Z, fol. 116r; immediately follows gloss 2 at the end of Operation 3:
Nota recipe quintam essenciam mercurii ii lotes 4 lotes lune tenuissime foliate et pone stratum super stratum in fiola lata et ampla sublima et ascendet luna in altitudine pauca. Iterum recipe illam lunam ascensa et sentissima dicto spiritu et sublima ut prius et iterum ascendet alcius hoc fac tociens donec in fundo maneat hoc separa et invenies solem.
(Note: Take 2 Loth of the quintessence of mercury and 4 Loth of silver in very thin leaves, and place them layer upon layer in a wide and ample phiol; sublime, and the silver will ascend to a short height. Again, take that silver, ascended and very rough by means of the said spirit, and sublime as before, and again it will ascend higher. Do this many times until it remains at the bottom. Separate this, and you will find gold.)

Ge, fol. 100v; following gloss no. 2, but the copyist broke off and struck through the words he had written:

Recipe 5am essentiam [mercury]

Bg_d, fol. 16r; following Operation 6:
Nota in illo opere invenies Solem.Merkch. nimb das fünfft wesen des Mercury / 20 / loth / und 4 loth cac[117] / in gar dünne blech / und leg mit den zwain ye ain legk auf die andern / in ain glas / als ain urinal höcher und weyt unden / und der cac steigt ain wenig / denß nimb aber ab / und legs aber miteinander alsvor / und sublimiers / so steigt daz cac / aber höcher / das anderwaid alslang / untz ez sich nicht mer sublimier und bleib am grünt / das leg in ain glas / und

[117] Throughout this recipe, the letters *cac* are anomalously used for *silver*. This rendering is probably a misreading of an ill-formed crescent moon symbol for silver as two c's.

verlutier das gantz / und Reverberiers / biz ez am grünt des glass fliezz / das schaid dann so vindestu gold.

(Note, in this work, you will find gold. Note: take the quintessence of mercury, 20 Loth, and 4 Loth of silver in very thin leaves and put them together one upon another in a glass like a urinal high and broad, and the silver rises a little. Then take it out and put it together as before, and sublime, thus the silver rises, but higher. Do this again until it sublimes no more and remains at the bottom. Put it in a glass and lute it all over and put it into a reverberatory fire until it flows at the bottom of the glass. Separate this, and you will find gold.)

$Wb^2{}_d$, fols. 126v–127r; following gloss 2 at the end of Operation 3:
Nim diese quinta Essentia solutum 20 lot, der dünn [luna] pletten 40 lot stratificirs in eine weiten und praiten Viole so wirdt die Luna ein wenig in die hohe steigen. Nim dise aufgestegene [luna] stratificiers mit vorgeflugten Spiritus und sublimirs zuvor, so wirdt es höher steigen. solches thue so offt biß das es im grundt liegen thuet, und fleist. dises soltu scheiden so wirstu [gold] finden."

(Take this dissolved quintessence, 20 Loth, and 40 Loth of silver in thin leaves and stratify it in a large and broad phiole, thus will the silver rise a little in the height. Take this arisen silver, stratify it with the previously sublimed spirit and sublime as before, thus the silver will rise higher. Do this so often until it remains at the bottom and melts. You should separate this, and you will find gold.)

Gloss 4

Z, fol. 116r; immediately follows gloss 3 at the end of Operation 3:
Autor noviter ponit calcinacionem mercurii fieri cum lento igne et ideo necessarius est hec calcinacio mercurii sublimati in isto opere preuti. Et ignis in calcinacione seu digescione ista debet fieri duum carbonum per 8 dies et in calida stuba et sicca. Post extrahe et tere tuam materiam [*illeg.*] donec sit sine tactu et hoc dicit autor prius f g [*sic*] que terra ponitur ad dissolvendum.

(The author finally posits that the calcination of mercury is to be done with a gentle fire, and therefore, this calcination of the sublimed mercury is very necessary first of all in this work. And the fire for this calcination or digestion ought to be made of two coals for eight days, and in the hot and dry oven. Thereafter, take out and grind your material until it is impalpable, and the author says this before ... which earth is put to be dissolved.)

Bg_d, fol. 16r–17r; immediately follows gloss 3 at the end of Operation 6:
Und der auctor nemlich schaft / die calcinacio oder kalchung dez gesublimierten Mercuri mit lindem fewer ze arbaitten / und das setzt Geber offenwarlichen in der Capitel der ausweysung der erhöchung / do er spricht der dritt gradus ist / das mon im zaig ain krankch linds fewer / und daz von den feceß aufsteigt / und das da gelewttert ist / das kawm ettwas von demß aufsteigt / und das da aufsteigt das wirt / das allersubtilist guet / das dem Werckch nichtsnicht bezimbt / wenn das ist ain ding dez mittel / damit mon würcht das verprennlich in der erhöchung / und darumb so ist nottürft des Mercury / kalchung id est calcinacio / in dem gegenwurtigen werkch / und das fewr in der kalchung / oder pratung / schol sein / als von klainen oder wenig kolen / und in der warmen stuben / Athanor / trükchen

darnach nimbs aus / und reibs auf ain Merbelstain untz ez wirt / das mon sein am griff nicht bephinden mag / und also spricht der auctor das leg dann zu dissolvieren / und ob du ain trophen derß Jünkchfrawmilich auf ain gluend blech würffest / so weisset ez innen und aussen / als du gesehen hast / als vorgeschrieben ist / etc.
(And the author specifically orders that the calcination of the sublimed mercury be done with a gentle fire. And Geber posits this openly in his chapter on the explanation of sublimation where he says 'the third degree is that a weak, gentle fire is given to it and what rises up from the feces and is purified, so that scarcely anything of it arises, and what does then arise will be the most subtle thing, which is not at all suitable for the work, for it is a thing of the means whereby the flammable is produced in the sublimation.'[118] And thus the calcination of the mercury is absolutely necessary in the present work, and the fire for the calcination, or roasting, ought to be from small or few coals, and dried in the warm oven athanor; then take it out and grind it fine upon a marble stone and it will become such that it cannot be perceived by touch and so the author says to put it to dissolve and if you throw a drop of this virgin's milk upon a glowing metal plate thus it whitens it inside and out as you have seen it written above, etc.)

Wb$^{2}_{d}$, fol. 128r–v; following Operation 4:
NB. Außerhalb daß texs ist zu mercken, der Author sizt, die Calcination [mercury sublima]ti solle geschehen in ein linden [fire] und dises thuet er nach der Meinung Gebri offenbahr, in den Capiten von underscheidt der Calcination:der dritte gradt ist, das ime dz allerlindestens feyer gegeben werde, was zu sublimiren ist von seinen fecebus, und dz welchs gereiniget ist, welches kaum von ime etwas aufsteige und was von ime aufsteiget, ist daz allersubtilleste ding welches inn dises werk nichts solle, das dises ist ein solches ding vermittels welches in einer verbrenung vernichtet würdt in den [sulphur], derenthalben ist dise Calcination [mercury sublima]ti Vernötten in dises gegenwertigen werk und dz [fire] inn der Calcination oder digestion sollen sein zwen kolben uf 8 tage in einer warmen trüken stuben, hernach nimbs auß und reibe dise Materia uf ein marbelstein, bis sie glein unempfindlich ist, und dises sagt der gegenwerdiger Author. Zu merken ist, das halbes die 4ten Opperation der quinta [essentia] sagt der Author, das die Materia, dz Wasser bey 2 fingern nicht berühren solle, wie er derselbsten sagt, dan es ist wol Calcinirt und geriben, den <*illeg.*> dir ein <*illeg.*> das beides M[ariae] B[alneum] ins doppelten geschirn, das du die wermen regeren kanst.
(Note well: beyond the text it is to be noted, the author sets down that the calcination of the sublimed mercury is to be done in a gentle fire, and he does this clearly following the opinion of Geber in the chapter on the differences of calcination: "the third degree is that the gentlest fire is given to that which is to be sublimed from its feces and which is purified, which scarcely anything arises from it, and what does arise from it is the most subtle thing which ought not to be in this work, that this is the thing by the means of which the flammable is destroyed in sulphur." Wherefore, this calcination of the sublimed mercury is absolutely necessary in the present work, and the fire in the calcination or digestion ought

[118] Geber, *Summa perfectionis*, bk. 1, ch. 40. The German translations here and in **Wb**$^{2}_{d}$ are cumbersome and somewhat faulty ; see Newman, *Summa perfectionis*, 363.

to be of two coals for eight days in a warm and dry oven. Thereafter, take it out and grind this material upon a marble stone until it is impalpably fine, and this the present author says. It is to be observed , the fourth operation of the quintessence, the author says that the material ought not touch the water by two fingers, as he himself says, for it is well calcined and ground <*illeg.*> a bain-marie in a double vessel, so that you can regulate the heat.)

Gloss 5

Z, fol. 117r; following Operation 4:

Notandum quod balneum marie sic fit id est olla super furnum et super illam ollam unam aliam ollam que in fundo habeat multa foramina et inferior olla habeat collum gracile ubi infundatur aqua et conjunge ollam preoptime luto sapientie et luta ollam superiorem luto sapientie undique et impone cum omni medicina et perluta bene ne vapor exeat et infunde in ollam inferiorem acetum vel aquam et obtura ei collum et fac ignem levem aut mitte in furnum et balnea medicinam ipsam sicud vis donec solvatur. Ego autem illud balneum feci et inveni quid quesivi.

(It is to be noted that the bain-marie is made thus, that is, a pot upon the furnace and above that pot another pot which should have in its bottom many holes and let the lower pot have a thin neck where the water may be poured in, and join the pot extremely well with *lutum sapientiae* and lute the upper pot with *lutum sapientiae* everywhere, and fill it with every medicine and lute very well so that no vapor escapes and pour vinegar or water into the lower pot and stopper its neck, and make a gentle fire or put it in the furnace and bathe the medicine itself as you wish until it is dissolved. I also made that bath and found what I sought.)

6.3 Germanic Family γ^2

Ko$^{\mathbf{I}}$, **J**, and **W**$^{\mathbf{6}}$; and an ancillary source for **X**$^{\mathbf{I}}$

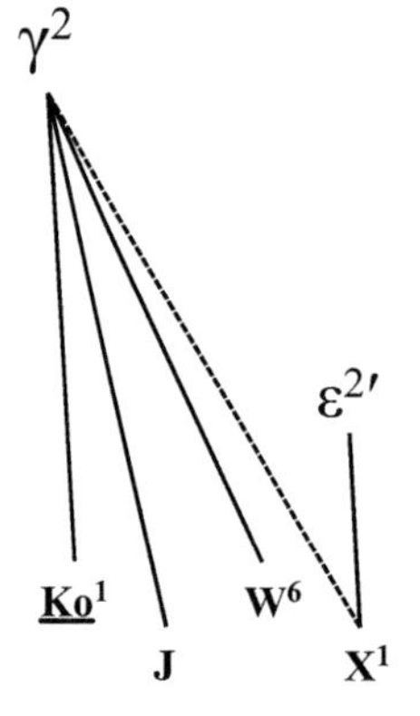

This small family contains three witnesses of North German and Netherlandish origin whose many unique textual variants indicate their descent from a common hyparchetype designated as γ^2. **Ko**$^{\mathbf{I}}$ has the most complete text; **J** both edits John's text and adds supplementary material; **W**$^{\mathbf{6}}$ is closer to **Ko**$^{\mathbf{I}}$, but has several substantial lacunae. The copyist of **X**$^{\mathbf{I}}$, working at Oxford in the 1450s, drew upon an unknown member of this family for the prologue and Operation 8, both of which were absent from his main exemplar descending from $\varepsilon^{2\prime}$ (see below). The characteristic readings of family γ^2 include:

- Several operations begin with descriptive headers. For Operation 3, **Ko**$^{\mathbf{I}}$ and **W**$^{\mathbf{6}}$: *Digestio lapidis et ejus calcinacio*; **J**: *Cuius digestio et calcinacio*; For Operation 5: *Fixacio lapidis et lactis virginis pulverisacio.*

- l. 5: *in successu temporis* rather than *in brevi*.
- ll. 13–14: *probacione ac recitacione ac impugnacione probare* instead of *probative aut recitative aut impugnative procedere*.
- l. 170: *amphoram intus rotundam* in place of *amphoram vitri rotundam*.
- l. 210: *acuto* rather than *aucto*.
- ll. 222–24: reworked as *deprecati fuerunt deum ne hoc archanum veniret ad manus virorum indignorum et quod maledictiones incurrent horrendas quicumque revelarent hoc alicui magisterium hujus artis et veritatem*.[119]
- l. 277: *veritate* instead of *utilitate*.
- l. 283: *juxta terram ubi immittantur carbones* (borrowed from *De famulatu*) added after *competenti*.

Ko[I] Copenhagen, Kongelige Bibliotek, GKS 1712 kvart, fols. 111r–18v (2nd q. 15th c.)[120]

Written in a carefully formed *gothica rotunda cursiva* bookhand on paper with wide margins, **Ko**[I] is part of a collection of alchemical treatises and recipes for transmuting agents, medicines, and pigments; it includes *De famulatu* (fols. 5r–38v). The texts are mostly Latin, but with some recipes in German. Several recipes are written in the first person, and several mention Cologne, possibly indicating an approximate locale for the manuscript's production.[121] **Ko**[I] contains the prologue and all eight operations. It is titled *Liber lucis magni magisterii*; the explicit repeats this title: "Here ends the *Book of Light of the Great Magistery* of Brother John of Rupescissa, Franciscan; thanks be to God."[122] Space was left for an illustration of the *lamina* in Operation 8, but it was not executed.

J Jena, Thüringer Universitäts- und Landesbibliothek, Ms. El. q. 21, fols. 151r–54v (mid-15th c.)[123]

J is written in a rough and highly abbreviated *gothica cursiva* on paper as part of a collection of alchemical materials in Latin and German. Headed simply *Incipit rupicissa*, **J** contains the prologue and Operations 1–7. There is considerable intervention in the text; sections are omitted or summarized and glosses added. The copyist was clearly interested in comparing the practical details of the *Liber lucis* with parallel sections of *De famulatu*. He imported material from the latter work, made references to it, and sometimes used it to replace text. He replaced,

[119] The reading given here is from **Ko**[I]; in **J**, this line is slightly condensed; in **W**[6], a lacuna begins following *indignorum*.

[120] Jørgensen, *Catalogus*, 452–53.

[121] E.g., fol. 105v: "Inveni in quodam scripto theutonico…"; fol. 108r: "Nota quod Colonie erat quidam fantasticus semifatuus quasi et dixit quod vellet facere unum bitumen ex homine cum quoquidem bitumine velle infixare animam vitalem humano corpori…."

[122] Fol. 118v: "Explicit liber lucis magni magisterii fratris Johannis de Rupescissa cordigeri Deo gracias."

[123] Bernhard Tönnies, *Die Handschriften der Thüringer Universitäts- und Landesbibliothek Jena*, vol. 1: *Die mittelalterlichen lateinischen Handschriften der Electoralis-Gruppe* (Wiesbaden: Harrassowitz, 2002), 275–88; https://collections.thulb.uni-jena.de/receive/HisBest_cbu_00015962

for example, the "demonstratio" (ll. 65–77) with "And this can be demonstrated visibly by separating the spirit of vitriol from the mercury in distilled vinegar, but these things are in the *Book of the Quintessence* in the chapter on minerals."[124] Likewise, in Operation 2, where only one-quarter of the usual text is present, the copyist summarized the rest. In Operation 3, **J** refers to "the furnace described in the aforementioned book, in the chapter on the reduction of sublimed spirits into water," and in Operation 4, it instructs the reader to proceed "as it is taught in the aforementioned place."[125] Operation 8 is summarized in a single sentence with another reference to *De famulatu*.[126] **J** has several marginal glosses providing practical information or explaining terms, such as defining *ovum philosophorum* in German. A marginal gloss to Operation 1 adds an interesting warning to practitioners: "And note that when you mix the quicksilver or mercury with the vitriol and salt, hold at that time some cold water in your mouth for the sake of your teeth because the fume of quicksilver very greatly damages teeth since it makes them become loose. But the cold water then held in your mouth greatly restrains the fury of the mercury fumes."[127]

W^{6} London, Wellcome Library, MS 517, fols. 33r–36v (2nd q. 15th c.)[128]

This witness is inserted between the two books of an abridged version of *De famulatu* and labeled as its second book (cf. **L**^{2}). It is written in a neat *gothica cursiva* on coarse paper in a collection of mostly alchemical and medical treatises and recipes in Latin and Dutch. Moorat dates the collection to the late fifteenth century on the basis of a list of dated events in the Netherlands ending in the late fifteenth century; however, this list is a later insertion by a different hand and the first date listed is actually 1580, indicating that this insertion dates from the late sixteenth century.[129] The paper, on the other hand, bears the watermark Briquet 12976, which is found in a Dutch manuscript dated 1430; this date is consistent with the paleography, hence my redating to the second quarter of the fifteenth century. The text contains several extensive lacunae in Operations 3–5 such that these sections are run together as one, and consequently Operation 6 becomes *quarta operacio*, which also omits most of the "Credas." As in **Ko**^{I}, space was left for an illustration of the *lamina* in Operation 8, but it was not executed.

[124] Fol. 151v: "Eciam potest demonstrari ad oculum separando spiritum vitrioli a mercurio in acceto distillato sed hec in libro quinte essencie capitulo de myneris." See Appendix 1.

[125] Fol. 152r: "furnus in libro prescripto capitulo de reduccione spiritum sublimatorum in aquam descriptus," and "ut in loco preallegato docetur." See Appendix 1.

[126] Fol. 154v: "Octava pars hujus operis continet modum athanar id est furnum phylosophici qui tum post in libro 5° essencie est descriptus cum lamina sua lata supra quam potes ponere fiolas multas vel modicas secundum quod diversa volueris temptare et in quamlibet fiolam ovum phylozophicum ponere et juncture clause sint et lutate in portis et coopertorii contactibus ne calor exalet et adhibe cautelam in ignis temperie." The reference is again to *De famulatu*, see Appendix 1.

[127] Fol. 151r: "Et nota quando commisces argentum vivum sive mercurium cum vitriolo et sale tunc tene aquam frigidam in ore propter dentes quia fumus argenti vivi permaxime nocet dentibus quod facit eos mobiles sed aqua frigida tunc in ore contenta maxime compescit ardorem<?> fumi mercurii."

[128] Moorat, *Catalogue*, 1:359–62.

[129] This date refers to a "great earthquake in Holland" [*grote ardtbeving in Hollant*], which is clearly a reference to the Dover Straits Earthquake of 6 April 1580.

6.4 Germanic Family γ^3

Ka^{2}, **Oth**^{I}, and **Wb**; in German translation, **Ka**$^{5}_{d}$ and **L**$^{8}_{d}$

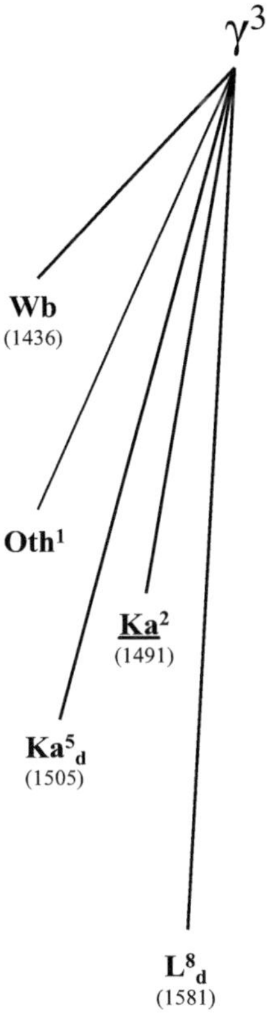

Three fifteenth-century Latin witnesses and two sixteenth-century German translations form a family of related texts descending from a common hyparchetype denoted γ^3. Two of the Latin texts and both German translations are of North German origin; **Oth**^{I}, a short extract, is northern Italian. The textual characteristics of this family include:

- l. 34: *et parum move et quando fuerit bene (com)mixta* added after *in ea*.
- l. 117: *lutata et fortissima taliter quod mercurius exire non possit* added after *vitrea*.
- ll. 126–27: *da ignem competentem ut ascendat et distilla rorem madii*.
- l. 128: *parum acetosa* rather than *acetosa*.
- ll. 229–30: *mentem/mente* in place of *librum* (**Wb/Ka**^{2}: *non habet mentem veri luminis*; **Oth**^{I}: *caret mente veri luminis*).
- l. 251 concludes *ut finaliter ab hac calamitate liberetur* (also in **Br**).
- l. 252–53: *iteratis virtutibus elixir hoc est de medicinis prefate medicine perfecte virtutis* between *praxis.... autem medicine*.

Wb Wolfenbüttel, Herzog August Bibliothek, Cod. Guelf. 444 Helmstedt, fols. 241r–42v (1436)[130]

Written in a highly abbreviated *gothica cursiva* bookhand on paper, **Wb** follows a copy of *De famulatu* (fols. 227v–41r) in the same hand. The volume contains medical, astrological, and alchemical treatises, mostly Latin, but some German. **Wb** is headed "Incipit opus lapidis per Johannem de rupcissa satis expertum." It contains the prologue and seven operations, omitting the furnace chapter.

[130] Otto von Heinemann, *Die Handschriften det Herzoglichen Bibliothek zu Wolfenbüttel: Die Helmstedter Handschriften*, 3 vols. (Wolfenbüttel, 1884–88), 1:346–50. The date I have given for this witness comes from the colophon to pseudo-Arnald's *Rosarius* that immediately precedes the Rupescissa material and is written in the same hand on the same paper: "Explicit Rosarius Arnaldi de villa nova de compositione lapidis preciosi anno domini 1436 ipso die elysabeth" (fol. 227r).

Ka^{2} Kassel, Universitätsbibliothek Kassel-Landesbibliothek und Murhardsche Bibliothek der Stadt Kassel, 4° Ms. chem. 10, fols. 25r–30r (1491)[131]

Ka^{2} is part of a short collection of alchemical treatises and recipes mostly in Latin but with some in German, written in northern Germany, apparently in a monastic context in Schleswig.[132] It is written on paper in a rapidly formed and highly abbreviated secretary hand and headed simply *Incipit rupssissa* [*sic*]. It contains the prologue and all eight operations and ends with a dated colophon using an idiosyncratic form of John's name: "Here ends the *Liber lucis magni magisterii*, written by Brother John of Petrescissa of the Order of Friars Minor, on Good Friday in AD 1491."[133] As the most complete and reliable witness of this family, it is included in the critical edition as the representative of γ^{3}.

Oth^{1} Philadelphia, Science History Institute, Othmer MS 2, fols. 1r–2r (extr.; 3rd q. 15th c.)[134]

This witness contains only the "Credas" (ll. 220–51) and the last lines of the text, namely the "mira ingenia" and doxology (ll. 324–30). **Oth^{1}** functions as a preface to an alchemical miscellany, being used not for practical information but as a warning about unrestricted communication of alchemical knowledge. The collection contains works by Franciscan and other clerical authors, mostly in Latin but with some Italian. It is written in a fine *humanistica* bookhand on parchment with rubrics. **Oth^{1}** is headed "Chapter eight<?> on the entrusting of the revelation of this art and the arcanum of the magistery."[135] The explicit assigns the text to the correct author but the wrong text: "Here ends the book of the quintessence of master John of Rupescissa, a great and prophetic man of the Order of Friars Minor."[136]

Ka$^{5}_{d}$ Kassel, Universitätsbibliothek Kassel-Landesbibliothek und Murhardsche Bibliothek der Stadt Kassel, 4° Ms. chem. 73, fols. 27r–36v (1505)[137]

Ka$^{5}_{d}$ contains the prologue and seven operations (omitting Operation 8, like **Wb**), and displays all the readings unique to this family.[138] The copyist identifies

[131] Broszinski, *Handschriften*, 40–42. https://orka.bibliothek.uni-kassel.de/viewer/image/1530004140628/51/

[132] Hartmut Broszinski, "Spuren klösterliche Alchemie in Kasseler Handschriften des 15. und 16. Jahrhunderts," 823–47 in *Nova de veteribus: Mittel- und neulateinische Studien für Paul Gerhard Schmidt*, ed. Andreas Bihrer, Elisabeth Stein, and Paul Gerhard Schmitt (Munich: Saur, 2004), 840–47.

[133] Fol. 30r: "Explicit liber lucis magni magisterii editus a fratre johanne de petrescisa ordinis minorum anno domini m cccc xci in bona sexta feria." The same spelling of his name is found in BNF, MS Lat. 7371, fols. 115r–19v.

[134] Wilson, "Catalogue," 30–39.

[135] Fol. 1r: "Capitulum Viii^{m}<?> de recommendatione revelationis hujus artis et archani magisterii." The numeral has been altered, making the reading uncertain.

[136] Fol. 2r: "Explicit liber quintae essentiae lapidis magistri Johanis de Rupecissa viri magni et prophetici de ordine minorum."

[137] Broszinski, *Handschriften*, 329–35. https://orka.bibliothek.uni-kassel.de/viewer/image/1550147333249/

[138] E.g., l. 34: "unnd wan sie zu hauf gemenget sein"; l. 117: "wol vercleibt das nicht daraus moge"; ll. 126-27: "unnd gib behörlich feuer so steiget das uf und distillier das dawe"; and ll. 229-30 "nicht hat den syn des warn lichtes."

himself in the colophon (fol. 36v): "written by the hand of the priest Valentine [of Ottrau], in the year of our Lord Jesus Christ 1505, on the Saturday after the feast of Mark the Evangelist (i.e., 26 April)."[139] Valentine showed special interest in John of Rupescissa: besides **Ka^5_d**, his collection also contains an excerpt from *De famulatu* (fols. 9v–10v), a reference to John about a composite *aqua vitae* (fols. 12v–13v), and a work attributed to John entitled *De lapide minerali* which incorporates parts of the *Liber lucis* verbatim (fols. 115r–18v; see Appendix 2). **Ka^5_d** is titled simply *Rupicissa*, opens "Here begins the book of John of Rupescissa about the secret natural art," and concludes "Here ends the book of John of Rupecissa of the Order of [Friars] Minor about the true stone of the wise."[140] Between the explicit and the colophon, a "parabola" and its "exposicio" with no evident relationship to the *Liber lucis* are inserted.

L^8_d London, British Library, Sloane MS 2701, fols. 142v–51r (1581)

This witness provides a Niederdeutsch translation independent of the German **Ka^5_d**. It contains the prologue and seven operations (Operation 8 is omitted, as in **Wb** and **Ka^5_d**), and despite omissions and rewordings, it nevertheless displays the readings characteristic of this family.[141] It also provides unique introductory material giving new, if dubious, titles to the author: "John of Rupescissa, professor of the Sacred Page, licenciate of canon law, and also bachelor of medicine."[142] The opening section (fols. 142v–43v), although severely damaged, tells how John was visited in prison by a "spiritus intelligentiae" from whom he learned the secrets of nature. The account is a considerably elaborated and theologically misunderstood version of John's remark at the end of book 1 of *De famulatu* that the secret of transmutation was "intellectually unsealed" for him during his confinement.[143] L^8_d is part of an acephalous bundle of Latin and Niederdeutsch alchemical treatises and recipes, dated March-May 1581; the colophon identifies the copyist as Heinrich Gronen.[144] This bundle also includes a Niederdeutsch translation of part of *De famulatu* (fols. 189v–93v). It currently resides among the papers of Peter Moritz (1638–1702), a Saxon alchemist and religious dissenter who in the late

[139] Fol. 36v: "Finitur scriptum per manum valentini prespiteri. Anno domini nostri jhesu christi xv^c v sabatho post marci Ewangelisti." His residence in Ottrau is given in another colophon (fol. 98v).

[140] Fol. 27r: "Hyr beginnet das Buch Johannis vonn der steinruszen vonn der heimlichen naturlichen kunst"; fol. 35r: "Hirmit endet sich das buch Johannis de Rupecissa von den myneren orden in den waren steyn der Weissen."

[141] E.g., l. 117: "vormake woll mit luto sapientiae dat dar nicht <*illeg.*> moge"; l. 126-27 "unde distilir dan dow der fuchticheit," and "ein wennich suerlick"; and ll. 229-30: "nicht hedde den sin dess waren lichtes."

[142] Fol. 142v: "Johannes De Rupecissa sacrae pagine professor Canonicum licentiatus nec non medicinae Baclarius."

[143] See *De famulatu*, "Excusacio": "magisterii veritas est michi in tribulacionibus carcerum, volente deo, intellectualiter reserata."

[144] Fol. 196r: "Complevi in hora matutina in die trinitatis Anno 1581 [21 May], et incepi ejusdem anni in pascalis festo [26 March]: dis alles vonn heinrich gronen aussgeschriebenn." The colophon continues "dasgleichenn da im brittenn<?> biss uff Bastian schultzen ausschrei[ben]."

1680s lived for a time in London, where his transmutational promises were briefly financed by Robert Boyle (1627–1691).[145]

6.5 Italian Family ι

O and **Fl**; in German translation **Le**$^{4}{}_{d}$; descendant subfamily ι′: **L**^{5}, **Oth**^{2}, and **Pd**

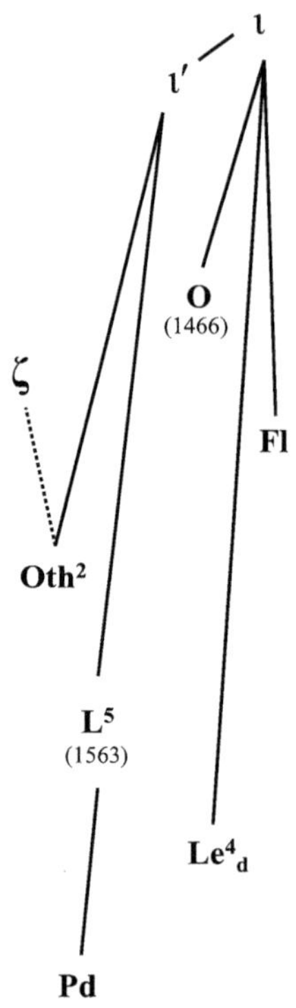

This family contains two distinct clusters of witnesses descending from two closely related hyparchetypes designated ι and ι′. These six witnesses share many characteristic variants which include:

- l. 21: *in cavernis* (chosen as the correct reading, see Section 4.3).
- l. 31: *panno lineo duplicato* (shared by **Y**^{I}/**G** and **W**^{I}).
- l. 33: *per partes* instead of *per poros*.
- ll. 108–9: *dissolutio mercurii in aqua forti vitrioli romani fit in 12 horas. Et in illa dissolutione bibit quantum bibere potest* between *quia* and *de sulphure*.

O and **Fl**, both mid-fifteenth century, and the sixteenth-century German translation **Le**$^{4}{}_{d}$ descend from a common ancestor ι, which was apparently badly damaged beginning at ll. 170–71. **O** ends there in the middle of a page, breaking off abruptly in mid-sentence. **Fl** continues beyond this point, but the rest of Operation 5 is very lacunose. I suggest that the copyist of **O** gave up at the word *vitri* due to illegibility, while the copyist of **Fl** endeavored to extract as much of the damaged text as possible, recovering about half of the rest of Operation 5. **Le**$^{4}{}_{d}$ follows **Fl** closely. All three lack the prologue.

O Ottobeuren, Benediktinerkloster Ottobeuren, MS.O.86, fols. 246v–48v (unfinished, Operations 1–5; ca. 1466)[146]

This witness is part of a quarto collection of alchemical and astrological texts, mostly Latin but with some German. It is written by a German hand on paper

[145] Lawrence M. Principe, *The Aspiring Adept: Robert Boyle and His Alchemical Quest* (Princeton: Princeton University Press, 1998), 134–36, describes the interactions of Boyle and Ambrose Godfrey Hanckwitz with an unnamed alchemist without having been able to identify him; Mike A. Zuber and Leigh T. I. Penman identified him as Moritz in "Robert Boyle's Anonymous 'Crosey-Crucian' Identified: The German Alchemist and Religious Dissenter Peter Moritz," *Notes and Records of the Royal Society of London* 74 (2020): 95–103, and Mike A. Zuber, "Alchemical Promise, the Fraud Narrative, and the History of Science from Below: A German Adept's Encounter with Robert Boyle and Ambrose Godfrey," *Ambix* 68 (2021): 28–48.

[146] Hermann Hauke, *Die mittelalterlichen Handschriften in der Abtei Ottobeuren* (Wiesbaden: Harrassowitz, 1974), 84–86.

in an idiosyncratic *gothica cursiva*. **O** follows excerpts from *De famulatu* (fols. 243v–46r), and is headed simply "Practice of the same philosopher Rupescissa, minorite brother, in the greater art and it is a most notable thing."[147] It lacks the prologue and breaks off abruptly in mid-sentence at the end of l. 170 about one-third of the way down fol. 248v.

Fl Florence, Biblioteca Laurenziana, Gaddi MS 174, fols. 10r–14v (2nd half 15th c.)[148]

Titled *Octo operationes mercurii Magistri Raymundi Lulli*, **Fl** is written in *gothica hybrida cursiva* by an Italian hand on paper. It is part of a collection of alchemical treatises attributed to Lull. Like **O**, it lacks the prologue, but, as the title indicates, it contains all eight operations. Operation 8 refers to the parallel section in *De famulatu*: "which [furnace] we sketched out to you in the first book of the quintessence" (*quem tibi designavimus in primo libro quinte essentie*), but ends abruptly at l. 292. It contains two simplified drawings of the furnace (fols. 13r and 14v).

Le^4_d Leiden, Leiden University Library, Vossianus Chym. F. 6, fols. 397r–403v (last q. 16th c.)[149]

This witness is part of a collection of alchemical works in German, produced in the vicinity of Dresden, and was probably owned by the alchemist Sebald Schwertzer (1552–1598).[150] Its title and attribution follows **Fl**: *Reimundi Meisterschafft von Mercurio*. It provides German equivalents for all the textual variants of this family.[151] Operation 8 is no more than a slightly expanded version of the same reference to *De famulatu* as found in **Fl**: "It [the furnace] is described in the first book of the quintessence; I will let it remain there. You may seek for it there as you wish."[152] **Le^4_d** repeats many other readings unique to **Fl**.[153] Where **Fl** has lacunae in Operation 5, **Le^4_d** often has corresponding lacunae, but sometimes provides text that is not present in **Fl** but which is often garbled, thus corroborating the hypothesis of a damaged hyparchetype that was difficult to decipher in this section.

[147] Fol. 246v: "Pratica ejusdem philosophi de rubescissa fratris minoris in arte majori et est res notabilissima."

[148] Angelo Maria Bandini, *Bibliotheca Leopoldina Laurentiana*, 2 vols (Florence: 1792), 2:172–73; Thorndike, *HMES* 4:645.

[149] Boeren, *Codices*, 22–26.

[150] On Schwertzer, see Vladimir Karpenko and Ivo Purš, "Sebald Schwertzer: Between Metallurgy and Alchemy," 617–90 in *Alchemy and Rudolf II*.

[151] E.g., l. 21: Mercury is generated "in den grübenn und hölen der Erden"; l. 31: "zweyfachen Leinwandt"; ll. 108-9 "der Mercurius mag innerhalb 12 stunde im wasser des Römischen Vitriols auffgelöset werden, und trincket auch in der aufflösung so Viel er mag."

[152] Fol. 403v: "...er aber in Ersten Buch vom funfften wesen beschrieben ist, will ichs dabey lassen bleyben du magst nach deinem gefallen da suchen."

[153] At l. 167: "cum benedictione dei et domini nostri yhesu christi" (**Fl**) / "im Nahmen Gottes unnd Jesu Christi" (**Le^4_d**); l. 184: "gaude et laude deum" (**Fl**) / "freue dich und sage danck dem allmachtigen" (**Le^4_d**). **Le^4_d** also has some unique readings, at l. 19 it states that mercury "auß dem wachs gezogen wirt," a result of misreading *terra* as *cera*.

6.5.1 Subfamily ι′
Oth², L⁵, and Pd

These three witnesses date from the sixteenth or early seventeenth century and were all produced in northeastern Italy—one around Gubbio, the other two in Padua. **Pd** is copied from **L⁵**. The copyist of their common ancestor, designated ι′, attempted to "restore" the damaged parts of ι, resulting in an extremely distorted text. The text jumps suddenly from the word *vitrea* at l. 117 (Operation 3) to the word *rotundam* at l. 171 (in Operation 5, the same place where **O** breaks off in mid-sentence), thus giving the appearance of an eye-skip (*vitrea...vitri*) leaping across more than 600 words. After continuing with ll. 171–75, the text returns to where it left off at l. 117, continues to *dies* in l. 119, then jumps back to l. 175 (*ubi vas*) until 182 (where *materia quinte essentie* replaces *materia lactis virginis*), and finally returns yet again to pick up l. 119 where it had previously been broken off (*vel ultra*). Thereafter, the text progresses in its usual order through Operation 4 and again into Operation 5, repeating all of the text of that Operation that had already been transposed into Operation 3, yet with some variants and an instruction added perhaps to account for the repetition: "by doing everything as was said in the third chapter or third operation" (*per omnia faciendo ut dictum est in capitulo tercio sive tercia operacione*).

While it is difficult to explain how such severe muddling came about, it is clear that the corruption began in ι at ll. 170–71 between the adjacent words *vitri* and *rotundam*. The copyist of **O** gave up at that point; **Fl** and **Le⁴d** become lacunose there; **Oth²**, **L⁵**, and **Pd** jump to exactly that point from l. 117. **Oth²** places a period nonsensically at that juncture, as well as an indecipherable symbol which is perhaps the relic of a marker present in ι′ that signified how the text had been modified or how it was to be read. It cannot be a mere coincidence that all six witnesses show anomalies at the very same point in the text. Presumably, the compiler of ι′ used another witness when attempting to restore the text, from which he also took the prologue (absent from **O**, **Fl**, and **Le⁴d**, but present in **Oth²**, **L⁵**, and **Pd**). The family of this supplemental witness is not identifiable. The Italian translation **Fn³i**, although descending predominantly from π^2 takes its prologue and Operation 2 from subfamily ι′, specifically from **L⁵** (Section 6.1.2).

While the unique rearrangement of the text suffices to identify members of this subfamily unambiguously, they nevertheless share other unique features, the most obvious of which include:

- l. 13: *Continet autem liber iste octo capitula sive octo operationes* added after *celaverunt veritatem*.
- l. 122–23: text following *est* rewritten as *deinde diligentissime tritura materiam predictam donec sit sine tactu et post hoc pone in distillatorio vitri quod colloca in balneo marie....*

- l. 124: Situating the flask in the bath is specified with a line lifted from the parallel text in *De famulatu*: *Et per duos digitos...bene triturata.*

Oth^2 Philadelphia, Science History Institute, Othmer MS 14, fols. 23r–32v (1520s)

When the manuscript bearing this witness was owned privately by the Ubaldini family in Urbino, Andrea Aromatico and Marcella Peruzzi studied and published images of **Oth**^{2} with a transcription and Italian translation.[154] Leah DeVun used this publication as the text of the *Liber lucis* in her study.[155] The manuscript was purchased in 2016 by the Chemical Heritage Foundation (now Science History Institute) in Philadelphia. **Oth**^{2} is bound in the middle of a manuscript book of secrets written by a different hand on different paper that dates from the 1540s.[156] The first page of **Oth**^{2} is more soiled than the rest, indicating the independent existence of these leaves before their incorporation into the present manuscript. Aromatico and Peruzzi claimed **Oth**^{2} to date from "the end of the fourteenth century," which would make it the earliest known witness, but their dating is incorrect.[157] **Oth**^{2} bears watermarks dating from the 1520s (Briquet 6305 and 748). Its *gothica rotunda* bookhand would suggest an earlier date, especially for an Italian manuscript, but the copyist seems to have used an antiquated script (and not correctly formed) to make the manuscript appear more venerable. The copyist's Latin was very poor; he made rudimentary grammatical and orthographic gaffes frequently throughout. He also had difficulty reading his exemplar, which, coupled with his poor Latin and the already garbled text of descendants of ι', resulted in frequent nonsensical readings.

Oth^2 contains the prologue, eight operations, and (uniquely for this family) Operation 9. The explicit names John as a "doctor of Paris" ("Explicit liber lucis fratris Johannis de ripa cissa gallici doctor parisiensis"), while the incipit garbles both author and title: "Incipit liber fratris Johannis de Ripasis qui liber lucus nuncupatur." Operation 9 is prefaced with "in this same chapter and last part concerning projection even to infinity, master Arnald says 'and because a grain is a thousand of parts of a thousand...'."[158] The presence of Operation 9 indicates contamination from another line of descent; in fact, at about l. 286, **Oth**^{2} ceases to have close affinity with other members of the subfamily, and instead bears numerous readings found only in descendants of $\pi^{1'''}$, which, coupled with the presence of Operation 9, implies that the ending material came from a descendant of ζ (its descendants **G** and **T** circulated in Italy). Thus, the exemplar for **Oth**^{2} was probably a descendant of ι' whose last folio was damaged or missing, hence the need to appeal to a descendant of ζ for the conclusion of the text.

[154] Andrea Aromatico with Marcella Peruzzi, *Il libro della luce* (Venice: Marsilio, 1997).

[155] Leah DeVun, *Prophecy, Alchemy, and the End of Time: John of Rupescissa in the Late Middle Ages* (New York: Columbia University Press, 2009).

[156] The surrounding Italian text was also treated by Aromatico and Peruzzi, *Medicamenti, pozioni e incantesimi del ricettario magico urbinate* (Milan: Roche, 1994).

[157] Aromatico with Peruzzi, *Libro della luce*, 116.

[158] Fol. 32r: "In eodem Capitulo et ultima parte de projectione usque ad infinitum dicit magister Arnaldus, et quia grana est mille milium partes..."; *grana* is an error for *grave*, typical of this copyist.

L^5 London, British Library, Sloane MS 2006, fols. 53r–62r (1563)

The quarto manuscript containing L^5, written by an Italian hand in a hurried *humanistica cursiva* and containing over two dozen complete or excerpted alchemical treatises was compiled in Padua and Venice in 1563–64. L^5 contains the prologue and eight operations. It is followed immediately (fols. 62v–64v) by parallel excerpts from *De famulatu*. These excerpts end with "I, master Nico Rudianus, although unworthy, from inspiration by divine grace found the true path of the secret art, and perfect in all examinations, for making gold and silver."[159] Two excerpts from an unidentified alchemical work follow (fol. 68v), prefaced by "on 13 November 1563, I saw in the shop of a certain master Simon, bookseller of Padua, a manuscript book from which I copied down these two things."[160] This superscription both dates and localizes the copying of the preceding L^5. There are three crude sketches of the furnace, its cover, and the position of the *lamina* (fol. 62r). The drawing is endorsed "I think this is the description of the furnace of John of Rupescissa" ("Hanc ego arbitror esse descriptionem furni Jo. de Rupescissa"). The similar shape and identical labeling suggests dependence upon the more carefully-rendered drawing in $Mp^I{}_a$ (Figure 2.6), although there are no commonalities of textual variants.

Pd Padua, Biblioteca Universitaria, MS 1380, fols. 157r–68r (ca. 1600)[161]

Written on paper by a Germanic hand in a clear, flowing cursive with red marginal notes, this witness is part of a collection of nine alchemical treatises, in Latin and Italian, including a complete *De famulatu* (fols. 27r–110v) in a different hand. **Pd** was copied directly from L^5. This relationship is attested by their nearly complete identity of readings and by **Pd**'s smooth incorporation into its text of a marginal note added at the end of L^5.[162] Additionally, both L^5 and **Pd** are immediately followed by identical excerpts from *De famulatu* (fols. 168r–70v), indicating that the copyist of **Pd** took both Rupescissan works as a unit directly from L^5. Since both manuscripts were in Padua—**Pd** was owned by the Benedictine abbey of Santa Giustina—this close linkage is fully plausible. **Pd**'s copyist modernizes the orthography and omits the illustrations of L^5.

6.6 Italian Family ι^2

Fn², **L³**, **Mp^I** and, with significant contamination, the published Cologne version and its forerunner **Ua** (see Section 6.1.1.4).

[159] Fol. 64v: "Ego Magister Nico rudianus[?] licet indignus ex inspiratione divina gratia Artis archanae viam inveni veram et in omnibus examinationibus perfectam ad solem et lunam."

[160] Fol. 68v: "Anno domini 1563 die 13 mensis novembris vidi in officina cujusdam magistri Simonis Librarii patavii librum manuscriptum ex quo duo notavi."

[161] Biblioteca universitaria di Padova, *Catalogo Gloria dei manoscritti*, fols. 196r–v; http://cataloghistorici.bdi.sbn.it/dett_catalogo.php?IDCAT=230

[162] Fol. 61v of L^5 and fol. 168r of **Pd**: *In centum diebus totum opus perficitur* ("The work is completed in one hundred days"); the same line appears in $Fn^3{}_i$.

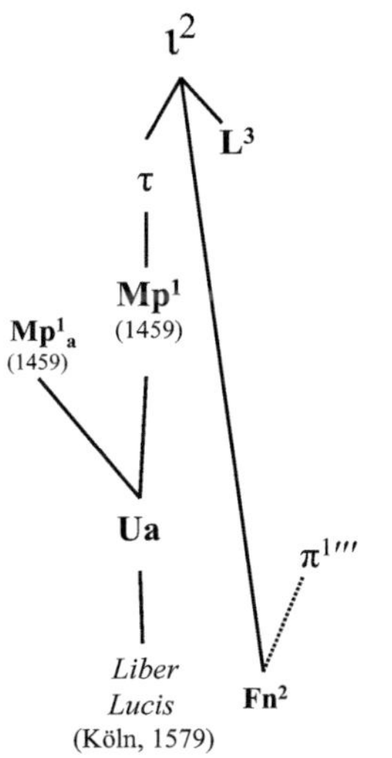

Three manuscript witnesses of Italian origin form a family descending from a common ancestor designated as ι^2. They show significant similarities with descendants of π^3 (e.g., parallel versions of Operation 9, common textual variants, and similar illustrations) demonstrating a genetic link between ι^2 and π^3. Yet **Fn**^{2}, L^3, and **Mp**^{I} diverge sufficiently uniformly from the other descendants of π^3 to imply a common intermediate ι^2, possibly of Italian origin. This ancestor ι^2 may have been damaged in the area of the prologue since that section is uniquely abbreviated in L^3, substantially shortened in **Mp**^{I}, and borrowed from a different family in **Fn**^{2}; the damage probably began in π^3 since its descendants omit the prologue entirely. This intermediate ι^2 is also the source for English family ε^2, as indicated by numerous unique readings shared by descendants of both ι^2 and ε^2 (Section 6.8). Major characteristics of descendants of ι^2 include:

- l. 54: pseudo-Arnald's book is called *tractatu parabolico de* lapide *majori.*
- ll. 65–66: *ascensum per ingenium* in place of *ad sensum per ingenium.*[163]
- l. 76: *portat spiritum et communiter* (**Mp**^{I}: *commovens*) *sulphur invisibile.*
- ll. 136–37: *quasi christallus* instead of *quasi christus.*
- l. 205: the phrase *ascendit de sepulchro rex* is expanded to *ascendit de sepulchro clauso rex.*
- l. 229 *montem* instead of *librum*, as in descendants of π^3.
- ll. 281–82: hair, salt water, and vinegar omitted.
- ll. 296: *quinque digitos aut quasi* (omitted in **Mp**^{I}) in place of *quinque digitos aut minus.*
- Operation 9 included with the reading *rosas bene odoriferas* (cf. family γ').

L^3 London, British Library, Sloane MS 276, fols. 75r–77v (1st q. 15th c.)

This witness is part of a small collection of Latin alchemical texts written on paper in double columns by two alternating hands—an idiosyncratic, rapid, and highly abbreviated *hybrida cursiva*, and a more carefully formed *gothica cursiva* containing elements of *bastarda.* The two hands sometimes switch in the middle of a sentence. The first hand is certainly Italian, while the second hand shows Germanic elements and orthographies such as *diwulgeret*, even though it too uses the Italianate crossed q for *qui.* The watermark (Briquet 7680) suggests an origin in the Veneto in the first two decades of the fifteenth century. L^3 is headed "Here

[163] In **Mp**^{I}, the sentence maintains the sense of the misreading but is reworded as "et ascendit spiritus vitrioli propter ascensus ingenium."

begins the *Book of Light of the Great Master*" (*Incipit liber lucis magistri magni*) but without John's name. It contains an abbreviated prologue and Operations 1–9. A set of seven "precepts" for the alchemist (*Primum preceptum est ut artifex hujus sit secretus...*), cribbed from pseudo-Albert's *Semita recta*, intervenes before the explicit.[164] There are three small illustrations of furnace parts inserted in the lines of text.

Fn² Florence, Biblioteca Nazionale Centrale di Firenze, MS Palat. 885, fols. 154r–60v (end of 16th c.)[165]

This witness is part of a miscellany of alchemical treatises and recipes in Latin and Italian of widely differing dates (fourteenth to sixteenth century). **Fn²** contains the prologue and Operations 1–9. Like **L³**, it is headed *Incipit liber lucis magistri magni* without attribution to John, and contains the same three illustrations. Unlike **L³**, it has a full prologue, but displays variants found otherwise only in the descendants of $\pi^{I'''}$. The three illustrations, although larger than those in **L³**, are identically stylized.

Mp¹ Montpellier, Bibliothèque Universitaire Historique de Médecine, MS H 493, fols. 236v–42v (1459)[166]

Mp¹ is part of a collection of Latin alchemical and medical materials written in *gothica rotunda cursiva* on paper in Naples by a single hand. The collection includes excerpts from *De famulatu* (fols. 267r–70v), including the sections that parallel the *Liber lucis*, as well as the abbreviated version **Mp¹ₐ**. **Mp¹** is headed "Here begins the treatise of the light of the work of the philosophical stone according to Brother John of Rupescissa of the Order of Friars Minor."[167] It contains a short version of the prologue and Operations 1–9. Short and unique insertions occur throughout: for example, noting that some authors call vitriol "magnesia" and the quintessence of quicksilver "terra foliorum" or "terra foliata," and adding Ramon Lull to the list of authorities at l. 131. As noted in Section 3.3, **Mp¹** is a back-translation into Latin from a presumed vernacular witness which I designate τ. A lengthy marginal gloss, in the same hand as the text, questions the workability of John's method: "But I judge indeed that this medicine made by this method will not work because it is without a ferment." It then quotes *in extenso* from the *Rosarius* of pseudo-Arnald ("the imitator of Geber and Albert the German").[168] There are two rough drawings of the furnace and the *lamina*.

[164] Pseudo-Albert, *Semita recta* in *Verae alchemiae*, 2:72–101, on 81–82; see also Vaticanus Palatinus MS 978, fols. 34v–35r. The list breaks off in mid-sentence near the end of the seventh precept, which may indicate that the copyist realized his error in copying past the *Liber lucis*.

[165] Luigi Gentile, *I codici Palatini*, 3 vols. (Rome, 1890), 2:394–96.

[166] Corbett, *Catalogue*, 2:120–33; Alessandrini, *Cimeli Lincei*, 287; and Principe, "Missing Link."

[167] Fol. 236v: "Incipit tractatus lucis operis de lapide philosophico secundum Johannem de Rupe cisa ordinis fratrum minorum."

[168] Fol. 241v: "Sed judico immo, ista medicina hoc modo non valet, quia est sine fermento" and "ymitator yeberis et alberti theotonici." The *Rosarius* quotations are from "de quarto regimine," "de utilitate magisterii," and "de inceratione elixiris albi"; see Calvet, *Oeuvres alchimiques*, 328, 332, and 346.

6.7 English Families

Ccc, Cu^{I}, Cu^{2}, Ct, H, L^{I}, L^{4}, X^{2}, X^{3}, X^{3b}, X^{4}, and **Y^{2}**; with contamination, **X^{I}, L^{2},** and **L^{6}**; in English translation, **X$^{5}_{e}$, X$^{6}_{e}$,** and **X$^{7}_{e}$**.

Eighteen witnesses are of English origin; twelve are complete, and four are fragments or excerpts; fifteen are Latin, and three are English translations. All but one (**Ct**) fall into two distinct and well-defined families—ε^{I} and ε^{2}—that descend from two different Continental families, π^{2} and ι^{2}, respectively, implying at least two separate importations of the text into England, in both cases from Italy.[169] Fourteen date from the fifteenth century, three from the sixteenth, and one from the seventeenth, indicating a sharp peak of popularity for the text in the mid-1400s.

6.7.1 English Family ε^{1}

Cu^{2} and **X^{4}**; with contamination, **L^{2}** and **L^{6}**

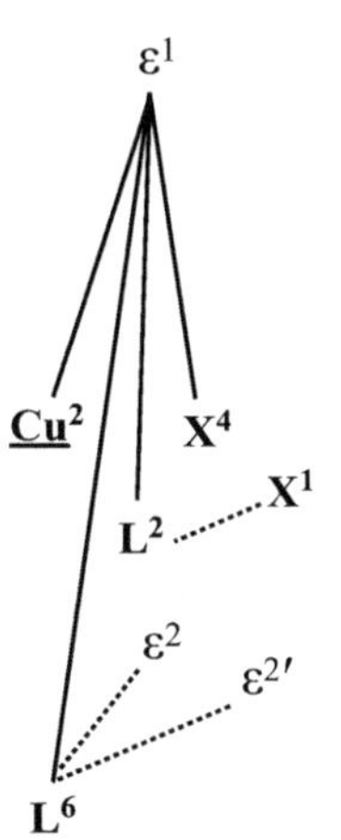

Cu^{2}, L^{2}, X^{4}, and **L^{6}** form a coherent family descending from a common ancestor designated as ε^{I}. Their texts differ significantly from all other English witnesses, being considerably more complete and reliable; consequently, one witness from this family (**Cu^{2}**) has been used in preparing the critical edition. They share many significant textual variants with descendants of π^{2} of the Provençal-North Italian family, indicating the genetic link between ε^{I} and π^{2}. The variants common to descendants of *both* ε^{I} and π^{2} include:

- ll. passim: The Deckname *leo* or *leo viridis* used for vitriol.
- l. 28: *libram i* rather than *libras ii.*
- l. 131: The title *vir magne reputationis* is given to Arnald, followed by *oportet a terra exaltari filium hominis*, erroneously transferred from l. 133 due to an eye-skip (and repeated there).
- l. 227: *materiam* instead of *maculam* (corrected in **L^{2}**).
- l. 229: *librum* rather than *montem* or *mentem/mente.*

Cu^{2}, L^{2}, X^{4}, and **L^{6}** nevertheless diverge sufficiently from π^{2} to demonstrate the existence of a common intermediate ε^{I}. The major characteristics of descendants of ε^{I} (*not* present in those of π^{2}) include:

[169] **Ct** does not appear to be a member of either family and may thus signal a third importation.

- l. 1: *Ego frater Johannes de Rupescissa consideravi...* (These are the only fifteenth-century English witnesses that attribute the work to John, and the only witnesses of any sort to include his name in the text.)
- ll. 22–23: *ipsam in latrina stercorum quesierunt* rather than *ipsam ad literam in stercoribus quesierunt.*
- ll. 53–54: *in quodam suo tractatu* rather than *in tractatu parabolico.*
- l. 174: *et cum collum amphore bene fuerit strangulatum cum forpicibus tunc* inserted after *destruatur ab igne.*
- l. 243: *evangelistarum ac eciam pauperibus hominibus* rather than *evangelicis pauperibus.*

Cu² Cambridge, University Library, Kk.6.30, fols. 62v–68v (mid 15th c.)[170]

This witness is part of a collection of alchemical treatises and recipes in English and Latin written by multiple hands dating from the fifteenth and sixteenth centuries. **Cu²** is written on paper in a neat *gothica cursiva* with red and blue capitals and paragraph markers. It contains the prologue and all eight operations, and is titled *Magisterium lucis*. It is immediately preceded (fols. 62r–v) by two short excerpts in the same hand from *De famulatu* that deal with antimony and gold. As the fullest and most reliable witness of this family, it has been included in the critical edition.

X⁴ Oxford, Corpus Christi College, MS 226, fols. 27r–30r (15th c.)[171]

Written in a hurried and irregular cursive on paper (with corrections in *anglicana*), this witness is part of a miscellany of alchemical tracts and recipes in Latin and English by multiple hands. **X⁴** contains the prologue but omits Operation 8.

L² Sloane MS 75, fols. 114r–15v (frag., Operations 4–8; 3rd q. 15th c.)

The manuscript, a collection of alchemical materials in Latin and English, written in *gothica rotunda cursiva* on paper, presents **L²** as the second book of *De famulatu*, inserting it between Books One and Two. Several folios have been lost preceding fol. 114, thus the *Liber lucis* begins at l. 132; it closes with "Explicit secundus liber quinte essencie." Most of the text closely resembles other members of this family, but Operation 8 follows almost exactly the unique variants of **X¹**, and minor contaminations from the same source appear in earlier operations. **X¹** also inserts the *Liber lucis* into *De famulatu*, but as Book Three rather than Two. Likewise, both **X¹** and **L²** attribute the text to Roger Bacon, although **L²** doubts Bacon's authorship because *De famulatu* considers quicksilver hot in the fourth degree, while Bacon calls it cold and wet: "This little tripartite book of the

[170] *Catalogue of the Manuscripts Preserved in the Library of the University of Cambridge*, 5 vols. (Cambridge: Cambridge University Press, 1858), 3:726–28 (listed as item 2110; the foliations given therein are incorrect) and Anke Timmerman, "Alchemy in Cambridge: An Annotated Catalogue of Alchemical Texts and Illustrations in Cambridge Repositories," *Nuncius* 30 (2015): 345–511, on 496–501.

[171] R. M. Thomson, *A Descriptive Catalogue of the Medieval Manuscripts of Corpus Christi College Oxford* (Oxford: Brewer, 2011), 112–14.

quintessence is ascribed by some to Roger Bacon...It is ascribed by others to John of Rupescissa...I don't care whose it is since many things are well said in it."[172] **L^2** would thus seem to have originated from a descendant of ε^1, which, like **X^4**, lacked Operation 8 and therefore the copyist appealed to **X^1** (or a closely related manuscript) for the missing text.[173]

L^6 London, British Library, Stowe MS 1070, fols. 51v–56r (2nd half 16th c.)[174]

Given the title "Ad curiam romanam," which is repeated in the explicit as "a work written to the Roman Church" (*opus ecclesie Romane scriptum*), **L^6** was left anonymous. A later hand identifies it as Ripley's *Medulla alchymiae*, and a yet later hand adds, "Some consider it to be the *Liber lucis mercuriorum* of John of Rupescissa." The same hands made similar additions to the explicit. **L^6** contains all eight operations but lacks the prologue. Remarkably, it was compiled from *three* exemplars, one from each English family. For Operations 1–4 and 7, **L^6** shares almost all the variants of **Cu^2/X^4**, and therefore descends predominantly from ε^1. Operations 5 and 6, however, share their variants with **Y^2**, and many with **X^3/L^4**, indicating their origin in a descendant of $\varepsilon^{2\prime}$.[175] Like all descendants of $\varepsilon^{2\prime}$, **L^6** omits the "Credas" from Operation 6, but that omitted text appears at the end of Operation 7, suggesting that after Operations 5 and 6 were copied from a descendant of $\varepsilon^{2\prime}$, the copyist reverted to his descendant of ε^1 for Operation 7, whereupon he discovered the substantial "Credas" omitted from the exemplar he used for Operation 6, and then copied this additional material as part of Operation 7. Finally, the compiler took Operation 8 from a descendant of ε^2; it displays the unique variants of **Ccc** and **X^{3b}**. Thus, the compiler of **L^6** worked from (at least) three exemplars, one (or more) from each of the families ε^1, ε^2, and $\varepsilon^{2\prime}$. He chose what he considered to be the "best" version of each operation—taking 1–4 and 7 from one exemplar, 5 and 6 from another, and 8 from a third. John Dee (1527–1608) owned this witness and made marginal annotations to it.

6.7.2 English Family ε^2

Ccc, Cu^1, L^1, and **X^{3b}**; in English translation, **X^6_e** and **X^7_e**; descendant subfamily $\varepsilon^{2\prime}$: **H, L^4, X^1, X^2, X^3,** and **Y^2**; in English translation, **X^5_e**.

Witnesses of the English family ε^2 are less complete than those of ε^1. Shared textual variants indicate that ε^2 descended from ι^2. Its incomplete state was due

[172] Fol. 123r: "Libellus iste de 5ta essencia tripartitus ascribitur ab aliquibus magistro Rogero Bacon...Ab aliis ascribitur a Johanni de rupecissa...non curo cujus sit dum bene dicantur in eo multa."

[173] As noted below, **X^1** itself took Operation 8 from a descendant of γ^2, so **L^2** might have accessed the same source. However, **X^1** and **L^2** share identical lacunae (ll. 284–86 and 309) and minor readings that differ from descendants of γ^2, so **X^1** is the more likely source for **L^2**, especially given the shared layout and attribution to Bacon.

[174] *Catalogue of the Stowe Manuscripts in the British Museum*, 2 vols. (London: British Museum, 1895–96), 1:686–87.

[175] It could not, however, have been copied from **Y^2** since that manuscript was in Italy when **L^6** was compiled, and it does not contain any trace of the additional texts of **Y^2**.

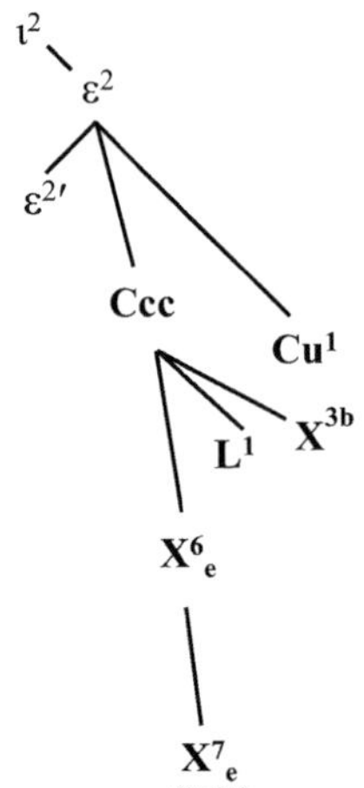

to the apparent damage to ι^2 plus additional editing. Unlike the Italian descendants of ι^2, the English descendants of ε^2 do not include Operation 9, suggesting that the copyist of ε^2 recognized it was not part of John's text and omitted it. Members of family ε^2 do, however, show nearly all of the other textual characteristics listed above (Section 6.6) for the Italian descendants of ι^2. Witnesses of family ε^2 omit the prologue—similarly to the highly edited or omitted prologue in descendants of ι^2—while those of subfamily $\varepsilon^{2\prime}$ contain many further omissions and include new materials (Section 6.7.2.1).

Within family ε^2 only the late **X**^{2} (1593) and its direct translation **X**^5_e attribute the text to John of Rupescissa; most witnesses leave it anonymous, **X**^{1} attributes it to Roger Bacon, and **Cu**^{1}, **X**^6_e, and **X**^7_e to Arnald of Villanova. Thus, John's name did not cross the Channel with his text as successfully as it propagated across Continental Europe. This may again be due to the presumably damaged state of ι^2, which apparently also lacked attribution to John, given that its Italian descendants **L**^{3} and **Fn**^{2} are also anonymous. Although the English witnesses of family ε^2 are valuable for understanding the dissemination, popularity, and modifications to the *Liber lucis* in England, none of them has been used in the critical edition due to their multiple omissions and other interventions.

Ccc is the most complete and probably the earliest Latin witness of the family, while **Cu**^{1} is heavily edited. **X**^6_e, a mid-fifteenth-century English translation, and likewise **X**^7_e, lack both the prologue and Operation 8, and omit the "Credas." **L**^{1} and **X**^{3b} are extracts (Operations 5–7 and Operation 8, respectively). **Ccc**, **L**^{1}, **X**^6_e, and **X**^7_e all uniquely indicate the title of the work as *Quinta essencia mercurii.* Likewise, the three small and simplified illustrations of the furnace found in descendants of ι^2 (e.g. **L**^{3}) recur in this family as tiny depictions inserted in the lines of text. Besides their exclusively English origin, the members of family ε^2 show many additional textual variants *not* found in family ι^2:

- The text, with the exception of **Cu**^{1}, is titled *Quinta essencia mercurii.*
- l. 26: *certissime* rather than *certe.*
- l. 71: *vastabitur* replaces *evapora.*
- l. 74: *mundissimum* in place of *marcidum.*
- l. 210: *educens de dicto vase quando volueris ad projecciones et ad augmentandum/augendam medicinam album ut inferius patebit* added following *inducta.*
- ll. 220–21: *Unde ego pauper evangelizo vobis secretissimum quod nullus homo ante me unquam in aperto scripsit. Scitote annuente deo et sacrosancta romana ecclesia quod philosophi....*

- l. 226: *sed fide dignis et hominibus deum pre oculis habentibus ac ministratoribus mistici/mistice sacramenti domini nostri jhesu christi* instead of *sed mistico sacrosancto corpori Jhesu Christi.*
- ll. 232–33: *reprobis et inquis ac eciam sibi qui tantum archanum adulteris reprobis et iniquis docebit et communicabit*, in place of *reprobandis... communicabit.*
- l. 283: following *parvum*, the exact measure *novem palmas* is given.
- ll. 295–96: *circulariter per parietes furni infra muros ejusdem per spacium unius digiti vel duorum* replaces *circulariter...minus et.*

Ccc Cambridge, Corpus Christi College, MS 99, pp. 249–54 (mid-15th c.)[176]

This witness is written in a formal *gothica quadrata* bookhand on parchment (one of only seven known witnesses on parchment, the others are **Fr$_a$**, **M**, **Oth^I**, **W^I**, **X^I**, and **Y^I**), forty-nine lines to the page. Space was left for the addition of initials; these were completed earlier in the manuscript, but not where this witness occurs. **Ccc** contains all eight operations but no prologue. It is entitled *Essencia 5^a Mercurii* without indication of authorship, but a later fifteenth-century hand added "per Arnaldum" in *anglicana*, which a seventeenth-century hand corrected to "per Johannem de Rupescissa, et vocatur Liber Lucis." Fifteenth- and sixteenth-century hands added marginalia throughout. In addition to the three drawings of furnace parts imbedded in the lines of text typical of this family, it also shows five small sketches of parts of the furnace in the lower margin of fol. 254r, possibly added at a later date.

Cu^I Cambridge, University Library, Dd.4.45, fols. 1r–4r (mid-15th c.)[177]

Written in *anglicana* on paper, this witness is titled simply "Arnaldus de Nona [*sic*] Villa." Space was left for an opening initial, but it was not filled in. Like **Ccc**, it lacks the prologue but contains all eight operations, although many are highly edited—Operations 1 and 2 lack roughly 50% of their usual content. It contains one rough illustration of the furnace *lamina*, inserted within the lines of text.

X^{3b} Oxford, Corpus Christi College, MS 185, fols. 80r–v (extr., Operation 8 only; mid-15th c.)[178]

This extract contains only Operation 8 and is headed *De fornace vel furno philosophorum* ("Of the furnace or oven of the philosophers"). It is part of the same collection of alchemical and medical texts described under the heading for **X^3** and written in the same hand as **X^3** that appears roughly forty leaves later in the collection and is instead a descendant of $\varepsilon^{2'}$ (see Section 6.7.2.1). The text of **X^{3b}** follows **Ccc** exactly.

[176] Montague Rhodes James, *A Descriptive Catalogue of the Manuscripts in the Library of Corpus Christi College Cambridge*, 2 vols. (Cambridge: Cambridge University Press, 1912), 1:185–88. Digital images available at "Parker Library on the Web": https://parker.stanford.edu/parker/catalog/canvas-4370ee785131e430f96cfea0846f0cd0

[177] *Manuscripts of the University of Cambridge*, 1:240–43 (where it is listed as item 220), and Timmerman, "Alchemy in Cambridge," 475–77.

[178] Thomson, *Catalogue*, 91–93, on 92, no. 9, here erroneously thought to be part of "De Quinta Essentia."

L^{I} London, British Library, Harley MS 3528, fols. 21v–22v (extr., mid-15th c.)[179]

This extract contains Operations 5 and 7 and the second half of Operation 6. It is part of a compilation of extracts from alchemical works made in the mid-fifteenth century and later incorporated into an alchemical miscellany by Giles du Wes (Aegidius de Vadis) in the early sixteenth century.[180] These snippets are generally run together without titles or identifications. Only a seventeenth-century marginal note identifies **L**^{I}: "Fifth chapter, John of Rupescissa" (*Quintum capitulum: Jo. De rupescissa*, fol. 21v) and then "the sixth operation, that is, one chapter, is missing" (*Deest sexta operatio, id est, unum capitulum*, fol. 22r). In fact, the latter half of Operation 6 (the "Credas") is present directly following this note, but the annotator did not recognize it as John's, possibly because he compared the manuscript with the Basel edition which omits this section. Operation 7 also appears on fol. 21v but is not identified as such by the marginalia, probably because it ends with the misleading attribution: "this is the chapter of the seventh operation extracted from the *Quinta essentia mercurii*" (*Hoc est capitulum de* 7^{a} *operacione extractum de* 5^{a} *essentia mercurii*)—using the same title for the *Liber lucis* found in **Ccc**, **X**$^{6}_{e}$, and **X**$^{7}_{e}$.

X$^{6}_{e}$ Oxford, Bodleian Library, Ashmole MS 1450, part 6, fols. 157r–76r (frag., third q. 15th c.)[181]

Written in *anglicana* on paper, this witness preserves one of the earliest vernacular translations. Like other members of this family, it lacks the prologue, but also omits Operation 8 and the "Credas." Lines 33–80 and 115–39 are missing due to the loss of the original folios 65 and 66 (between fols. 157 and 158 in the later foliation) and 70 (preceding fol. 169).[182] While the translation is often very literal, it also adds expanded explanations of some directions that were obscure in the original. **X**$^{6}_{e}$ is headed "In the virtue of allmighti god Jesu criste here I begyne the grett and Royall worke of the 5^{a} essense of mercurye acordyge [to] the worke of master Arnold of the newe towne," using the same title for the *Liber lucis* found uniquely in this family. The text starts with a short reference to Arnald regarding metals before beginning at l. 18: "the matter of the stone is of a vyle price...."[183]

[179] Robert Nares, Stebbing Shaw, Joseph Planta, and Francis Douce, eds., *A Catalogue of the Harleian Manuscripts in the British Museum*, 4 vols. (London, 1808–12), 3:38. On this manuscript and its owner, see Rampling, *Experimental Fire*, 141–42.

[180] On Giles du Wes, see Rampling, *Experimental Fire*, 137–54.

[181] William Henry Black, *Catalogue of the Manuscripts Bequeathed...by Elias Ashmole* (Oxford: Oxford University Press, 1845), col. 1238.

[182] These omissions remove the sections bearing most of the textual variants unique to this subfamily, but this witness does display the characteristic addition to l. 210: "thou shalt drawe oute the vessell unto the whyte and mark that the same vessell that thou ledist the whytenez whan that thou wolt goo unto the projection and to augement the same whyte medizyn as withinforyth it shall appere unto thee openly."

[183] The prefixed text reads "Master Arnolde of the newe towne the great philosopher declarith and seith by dyverse Autorez that the mynerall bodyes are great and grosse in substannce. And he seith that the sprites of the same bodyez are light and subtile. And therfore he seyth unyversally they may not be medled to gethers that is to sey eyther with other."

X^{7}_{e} Oxford, Bodleian Library, Ashmole MS 1490, fols. 143v–47v (1585)[184]

The witness was copied by the astrologer and physician Simon Forman (1552–1611) and is clearly of the same lineage as X^{6}_{e}, but with frequent minor textual variants and rewordings presumably from intermediates. Yet it bears the same heading ("In the vertue...Arnold of newtowne"), a shorter version of the added opening line about metals, and omits the same materials. It does preserve the text lost from X^{6}_{e} due to that witness's now-missing folia.

6.7.2.1 *Subfamily* $\varepsilon^{2\prime}$

H, L^{4}, and X^{3}; with contamination, X^{1} and Y^{2}; in English translation, X^{5}_{e}

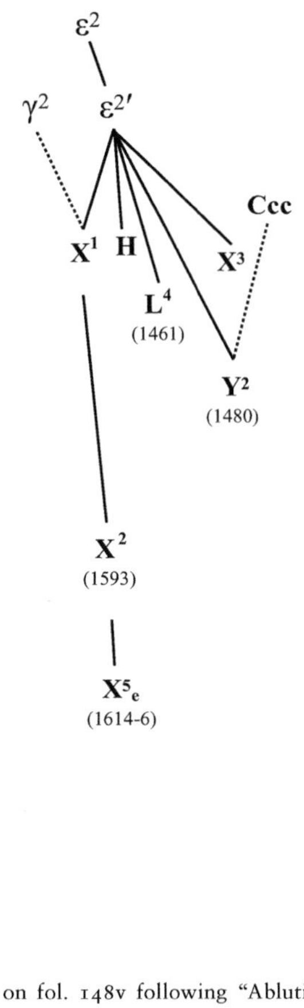

Six witnesses show numerous further textual variants that link them as derivatives of a common ancestor $\varepsilon^{2\prime}$. X^{1} and Y^{2} differ from the others due to clearly identifiable contaminations. Those contaminations aside, the witnesses of this subfamily are left anonymous and omit the prologue, like those of the larger family ε^{2}, but also omit Operation 8, the "Credas," most references to Arnald, and the litany of alchemical authorities at ll. 129–31.[185] Yet these witnesses, with the exception of Y^{2}, also contain multiple explanatory glosses and expansions, and append a new text about projection and the differences between transmutatory stones that references the *Liber de septuaginta* and *Liber de triginta verbis*. This addition concludes with two verse texts: the couplet *Est in mercurio quicquid querunt sapientes/Corpus ab hac anima spiritus tinctura trahuntur* and *Est fons in binis*.[186] "Est in mercurio..." appears frequently in alchemical literature, and dates from the fourteenth century; its first appearance in print occurs in the 1561 collection *Verae alchemiae* in the context of the

[184] Black, *Catalogue*, cols. 1354–68 on 1361, no. 50. The date 1585 appears on fol. 148v following "Ablutio sulphuris."

[185] H is a single-folio fragment containing only part of Operation 7 and therefore cannot display all the characteristics listed here.

[186] X^{1}, L^{4}, and H contain both couplet and poem; X^{3} omits the poem. The added text shows minor variations among these witnesses; the fullest version (X^{1}) reads: "Eciam qualiter in corporibus debet fieri projeccio patet. Pone super ignem corpus quo debet fieri projeccio donec liquifiat et si sit spiritus donec tepescat super cujus faciem proice elixir movendo bene usque elixir liquifiat et se cum corpore miscuerit et tunc remove ab igne et fac ut supradicitur quia secundum nobilitatem medicine patebit excessus tincture. Sed scias quod elixir lapidum plures tingit partes corporum quam elixir spirituum quia elixir lapidum tingit mille milia et duo mille milia sicut patet in libro de septuaginta et de triginta verbis. Elixir vero spirituum tingit mille partes et aliquando minus. Elixir vero corporum tingit minus quam spirituum quia tingit nisi centum."

pseudo-Arnald's *Practica ad quemdam Papam*.[187] The rhymed verses *Est fons in binis* also occur frequently, with considerable textual variation (e.g., *Est fons in limis* and *Est fons sublimis*), in fourteenth- to sixteenth-century texts, and were published alongside the works of Geber, first in 1486 and again in 1529; their origin is not known.[188]

Characteristics of $\varepsilon^{2\prime}$ not found in ε^2 include:

- l. 19: *dicitur aqua viscosa quia a sua specie de facili non mutatur in aliam speciem durabilem quin ad suam propriam speciem revertetur*, following *viscosa*.
- l. 27: *Et istud/illud nomen propter caritatem devulgo tradens hujus custodiam omnipotenti deo*, following *vitrioli romani*.
- l. 70: *patella frigida* instead of *patella terrea*.
- l. 72: *floris mundissimi* rather than *floris gaudii amenissimi*.
- l. 172: *Et hec dicitur sigillacio hermetis. Et sis securus quod quovismodo nil exire possit et taliter ingenia in vitri sigillacione*, in place of *ita quod nichil... et amphoram*.
- l. 190: *Postquam ut predicitur lapis albus perfectus est qui non est lapis nec habet naturam lapidis vase infrigidato aperi et* added following *remoto sic*.
- l. 267: *tripliciter* in place of *dupliciter*.
- ll. 277–78: *congelari ad aerem* instead of *congelari ad ignem*.
- text on projection and verses added at end.

X³ Oxford, Corpus Christi College, MS 185, fols. 121r–24v (mid-15th c)[189]

This witness is part of a quarto collection of alchemical treatises written in a fine English secretary hand by a single scribe on paper; it also contains *De famulatu*. Like the other members of this subfamily, **X³** lacks Operation 8; however, Operation 8 does exist elsewhere in the same collection on fols. 80r–v, far separated from **X³**, having been copied earlier and from a different exemplar. Since its text descends through a different family, I treat it as a separate witness designated **X³ᵇ** (Section 6.7.2). **X³** contains a gloss reading "The sayings of the *Rosarius* seem to agree with this foregoing little work," followed by chapter 22 of the *Rosarius philosophorum* attributed to John Dastin.[190] Dastin's *Rosarius* and his *Libellus*

187 Hans Walther, *Alphabetisches Verzeichnis der Versanfänge mittellateinischer Dichtungen* (Göttingen: Vandenhoeck & Ruprecht, 1959), #21054, and D. W. Singer, *Catalogue of Latin and Vernacular Manuscripts*, 3 vols. (Brussels: Lamertin, 1928–31), 1:228; *Verae alchemiae artisque metallicae*, ed. Guglielmo Gratarolo, 2 vols. (Basel, 1561), 2:74, I thank Didier Kahn for this reference.

188 Walther, *Verzeichnis*, #5670; Thorndike and Kibre, col. 509. For other copies, see Oxford, Bodleian MS 1384, fols. 74v–75r and Ashmole 1450 (which contains X^6_e), fol. 143r; Biblioteca Apostolica Vaticana, Palatinus MS 1328, fol. 39v; BNF Lat. MS 11202 (which contains P^2), fol. 138v (cf. fol. 3r); and Geneva, Bibliothèque de Genève, MS 82 (which contains Ge), fols. 131r–v. It is extended with further lines and given the title *Aurea carmina de lapide philosophico* in Munich, Bayerische Staatsbibliothek, Clm 2848, fol. 97v, which dates from 1531. The version in Bamberg, Staatsbibliothek, Ms. Nat. 6 (which contains Bg_d), fol. 130r–v, is intriguingly headed *Cardinalis parusius*.

189 Thomson, *Catalogue*, 91–93.

190 Fols. 124r–v: "Huic opusculo precedenti videntur dicta Rosarii consonare dicentis Quia nimius sermo intellectum... quod est super omne mundi preciosum preciosissimum. Hinc Rosarius in fine libri sui." See *BCC* 2:324 for the complete work. On John Dastin and his *Rosarius*, see Calvet, *Oeuvres alchimiques*, 128–30 and 147–52.

aureus appear elsewhere in the same volume (fols. 139r–55v and 119r–20r, respectively), although the *Rosarius* is there attributed to Arnald.

L⁴ London, British Library, Sloane MS 338, fols. 53v–56v (1461)[191]

Written on paper in a well-formed English *gothica cursiva* with decorated initials, **L⁴** bears many seventeenth-century marginal annotations. Its colophon dates it to 1461: "Explicit liber iste Anno domini 1461, ultimo die Maii." Although **L⁴** parallels **X³** very closely, the copyist of **L⁴** evidently had access to an additional witness, resulting in minor contamination: in several places, one- or two-word alternate readings are interlined, prefaced with *aliter.* These alternate readings are too slight to identify the source. There are also three minor insertions taken from *De famulatu*. The first adopts John's voice: "I described [the *stupha sicca*] in the preceding book, namely, the *Quintessence*, in the chapter 'All sublimed spirits...'."[192] A witness of *De famulatu* does precede the *Liber lucis* at fols. 2r–48v. At ll. 117–18 an alternate name is given to the *stupha sicca*: "and this furnace is called the athanor" (*et dicitur iste furnus Athanor*); a section from the "second remedy" of *De famulatu* is incongruously tacked onto the end of Operation 6 at l. 219.[193]

H San Marino (California), Huntington Library, HU 1051, fols. 32r–v (frag., mid-15th c.)[194]

This witness is a single-leaf fragment containing only Operation 7 and the verses characteristic of subfamily $\varepsilon^{2'}$. It is written in *anglicana cursiva* on paper and heavily damaged by damp. It is part of a miscellany of alchemical and medical manuscript fragments in Latin and English written in multiple hands. This miscellany also contains a single-leaf fragment from the end of *De famulatu* (fol. 33r–v) immediately following **H** and in the same hand. It seems likely, therefore, that the original manuscript from which only these two leaves survive contained both of John's alchemical works.

X¹ Oxford, Bodleian Library, MS e Museo 155, pp. 497–507 (ca. 1450s)[195]

Written in a fine English *gothica cursiva* bookhand on parchment, **X¹** is appended to *De famulatu* as its "third book" and is part of a collection of texts attributed to Roger Bacon. A colophon identifies the copyist: "Here ends the third

[191] *British Library Finding Aid: Sloane Manuscripts*, 193.
[192] Fol. 54v: "descripsi in libro precedenti scilicet 5ª essencia in capitulo Omnes spiritus sublimatos, etc."
[193] Fol. 56r: "Et si vis in ictu oculi sanare desperantem de vita ita ut credatur illusio ex subito miraculo evidenti. Habeas ab hac herba celidonia cujus flores et fructus etc. Habetur alibi etc."
[194] C. W. Dutschke, *Guide to Medieval and Renaissance Manuscripts in the Huntington Library*, 2 vols. (San Marino, CA: Huntington Library, 1989), 2:739–47; Wilson, "Catalogue," 424; https://hdl.huntington.org/digital/search/searchterm/hu%201051
[195] *Summary Catalogue of Western Manuscripts in the Bodleian Library*, 7 vols (Oxford: Clarendon Press, 1895-1953); vol. 2, part 2, 735-37.

book of the *De consideracione quinte essencie* according to master Roger Bacon, corrected and written by John Cokkes with his own hands at Oxford."[196] This identification of the copyist renders the dating of this manuscript in the *Summary Catalogue* to "early in the fifteenth century" impossible, since John Cokkes (or Cokkys), known as a rector and physician, was not resident in Oxford until 1448. He earned degrees in canon law (1449) and medicine (1450) at Oxford and resided there the rest of his life (d. ca. 1475).[197]

Unlike other members of this family, **X**^{1} contains both the prologue and Operation 8. These sections were taken from a descendant of γ^2; such additions presumably constitute the "correction" that Cokkes mentions. Throughout Operations 1–7, **X**^{1} follows the text of subfamily $\varepsilon^{2\prime}$ very closely. Yet at a few points, especially in Operation 1, these readings are doubled with borrowings from the descendant of γ^2, and in a few cases lacunae characteristic of subfamily $\varepsilon^{2\prime}$ are filled with text closely resembling the readings of **Ko**^{1}. These contaminations indicate that a member of the Germanic family γ^2 made its way to England before 1450 but did not leave any complete extant descendants there. **X**^{1} also appends a further alchemical poem after Operation 8, which begins *Est fons ornatus vulcani monte creatus* and ends *Deridens stultum qui nescit solvere nodum*.[198]

X^2 Oxford, Bodleian Library, Ashmole MS 1423, section 5, pp. 68–77 (1593)[199]

This is the latest known English witness of the Latin *Liber lucis*. It is part of a collection of alchemical treatises and recipes collected by Thomas Mountfort, who dates the section containing **X**^{2} to 1593.[200] It contains the prologue and all eight operations, and the textual variants clearly indicate that it descends through **X**^{1}.[201] Nevertheless, the text has been cut to less than half its normal length. This witness is not a paraphrase; instead, the text is highly lacunose, e.g., only the middle third of the prologue is present, and what this witness labels as Operation 3 is a pastiche of the first two-thirds of Operation 3, the middle third of Operation 4, and the last third of Operation 5. The process is thus impossible to follow. This sad state may indicate a gravely damaged intermediate, perhaps the one that also provided a notable feature of **X**^{2}, namely, a date of composition: "Johannes de rupiscissa practica quarta die mensis octobris anno domini 1350," and in the colophon: "Explicit veritas hujus artis, per Johannem de Rupiscissa anno domini 1350 mensis octobris

196 Fol. 507r: "Explicit liber tertius de consideracione quinte essencie secundum magistrum Rogerum Bacon correctus et scriptus per Johannem Cokkes manibus suis propriis Oxonii."

197 C. H. Talbot and E. A. Hammond, *The Medical Practitioners of Medieval England: A Biographical Register* (London: Wellcome Historical Medical Library, 1965), 134–36 and A. B. Emden, *A Biographical Register of the University of Oxford to A.D. 1500*, 3 vols. (Oxford: Clarendon, 1957), 1:457. Some of Cokkes' medical lectures are preserved in Cambridge, King's College MS 16.

198 Another copy of this poem appears in Cambridge, Trinity College Library, MS O.8.25, fol. 71r.

199 Black, *Catalogue*, cols. 1150–54.

200 On Mountfort, see Rampling, *Experimental Fire*, 325–33.

201 The link to **X**^{1} is clear from the presence of the prologue and Operation 8 and from unique readings such as "dies et noctes" (l. 119) and the addition "juxta terram ubi immittantur carbones" following *competenti* at l. 283, characteristic of family γ^2.

quarta die." This date is identical to that provided by **G**, which may mean that a witness intermediate between **X¹** and **X²** was contaminated by witness related to **G**, but textual variants do not supply corroborating evidence.

X^5_e Oxford, Bodleian Library, Ashmole MS 1424, fol. 24v–26r (1614–16)[202]

Entitled *A worke of John de Rupesisa on mercury sublymatte*, this English translation in the hand of Thomas Robson was done directly from **X²**, which exists in the immediately preceding manuscript of the Bodleian's collection, Ashmole MS 1423.[203] Although **X²** was already heavily edited, X^5_e edits it yet further. The translator deleted the oddly shortened prologue, the colophon, the chapter divisions, and almost everything that is not strictly practical, further reducing the text into a recipe. Like its Latin exemplar, X^5_e contains a single drawing, that of the "lamina" for the furnace, and here, exactly as in **X²**, it is drawn with a hole "in the midle, wher is the letter E" (**X²**: *ubi lettera e est*).

Y² New Haven, Yale University, Beinecke Library Mellon MS 19, fols. 116v–120r (1480)[204]

Y² is part of a collection of alchemical texts in Latin and French written by a single scribe in a tiny, extremely neat, and highly abbreviated *gothica rotunda* bookhand on paper. The writing style and occasional Italianisms indicate an Italian copyist, while the colophons (fols. 25v, 32r, 52v, and 96v) record that he compiled this volume in London in June and July 1480. This duodecimo volume returned with its copyist to Italy. While the copyist generally followed the text of other members of this subfamily—including the extension that references the *Liber de septuaginta*—he omitted the verses and frequently took readings that match those of **Ccc** from another exemplar. The copyist also used capital letters throughout for the metals and other substances, as either simple ciphers or abbreviations, although he could have used the standard planetary symbols for the latter purpose.[205] He titled the text *De sulphure philosophorum* without authorial attribution. **Y²** contains a version of Operation 8, although highly paraphrased (bearing no lines in common with any other witness) and separated from the rest of the text by an insertion taken from pseudo-Arnald's *Rosarius philosophorum*.[206] There is one small drawing of the furnace with its cover removed, and bearing the caption: "This is the plate standing and being supported upon the iron teeth, and

202 Black, *Catalogue*, cols. 1154–57.
203 On Robson, see Rampling, *Experimental Fire*, 331–35.
204 Witten and Pachella, *Alchemy and the Occult*, 3:126–33.
205 A = gold, B = silver, C = mercury, D = copper (although sometimes G), E = lead (although sometimes used for iron), F = tin, and H = sulphur.
206 Bk. 2, ch. 23; see Calvet, *Oeuvres alchimiques*, 334–36; *BCC* 1:672–3.

the vessel with the material to be digested or calcined is placed upon the said plate and then covered with the round cap, hemispherical like a biretta."[207]

6.8 Witnesses That Do Not Fit Clearly into Any Currently Identified Family Ct, Go^{I}, Le^{5}_{d}, Lh_{cat}, Sa, W^{8}, and Wb^{4}_{d}

Ct Cambridge, Trinity College Library, MS O.8.36., fols. 51r–55r (ca. 1500)[208]

This is the only English witness that cannot be placed into either of the two known English families since it does not share any of their characteristic variants, nor those of any other known family. Written on paper in *gothica cursiva* with elements of *anglicana* and secretary hand, **Ct** contains the prologue and all eight operations, omits the "Credas," and is part of a collection of alchemical treatises and recipes which includes the first book of *De famulatu* (fols. 60r–90v). It is headed with the author's name, "De Rupe Scissa" (one of the few English witnesses to claim John's authorship), but no title. The copyist, clearly an alchemical practitioner himself, offered his evaluation of the text: "Here an incredible one and, as it seems, impossible to reason; I proposed to write it as I found it, even though in terms of nature it does not seem to be true."[209] It contains two feeble attempts at illustrations of furnace parts,but they are no more than four lines making a hash markset in the text lines.

Go^{I} Gotha, Forschungsbibliothek Gotha der Universität Erfurt, Chart. A 1017, fols. 76r–77r (ca. 1450)[210]

Produced in the Low Countries and written on paper in a compact and highly abbreviated cursive script, **Go**^{I} is titled *Opus lapidis philosophorum optimum*, although a marginal endorsement clarifies that "This work is by Rupescissa" (*Opus istud est Rupescisse*). It omits the prologue and begins at l. 28, but otherwise includes all the operations. The text is frequently rephrased and shows minor additions and omissions that are unique. It contains two small drawings of furnace parts in the text, and leaves space for a third. It does not belong to any identified Germanic family.

[207] Fol. 119v: "Ista est lamina stans et substentata super dentes ferreos et supra dictam laminam ponitur vas cum materia ad digerendum sive calcinandum et postea coperitur cum capello rotundo: semisperico sicut est biretum."

[208] Montague Rhodes James, *The Western Manuscripts in the Library of Trinity College, Cambridge*, 4 vols. (Cambridge: Cambridge University Press, 1900–04), 3:437–38; Timmerman, "Alchemy in Cambridge," 434–37.

[209] Fol. 51r: "Hic unum incredibile, et ut videtur racioni inpossibile, et ut reperi scribere proposui, licet naturaliter non videatur esse verum."

[210] Elisabeth Wunderle, *Katalog der mittelalterlichen lateinischen Papierhandschriften aus den Sammlungen der Herzog-von-Sachsen-Coburg-und-Gotha'schen Stiftung für Kunst und Naturwissenschaft* (Wiesbaden: Otto Harrassowitz Verlag, 2002), 165–73.

Le^5_d Leiden, Leiden University Library, Vossianus Chym. F. 27, fols. 239r–50r (last q. 16th c.)[211]

This German translation bears the Latin title *Opus fratris Johannis de Rupicissa de lapide* and is independent of the other known German translations. It contains Operations 1–7 (omitting the prologue, "Credas," and Operation 8). It displays the textual variants of no one family. It contains a few variants typical of family γ, but also scattered resemblances to family ι and some descendants of π.[212] Yet these occasional commonalities do not permit assignment to any known family. It contains added references to pseudo-Arnald's *Testamentum*, explanatory lines, and several extended insertions. Most oddly, one of these insertions is a German version of the same text inserted at the same point in Operation 4 that appears in the roughly contemporaneous Italian translation **Fn^3_i**.[213]

Lh$_{cat}$ Bethlehem, Pennsylvania, Lehigh University Library, MS 10, fols. 145r–49r (1472)[214]

This witness is part of a collection on paper containing alchemical treatises and recipes in Latin, Catalan, and Italian, compiled in Naples by Arnald of Brussels during the last three decades of the fifteenth century. **Lh$_{cat}$** bears a colophon indicating that it was finished on 31 December 1472 ("complida la obra en Napols per Arnau de lishout de bruxella Any 1473 pridie kal. Januarii"). An opening rubric introduces the text as "Here begins the work that Brother John of Rupescissa composed concerning the philosophical stone which he wished to show to the chosen of God of the holy Roman Church."[215] Then follows a prayer not found in other witnesses, the prologue, and Operations 1–6; Operations 7 and 8 are absent. The text concludes with a brief paragraph on multiplication (*La moltiplicatio dela dita obra*) of unknown origin. Catalan rubrics preface each operation, and these are given in Latin in the table of contents (fols. xiiv–xiiir). Most strikingly, the

211 Boeren, *Codices*, 77–82.

212 As is typical of γ, it squeezes mercury through hare's leather (*weltzen ihn umb in einer hasenhautt*) but shows none of the other characteristics; as in ι, it claims that mercury is generated in caverns (*in den hölen oder klüften der Erden generirt*). At l. 116, it doubles the alternate readings *minutissime* and *mundissime* (*woll klein undt sauber*) found in differing families, the former in π^2 and γ^2, and the latter in all the others.

213 Fols. 245r-v: "undt conjungir es in gleichem gewichte wie sich solches in einem Matrimonio gebürett mitt dem obgemelten Weibe dan dieses Mänlein (das ist das Elementum ignis) ist warmer oder hitziger Natur, das weib aber in welchem der Man wirrkett undt mitt ihr zu schaffen hatt, ist kalter Natur, undt allso wirdt sich der Man und das weib zusammen finden, Wan aber diese zwey allso mitteinander conjugirtt sein, (dan sie als unterschiedene, geteilt undt wiederwertige Naturen können ohne Medium nicht zugleich bey einander stehen oder bleiben) / So mus man derhalben den Lufft oder das Elementum aeris als ein Medium hinzu thun, das es die gemelten zwey Elementa fein friedlich mitteinander conjugirn undt verbinde, das sie beyeinander wohnen undt leben."

214 William Jerome Wilson, "An Alchemical Manuscript by Arnaldus de Bruxella," *Osiris* 2 (1936): 220–405; *Catalogue of Latin and Vernacular Alchemical Manuscripts in the United States and Canada*, *Osiris* 6 (1939): 1–836, on 473–514; and Michela Pereira, "Alchemy and the Use of Vernacular Languages in the Late Middle Ages," *Speculum* 74 (1999): 336–56, esp. 351–53.

215 Fol. 145r: "Açi comença la obra que composa frate Johan de roccatagliada dela pedra phyilosophal la qual volgue mostrar als elects de deu dela sancta sglesia Romana."

"Credas" ("Cresegnas per ver tu pobre evangelich...", ll. 220–51) is significantly expanded with new text about the need for secrecy and the punishments awaiting those who fail to maintain it ("Maledictio contra los revelants daquesta scientia"). **Lh$_{\text{cat}}$** ends at the same point as **Sa,** the only known Iberian witness, but the two do not share textual variants.

Sa Salamanca, Biblioteca Histórica General, MS 2108, fols. 59v–65r (mid 16th c.)[216]

This is the only currently known witness produced in Iberia. It is part of a collection of alchemical treatises in Latin and Catalan by pseudo-Lull and pseudo-Arnald written on paper in a humanistic bookhand. Titled *Ars Alkimica fratris Joannis de Ruppe scissa*, it immediately follows a witness of *De famulatu*. In fact, the two Rupescissa works are announced together (fol. 1v), where the *Liber lucis* is described as the "third volume" of *De famulatu*, and "concerning the practice of the tree of philosophy and of the transmutation of metals through their reduction into a quintessence, according to master Raymond Lull" ("Tertius est de practica arboris philosophie et de metallorum transmutatione per reductionem ipsorum ad quintam essentiam secundum Magistrum Raymundum Lulli"). **Sa** contains the prologue but only the first six operations, ending with the "Credas" and concluding "Deo Gratias et Virgini matri ejus Marie." *Amen.* Strikingly, Operation 6 shows all the unique textual variants found in γ^3; its readings are especially close to those of **Oth$^{\text{1}}$**.[217] The other operations, however, do not correlate with γ^3, and since **Oth$^{\text{1}}$** is an excerpt containing only the "Credas," it is not possible to make further comparisons.

W^8 London, Wellcome Library, MS 710, fol. 24v (extr., Operation 8; late 15th c.)[218]

This single sheet immediately follows a witness of *De famulatu* (fols. 1r–23r and 25r–26v) in the same *gothica cursiva* hand, probably of North Italian origin. It contains one drawing of the *lamina* within the lines of text, which is followed by a long line lifted from *De famulatu* about how to position the *lamina*. It shows no textual variants that could serve to place it in any known family.

Wb$^4{}_{\text{d}}$ Wolfenbüttel, Herzog August Bibliothek, Cod. 306.1 Extravagantes, fols. 1r–7v (1st half 16th c.)[219]

This witness is part of a hastily written collection of alchemical materials in small duodecimo format. **Wb$^4{}_{\text{d}}$** bears the title *About the singular work of the*

[216] *Catálogo de Manuscritos de la Biblioteca Universitaria de Salamanca*, 6 vols. (Salamanca: Ediciones universidades, 2002), *Part 2: Manuscritos 1680–2777*, 473–75; Guy Beaujouan, *Manuscrits scientifiques médiévaux de l'Université de Salamanque et de ses colegios mayores* (Bordeaux: Féret, 1962), 112–15; and J. Massó Torrents, *Manuscritos catalanes de la bibliotheca de S. M.* (Barcelona, 1888), 23–24.

[217] For example, "Sa ends ut finaliter ab hac calamitate liberentur," a phrase found only in **Wb**, **Ka^2**, and **Oth^1**.

[218] Moorat, *Catalogue*, 1:522–23; Wellcome MSS 443, 444, and 710 were originally one manuscript.

[219] Hans Butzmann, *Kataloge der Herzog-August-Bibliothek Wolfenbüttel 15: Die mittelalterlichen Handschriften der Gruppen Extravagantes, Novi und Novissimi* (Frankfurt: Klostermann, 1972), 167.

philosophers of Peter[!] *of Rupescissa* (*Super opus singulare philosophorum petri de rupe scissa*), and is prefaced with "A blessed father of the Franciscans called Peter of Rupescissa made the present little book about the stone that one calls the stone of the pagan masters, in Latin, the *lapis philosophorum* or *elixir.* The same stone is twofold, white and red, as stands written hereinafter."[220] It begins at l. 28, continues through Operations 2–7, but omits Operation 8. It is highly edited throughout, but especially in the later operations; Operation 6 is edited down to about fifty words. It does contain several short interpolations; the term *azoth* is used for mercury. Due to its highly paraphrased state, it bears no clear characteristics that allow for assignment into any family.

6.9 Abbreviated Versions

$\mathbf{Z_a}$, $\mathbf{Hb_a}$, and $\mathbf{Le^{I}{}_{a}}$; $\mathbf{Fr_a}$; $\mathbf{Ve_a}$; and $\mathbf{Mp^{I}{}_{a}}$

Six witnesses preserve abbreviated versions of the *Liber lucis*; they represent four independent abridgments made from exemplars from three different families.

$\mathbf{Z_a}$ - Venice, Biblioteca Nazionale Marciana, Lat. fondo antico 323 (= 1579), fols. 185v–86v (1470s–80s); $\mathbf{Hb_a}$ - Hamburg, Staats- und Universitätsbibliothek, Cod. Alch. 659, fols. 119v–20v (1468); and $\mathbf{Le^{I}{}_{a}}$ - Leiden, Leiden University Library, Vossianus Chym. Q. 2., fols. 122r–v (c. 1500)[221]

These three are witnesses of an abridgment produced prior to 1468. Sufficient text remains to identify the source text as a member of family γ.[222] The fullest witness, $\mathbf{Z_a}$, contains Operation 9, which $\mathbf{Hb_a}$ and $\mathbf{Le^{I}{}_{a}}$ do not. Thus, if this spurious operation was copied from the same source as the rest of the abridgment, which seems likely, then the original source must have come from subfamily γ', the only subfamily that includes Operation 9. The prologue and first half of Operatio 1 are absent, as are the "demonstratio," "Credas," explications of allegories, references to authorities, and theoretical considerations, the intent apparently having been to reduce the text closer to a recipe.

$\mathbf{Z_a}$, titled *Liber magisterii perfeccionis veri lapidis philosophorum*, provides short versions of Operations 1–6 and 9 (labeled as Operation 7); Operations 7 and 8 are absent. Its colophon is drawn from Operation 9: "Here ends of the little treatise of master John of Rupescissa bearing white and red roses extracted concisely

[220] This witness's unique mistake of giving John the first name of Peter may be the result of an earlier translation of Rupe as Petrus, as in $\mathbf{Ka^{2}}$ where John is called "Petricissa," or $\mathbf{Ka^{5}{}_{d}}$ where he is called "von Steinruszen." Fol. 1r: "Ain seliger vater der minder bruder genant petrus de rupe scissa, hat gemacht dez gegenwerttige puechelein van den staine man nent den stain der haydnischen maister in lateyn lapis philosophorum oder elixir. Der selbig stain ist zwayerlay weiss und rot alß hernach darvon geschrieben stadt."

[221] Valentinelli, *Manuscripta*, 5:164, on the dating, see note 76 above; for $\mathbf{Hb_a}$, see Paulus, "Katalog," 69–73; and for $\mathbf{Le^{I}{}_{a}}$, Boeren, *Codices*, 116.

[222] For example, the squeezing of mercury through leather, the addition of seven parts (rather than a seventh part) of sal ammoniac to the aquafort, and so on.

from the books of the philosophers, and the book is called the *Rose Garden of the Philosophers*."[223] Although the same manuscript volume contains the complete witness **Z** (fols. 114v–19r), that version and the abridgment do not seem to have traveled together since $\mathbf{Z_a}$ is separated from **Z** by about seventy folios, and hence was copied at a different time, presumably from a different source.

$\mathbf{Hb_a}$ is only slightly earlier than $\mathbf{Z_a}$, but it did travel together with a full version (**Hb**) that descends through γ'''. The third witness, $\mathbf{Le^{I}{}_a}$, which is likewise followed by a full version ($\mathbf{Le^{I}}$), was probably copied directly from $\mathbf{Hb_a}$; their texts are almost identical, as are the texts of the accompanying full versions $\mathbf{Le^{I}}$ and **Hb.** Neither $\mathbf{Hb_a}$ nor $\mathbf{Le^{I}{}_a}$ provides a title for the abridgment.[224] Like $\mathbf{Z_a}$, $\mathbf{Hb_a/Le^{I}{}_a}$ omit Operations 7 and 8, but do not contain Operation 9. Although it cannot be definitively decided whether $\mathbf{Z_a}$ added Operation 9 or $\mathbf{Hb_a/Le^{I}{}_a}$ omitted it, yet since $\mathbf{Hb_a/Le^{I}{}_a}$ omit several other sections of text present $\mathbf{Z_a}$, it seems likely that they are further edited and therefore, $\mathbf{Z_a}$ more closely represents the common ancestor of all three.

A transmutational recipe attributed to John immediately follows all three witnesses; see Appendix 2.

$\mathbf{Fr_a}$ Florence, Biblioteca Riccardiana, MS 439, fols. 45v–47v (mid-15th c.)[225]

This abridgment follows a full *De famulatu* and is titled *Lapis philosophorum.* It is written on vellum in a *gothica rotunda* bookhand and the manuscript is of Italian origin. It omits the prologue, the first part of Operation 1, the "Credas," and Operation 8. Its inclusion of Operation 9, its use of hare's leather for the addition of the mercury, and other readings indicate its descent through family γ', but it is independent of the previous abridgment ($\mathbf{Z_a}$, $\mathbf{Hb_a}$, and $\mathbf{Le^{I}{}_a}$). It concludes with references to *De famulatu* for details of the furnace and the *lac virginis.*

$\mathbf{Ve_a}$ Vienna, Österreichische Nationalbibliothek, Cod. Palatinus 5487, fols. 105r–6v (2nd half 15th c.)[226]

This abridgment is independent of the previous ones and does not descend through γ, although its origins cannot be further determined.[227] It is less condensed, retaining more of the theoretical and explanatory material than the previous abridgments. The prologue is absent (although its first line is correctly cited in the title), Operations 1 through 7 appear in condensed form; Operation 8 is

[223] Fol. 186v: "Explicit tractatulus Magistri Johannis de rubecisse ferens rosas albas et rubeas extractas breviter ex libris philosophorum et vocatur liber rosarius philosophorum."

[224] A later owner of the manuscript, Caspar Kegler, added the title "Tinctura ex solo mercurio" to $\mathbf{Hb_a}$ in 1515.

[225] Francesca Mazzatinti and Maria Luisa Tanganelli, *Catalogo dei manoscritti della Biblioteca Riccardiana di Firenze: Ricc. 421–520* (Rome: Istituto Poligraphico e Zecca dello Stato, 2022), 59–60, in which the abbreviated *Liber lucis* not recognized as such.

[226] *Tabulae codicum manu scriptorum praeter graecos et orientales in Bibliotheca Palatina Vindobonensi asservatorum*, vol. 4: *Cod. 5001–Cod. 6500* (Vienna, 1870), 133–34.

[227] The customary doubling of "materia et natura" is absent, the mercury is squeezed through a cloth (*per pannum*), and the correct "seventh part" of sal ammoniac is added to the aquafort, all of which indicate a source outside of γ.

omitted, as is the "demonstratio," but two modified sentences from the "Credas" are included. The title claims it to be "an abridgement of Raymond's *Liber lucis*, which begins 'I considered the tribulations of the elect, etc.,' compiled by John of Rupescissa."[228] Most of the texts in its collection are pseudo-Lullian. This abridgment calls Arnald of Villanova "my master" (*magister meus*), and refers to two suppositious books by John, a "treatise of furnaces" and a "treatise of vessels," as well as to pseudo-Lull's *Codicillus* and *Testamentum*.[229] It also contains practical specifications from other sources, such as the proportion of aquafort and sublimate to combine (3 to 1 by weight), a description of the heat for digesting the white stone (that of the sun or a fever), and a comparison of the white stone's color to camphor.

$\mathbf{Mp}^{\mathrm{I}}_{\mathrm{a}}$ Montpellier, Bibliothèque Universitaire Historique de Médecine, MS H 493, fols. 233r–36r (1459)[230]

This witness provides a fourth independent abridgment. $\mathbf{Mp}^{\mathrm{I}}_{\mathrm{a}}$ immediately precedes a full version of the *Liber lucis* ($\mathbf{Mp}^{\mathrm{I}}$) written in the same hand on the same paper; both are Neapolitan productions dating from 1459. $\mathbf{Mp}^{\mathrm{I}}_{\mathrm{a}}$ is headed "This was extracted from the *Liber lucis* of Brother John of Rupescissa of the Order of Friars Minor, abbreviated not at all in substance but only in words."[231] Its colophon repeats: "This is extracted from the *Liber lucis* concerning the work of the philosophical stone, the sense has not been altered, but I have abbreviated it by setting aside much prolixity."[232] It omits the prologue and the first part of Operation 1, but contains condensed versions of all eight operations and the spurious ninth. Fol. 236r contains a red and black drawing of the furnace (Figure 2.6). $\mathbf{Mp}^{\mathrm{I}}_{\mathrm{a}}$ is *not* an abridgment of $\mathbf{Mp}^{\mathrm{I}}$; $\mathbf{Mp}^{\mathrm{I}}$ descends through Italian family ι^2 while $\mathbf{Mp}^{\mathrm{I}}_{\mathrm{a}}$ descends through $\pi^{\mathrm{I}'''}$, as indicated by its identical lacunae in Operation 1 and several characteristic textual variants which persist despite the abbreviation of the text. It most closely resembles **G**, since they share Operation 9 and further unique variants.[233] This evidence suggests a common intermediate descending from $\pi^{\mathrm{I}'''}$ but contaminated with material taken from a descendant of π^3; this postulated hyparchetype is shown as ζ in the stemma. See also Section 3.3.

228 Fol. 105r: "Accurtata Libri Lucis Reymundi Qui incipit Consideravi tribulaciones electorum etc. compilata per Joanne de Rupescissa."

229 Fol. 109v: "hunc furnum demonstravi in tractatu furnorum et vocatur athanor"; fol. 110r, "vas sit rotundam sicut docui in tractatu vasorum."

230 Corbett, *Catalogue*, 2:120–33; Alessandrini, *Cimeli*, 287; and Principe, "Missing Link."

231 Fol. 233r: "Hoc fuit extractum a libro lucis fratris Johannis de Rupe cisa ordinis minorum in substantia nichil sed in verbis abbreviatum."

232 Fol. 236r: "Extractum est hoc a libro lucis de opere lapidis philosophici non variata sententia sed abbreviavi dimittendo plura prolixa."

233 For example, a shared lacuna at l. 183: "si nimis tardaret (*om.* adauge ei ignem et cum videris materiam denigrari) gaude" and at l. 188 "*notabile(m)* perfectum ad album."

6.10 Witnesses Based Upon Printed Editions
E, Go², L⁷ₑ, Lu, Psgₛ, Ua²ₑ, W⁵, and Zcᵢ

It is worth noting that of the eight witnesses based upon printed editions, only three of them are direct copies. The others are either translations (into three vernaculars) or, in the case of **Go²**, a compilation of elements from both the Basel and Cologne versions.

E El Escorial, Real Biblioteca del Monasterio de San Lorenzo, Cod. Lat. H-1-9, fols. 181v–83v (18th c.)[234]

This very late manuscript contains four alchemical treatises "extracted from the second volume of the *Theatrum chemicum*" (fol. 170r) including the Cologne version of the *Liber lucis* found there.[235]

Go² Gotha, Forschungsbibliothek Gotha der Universität Erfurt, Chart. A 519, fols. 22r–34r (mid-1580s)

The copyist of **Go²** reproduced the entirety of the 1579 Cologne edition, including even the layout of the title page and the furnace illustrations, yet also appended the three supplemental texts added by Gratarolo to his 1561 Basel edition, indicating that copyist had compared the editions even if he did not modify the Cologne text.

L⁷ₑ London, British Library, Harley MS 4724, fols. 54v–61v (1st half 17th c.)[236]

This *Boke of Light of John De Rupecissa of the order of the Friers Minors* was translated directly from the Cologne version and the manuscript includes translations of the two other texts in that publication: *A Treatise of Thomas of Aquine* and *The Keye of Raymond Lully.*

Lu Lübeck, Stadtbibliothek, MS. Med. Quarto 12, fol. 40r–53r (last q. 16th c., after 1579; lost)[237]

Described by Sudhoff in 1899 and Hagen in 1932, this manuscript is no longer in the Stadtbibliothek Lübeck.[238] Nevertheless, the incipit and explicit given by Hagen identify its text as that of the Cologne version. The lost manuscript also

[234] Guillermo Antolín, OSA, *Catálogo de los códices latinos de la Real Biblioteca del Escorial,* 5 vols. (Madrid: Imprenta Helénica, 1910–23), vol. 2 (1911), 413–16, on 415.

[235] The four treatises copied here actually come from the *third* volume of the *Theatrum.*

[236] Nares, et al., *Catalogue of the Harleian Manuscripts*, 3:196.

[237] Karl Sudhoff, *Versuch einer Kritik der Echtheit der Paracelsischen Schriften, Theil 2: Paracelsus-Handschriften* (Berlin, 1899), 118–22, on 122; and Paul Hagen, *Naturwissenschaftliche Handschriften*, Cat. manuscr. 106, Lübeck 1932, fols. 68r–72r (in manuscript, edited by Anne Harnisch, 2018: http://diglib.hab.de/?db=mss&list=ms&id=hl-sb-med-4f-12&catalog=Cat-manuscr&hi=C.%20S

[238] I wish to thank Britta Lukow of the Stadtbibliothek Lübeck and Kathrin Hoffmann of the Institut für Medizingeschichte und Wissenschaftsforschung at the University of Lübeck for looking into this matter at my request in 2019 and confirming that the manuscript is no longer in either of their collections.

contained the same two alchemical works falsely attributed to Thomas Aquinas that appear alongside the *Liber lucis* in the Cologne edition.[239]

Psg$_s$ Paris, Bibliothèque Sainte-Geneviève, MS 3438, fols. 43r–49r (2nd half 17th c.)

This only known Castilian translation occurs in a large collection of alchemical and astrological treatises and notes, the majority in Latin, but fols. 1r–57v are in Spanish and a few later pages are in French. The text of **Psg$_s$** is translated from the Cologne version, but with alternate readings drawn from the Basel version. Both the prologue and the furnace description are omitted. The library's catalog dates the manuscript to 1768 based on the appearance of that number in the upper margin of fol. 1r, but this is not actually a date, and the hand of the text is distinctly seventeenth century. The latest material is a Latin text "ex Guillelmo Trogniano," (fols. 58r–v) that is, Guillaume de Trougny, an alchemist from Orléans in the circle of Théodore de Mayerne during the first quarter of the seventeenth century, and most of the Latin treatises on fols. 59r–69r appear in the same order as they do in volume 6 of the *Theatrum chemicum* (1661), from which they were almost certainly copied.

Ua^2_e Amsterdam, Universiteit van Amsterdam, Allard Pierson Depot, MS PH177, fols. 101r–06r (ca. 1800)[240]

This very late witness ("The book of Light of Joannes de Rupescissa a Minorit Friar") is translated from the Cologne version as part of a collection of English translations of twelve Latin alchemical texts.

W^5 London, Wellcome Library, MS 512, pp. 49–60 (1660s)[241]

This witness is part of a collection of alchemical treatises compiled in France in the 1660s, that also contains two pseudo-Thomas Aquinas tracts and pseudo-Lull's *Clavicula*, the same works in the same order they appear in the Cologne edition. The same tracts and order appear also in volume 3 of the *Theatrum chemicum*, which might be the source for the translation, but since no other treatises from the *Theatrum* appear here, the Cologne edition seems more likely.

Zc$_i$ Venice, Biblioteca del Museo Correr, MS Correr 106, fols. 125r–34r (c. 1605)

This second Italian translation is roughly contemporaneous with that of **Fn^3_i** but was clearly done from the Basel version, given the unique readings that they share; the concluding "annotations" have been omitted.

[239] The pseudo-Lull's *Clavicula*, the fourth work present in the 1579 printing, was not included. The date 1579 appeared on the original binding of the manuscript.

[240] https://lib.uva.nl/discovery/fulldisplay?docid=alma994050053450513 1&context=L&vid=31UKB_UAM1_INST:UVA&lang=en&search_scope=DN_and_CI_and_PURE&adaptor=Local%20Search%20Engine&tab=LibraryCatalog&query=any,contains,ph177&facet=rtype,include,manuscripts&offset=0

[241] Moorat, *Catalogue*, 1:352–53.

APPENDICES

Appendix 1: Sections from *De famulatu* That Parallel the *Liber lucis*

For purposes of reference and comparison, I have included here the two sections of John of Rupescissa's *De famulatu philosophie evangelio* that parallel material in his later *Liber lucis magisterii magni.* These sections make it clear that John already had his theory and practice of preparing the virgin's milk, the immediate precursor of the philosophers' stone, in mind (and presumably worked out at least in part) roughly two years before writing the *Liber lucis.* Nevertheless, he did not at that time connect the process explicitly to chrysopoeia, beyond the (rather obvious) hint at the end of the second section that the product contains the "secret of secrets."

The Latin texts I present here are *provisional.* I have compiled them from no more than a preliminary collation of twenty selected witnesses (of the over 250) of the *De famulatu.* The critical edition now in preparation will replace these provisional texts, which are not to be considered in any way critically edited. I have purposely suppressed the critical apparatus here in order to avoid any confusion about the provisional status of both the Latin texts and their English translations.

Scientia ad extrahendum quintam essenciam ex argento vivo et vitriolo romano et cuperosa.

Gloriosus et sublimis deus ita ordinavit ut quinta essencia vitrioli et cuperose cum argento vivo invisibiliter extrahatur, quoniam quando argentum vivum sublimatur cum eis et cum sale communi tunc ascendit argentum vivum albissimum et quintam essenciam vitrioli et cuperose sine dubio trahit secum. Et cum habes argentum vivum sublimatum cum istis, habes invisibiliter quintam essenciam eorum absconsam infra albedinem argenti vivi predicti. Et illa quinta essencia invisibilis vocatur sulphur philosophorum quod est invisibile. Ideo dicunt quod sulphur eorum non est sulphur vulgi seu commune, et paucissimi hominum ad hanc noticiam pervenerunt. Et ipsum est quod statim congelat mercurium in album supra albedinem nivis et quia sunt ejusdem nature ideo secum eciam sublimantur. Et scito quod argentum vivum sublimatum habet quasdam partes combustibiles et habet quintam essenciam que corrumpi non potest per ingenium nostrum. Et nos videmus quod si infra aquam corrosivam factam ex vitriolo et sale petra immittas argentum vivum sublimatum laniabitur totum et in athoma convertetur. Immitte ergo argentum vivum sublimatum cum cuperosa vel vitriolo et sale communi albissimo infra aquam corrosivam predictam donec totum argentum vivum dissolvatur in aquam. Consequenter totum illud pone ad distillandum, ascendet enim primo aqua corrosiva deinde sequetur quinta essencia mercurii et vitrioli simul ita alba ut albedo nivis ei non valeat adequari. Et remanebit in fundo pars mercurii combusta et turpis. Reitera dissolucionem hujus quinte essencie in aqua forti predicta et sublimacionem predictam ter vel plus donec post sublimacionem nulle feces remaneant in fundo vasis sed totum sublimetur in album et habebis, jussu dei, quintam essenciam mercurii et vitrioli simul perfectissime separatam. Si quis caperet misterium ejus maximum foret quia res grandes habet, ex quo ejus quinta essencia incorruptibilis est effecta et de natura incorruptibilis auri et incorruptibilis plusquam aurum. Et si cupis quintam essenciam vitrioli vel cuperose extrahere a quinta essencia mercurii infunde illud in aceto distillato vel alio et dimitte resolvi diu et argento vivo residente in fundo, quinta essencia vitrioli vel cuperose in aceto sine dubio remanebit. Quod si ad oculum probabis extingue pluribus vicibus ferrum in aceto et postea illud acetum misce cum illo in quo est quinta essencia vitrioli vel cuperose et habebis rem rubicundissimam et splendidam supramodum. Distilla per filtrum et pone ad lentissimum ignem et collige nebulas nigras que generabuntur in superficie et quinta essencia pulchrior omni auro tibi remanebit in fundo que valet ad vulnera sananda et morbos corrosivos sic et quinta essencia mercurii sanat fistulam et cancrum et mala alia plurima.

Translation

The knowledge for extracting the quintessence from quicksilver, Roman vitriol, and copperas.

The glorious and sublime God has so ordained that the quintessence of vitriol and copperas is invisibly extracted with quicksilver, because when quicksilver is sublimed with them and with common salt, the quicksilver then ascends very white and without a doubt carries up with itself the quintessence of the vitriol and copperas. And when you have quicksilver sublimed with these substances, you have their quintessence hidden invisibly beneath the whiteness of the aforesaid quicksilver. And that invisible quintessence is called the sulphur of the philosophers because it is invisible. Therefore they say that their sulphur is not the sulphur of the vulgar or common sulphur, and very few people arrive at this knowledge. And it is this same thing that immediately coagulates mercury into something white beyond the whiteness of snow, and because the two are of the same nature, they are likewise sublimed together. And know that sublimed quicksilver has some combustible parts and has a quintessence that cannot be corrupted through our techniques. And we see that if you put sublimed quicksilver into a corrosive water made from vitriol and saltpeter, it will be entirely torn to pieces and converted into atoms. Therefore, put quicksilver sublimed with copperas or vitriol and the whitest common salt into the aforesaid corrosive water until all the quicksilver is dissolved into water. Thereafter, put all of it to be distilled, for the corrosive water will ascend first, then the quintessence of mercury and vitriol together will follow so white that the whiteness of snow cannot equal it. And a part of the mercury will remain at the bottom, burnt and dirty. Repeat the dissolution of this quintessence in the aforesaid aquafort and the aforesaid sublimation three or more times until no residue remains at the bottom of the vessel after the sublimation, but everything is sublimed white, and you shall have, by the command of God, the quintessence of mercury and of vitriol together, separated most perfectly. If anyone should grasp its mystery, it would be the greatest because it contains great things, whereby its quintessence is made incorruptible, and of the incorruptible nature of gold, and more incorruptible than gold. And if you wish to extract the quintessence of vitriol or of copperas from the quintessence of mercury, pour it into distilled or other vinegar and set it aside for a long time to be resolved, and while the quicksilver settles to the bottom, the quintessence of vitriol or copperas will without a doubt remain in the vinegar. If you will prove this to your eyes, quench iron many times in vinegar and then mix that vinegar with the one that contains the quintessence of vitriol or copperas, and you will have something extremely red and beautiful beyond measure. Distill it by a felt, place it in a most gentle heat, and collect the black clouds that will be generated on its surface, and the quintessence, more beautiful than all gold, will remain for you at the bottom. This quintessence is good for healing wounds and corrosive diseases; so likewise the quintessence of mercury heals fistula and cancer and many other ills.

Scientia ad reducendum argentum vivum sublimatum vel quintam essenciam ejus separatam in aquam que apud philosophos lac virgineum appellatur.

Omnis spiritus sublimatus sicut est argentum vivum sublimatum aut eciam quinta essencia ejus, aut arsenicum sublimatum, aut sulphur, aut armoniacum, et maxime hoc intellige pro argento vivo sublimato, sic enim reduces in aquam vel in lac virgineum secundum nominationem philosophorum, et est modus iste. Recipe argentum vivum sublimatum vel quintam essenciam ejus, et pone in vase vitreo vel vitreato lato et amplo et constitue illud in tripoda athanor, id est in furno philosophorum, ubi est ingenium mirabile ut illud quod in eo ponitur digeratur equaliter ab omni parte, et fit sic: Construe parietes furni rotundi in altitudine medii pedis, et super parietes elevatos ingenia laminam ferream talem <*many manuscripts have here a small drawing of the plate*> et sic facies ut latera lamine non tangant parietes furni sed sustentetur in quatuor illis brachiis suis, et calor qui ascendet ab infra tantum per latera ascendere possit in circuitu lamine atque furni. Supponitur quod feceris hostiolum juxta terram unde immittantur carbones. Deinde super parietes a lamina et supra edificare incipies in rotundum per medium palmum vel citra et desuper testudinem rotundissimam ingenio figuli factam locabis, cum qua cooperies furnum tuum. Et in latere supra lamina foramen facies quod claudes cum hostio de terra aptato, et sic habes furnum tuum. Certum est tibi si intellectum habes quod iste furnus equaliter calefaciet ab omni circumferencia ejus quia calor propter laminam impeditur ab infra et projicitur ad latera et a lateribus reflectitur equaliter ad medium, ideo furnum fecimus scienter rotundum. Cum autem vas voluerimus in furno locare, in medio lamine constituemus tres dentes per duos digitos elevatos equaliter distantes et scutellam desuper apponemus ut in aere elevetur. Et infra scutellam terream alios tres dentes minores similiter erigemus et super hos constituemus amphoram vitream sigillatam quam cooperiemus cum alia scutella terrea ut in medio ambarum scutellarum amphora in medio intersticio aeris collocetur non tangens nec fundum nec latera undecumque. Et quia supponitur quod caput seu medietas furni elevari potest, aperiemus et cooperiemus quando volumus. Hic est furnus secretissimus philosophorum qui athanor appellatur et ad noticiam ejus paucissimi homines pervenerunt. In hoc fit digescio equaliter ab omni parte.

Cum ergo voles mercurium vel aliquem spiritum sublimatum convertere in aquam seu lac virgineum pone super laminam in tridentem scutellam cum re diligenter trita, et non cooperies eam sed aer calidus materiam nudam tangat et da ignem trium carbonum, quia modicus ignis sufficit tibi. Et continua ignem jugiter per octo dies et noctes et ibi in stupha calida et sicca spiritus calcinatur. Extrahe ipsum deinde et diligentissime tritura super marmor materiam donec sit sine tactu. Post hoc pone in distillatorio vitri quod colloca in balneo marie in caldario et de duobus digitis non tangat aquam sed in aere suspendatur. Fiat ignis sub balneo in stupha humida et calida. Licet non tangat aquam materia argenti vivi resolvetur in aquam. Hoc est quia fuit bene calcinatum et bene trituratum. Posito ergo desuper alembico distillabitur lac virgineum clarissimum, et est quinta essencia tam argenti vivi quam vitrioli simul mixta. Et est in ea secretum secretorum. Et mensura seu pondus ignis est in digestione omnium volatilium in nostro athanor, hoc est in furno, ut per hostium quod est factum supra laminam immittas manum in furnum quam tenere possis in aere sine lesione, alias nimis magnus est ignis; minue eum et cave. In fixione autem omnium rerum volatilium ideo necesse est rem infra amphoram vitream sigillare et ignem debilem dare. Intellige ergo extractionem quinte essencie in hac parte quia hec est res magna supra quam credi possit.

Translation

The knowledge for reducing sublimed quicksilver, or its separated quintessence, into a water that is called virgin's milk among the philosophers.

Every sublimed spirit, such as sublimed quicksilver or likewise its quintessence, or sublimed arsenic, or sulphur, or [sal] ammoniac—and understand this most of all for sublimed quicksilver—you will thus reduce into a water, or into the virgin's milk according to the nomenclature of the philosophers, and this is the method: Take sublimed quicksilver or its quintessence and put it into a wide and ample glass or glazed vessel and set that up on the tripod of the athanor, that is, in the furnace of the philosophers, where there is a marvelous technique such that what is placed into it is digested equally from all sides. It is made thus: Construct the walls of a round furnace half a foot in height and cleverly devise upon the raised walls an iron plate like this: <*some manuscripts have here a small drawing of the plate*> and you shall make it such that the edge of the plate does not touch the walls of the furnace but rather is supported upon its four arms. The heat that will ascend from below can ascend only around the edge at the circumference of the plate and furnace. It is supposed that you will make a little door near the ground where coals can be put in. Then, upon the walls upwards from the plate, you will begin to build in the round for half a palm or more, and you will place on top a very round shield cleverly made of clay with which you will cover your furnace. And in the walls above the plate, you will make an opening which you will close with a fitted door of earth. Thus you have your furnace. It is clear to you, if you have any intellect, that this furnace will heat equally from its entire circumference because the heat from below is impeded due to the plate and is projected toward the walls and reflected by the walls equally toward the middle; we have thus wisely made the furnace round. When we should wish to place a vessel in the furnace, we will set up three teeth raised two fingers high and equally spaced in the middle of the plate, and we will set a dish upon them so that it is raised up in the air. We will similarly raise up three smaller teeth within the earthen dish and upon those place the sealed glass flask that we will cover with another earthen dish so that the flask is located in the interior airspace in the middle of the two dishes, not touching either the bottom or the walls anywhere. And because it is supposed that the head, or middle part, of the furnace can be lifted up, we will open and cover it whenever we wish. This is the most secret furnace of the philosophers that is called athanor, and extremely few people come to the knowledge of it. In it, a digestion is made equally from all sides.

When, therefore, you wish to convert mercury or any sublimed spirit into water, or the virgin's milk, place a dish with the material diligently ground on the tripod above the plate and do not cover it, but let the hot air touch the naked material, and give a fire of three coals because a moderate heat is sufficient for you. And continue the heat without ceasing for eight days and nights, and there in the hot and dry oven the spirit is calcined. Then take it out and grind the material most diligently upon a marble until it is impalpable. Thereafter put it in a glass distillatory, which place in a bain-marie in the warm bath [*caldarium*] and let it not touch the water by two fingers but rather let it be suspended in the air. Let a fire be made under the bain-marie in the humid and hot oven. Although it does not touch the water, the material of the quicksilver will be resolved into water. This is because it was well calcined and well ground. Therefore, after having placed an alembic on top, let the most clear virgin's milk be distilled; it is the quintessence of both the quicksilver and the vitriol mixed together, and the secret of secrets is in it. And the measure or power of the fire during the digestion of all volatiles in our athanor, that is, in our furnace, is such that you may put your hand into the furnace through the door that is made above the plate and hold it there without any hurt; otherwise, the fire is too great, reduce it and take care. During the fixation of all volatile things likewise, it is therefore necessary to seal the thing within a glass flask and give a weak heat. Understand therefore the extraction of the quintessence in this part because this is a great thing beyond what might be believed.

Appendix 2: Spurious Witnesses of the *Liber lucis* and Other Alchemical Works Attributed to John of Rupescissa

The first part of this appendix lists and describes manuscript materials previously, but incorrectly, cited elsewhere as witnesses of the *Liber lucis*. This first part also includes several manuscripts uncovered in the course of this study that have not been previously cited by other scholars but which either contain sections of text coopted from the *Liber lucis* or, in some cases, endeavor to pass themselves off as the *Liber lucis*. The second part examines three alchemical texts that are not related to the *Liber lucis* but have nevertheless been attributed to John of Rupescissa by their authors. I have examined all the manuscripts listed in both parts.

Part 1—Spurious Witnesses of the *Liber lucis*

These texts fall broadly into four categories. The first and most trivial category contains manuscripts that have previously but erroneously been identified as the *Liber lucis*, probably only on the basis of a title given in a catalog without further investigation. A second group consists of documents that do contain a line or two quoted from the *Liber lucis*, often as an authoritative statement, but no more; such minor quotations do not render a document a witness in any meaningful sense of the word. More interesting cases fall into a third category, in which an author has coopted extended portions of the *Liber lucis*—one or more operations—and incorporated them seamlessly into his own composition. These documents indicate ways in which John's authentic text was coopted, modified, and redeployed by later authors for centuries after its composition, as was done most famously in the case of reworking John's *De famulatu* into the pseudo-Lullian *De secretis naturae*. These later compositions often borrow sufficient text to identify through which branch of the stemma the source text descended. Finally, a fourth category comprises intentional literary forgeries: Kassel MS 73, Leiden VCQ 54, Riccardiana MS 923/Lincei MS Rossi 238, and Bayerische Staatsbibliothek Clm 27000. In these cases, an unknown author consciously endeavored to disguise his composition as an authentic work by John of Rupescissa by explicitly attributing it to him and by including phrases or ideas characteristic of the Franciscan friar, such as appeals to the *viri evangelici* or references to the coming antichrist.

Bologna, Biblioteca universitaria, MS Lat. 500, fols. 218v–24r (15th c.)

Cited in Halleux, "Rupescissa," 282. The text has no relationship to the *Liber lucis*. It bears the title *Incipiunt septem operationes lapidis majoris secundum expertos per modum auctoritatum*, and a later hand has (mis)attributed it to "Ripleyus."

Bordeaux, Bibliothèque Mériadeck, MS 531, fol. 164v (ca. 1450)

Cited in Halleux, "Rupescissa," 282. This folio contains several quotations taken from Operation 4, prefixed with "Dicit de Rupecissa in libro qui intitulatur liber lucis magisterii." The text continued onto a following folio, now lost. Sufficient textual variants exist to identify the notetaker's exemplar as a member of family γ′. At fols. 2r–41r the collection contains a witness of *De famulatu*; see Corbett, *Catalogue* 2:21–22.

Chantilly, Musée Condé, MS 419, fols. 55r–56r (16th c.)

Cited in Corbett, *Catalogue*, 2:60–65 and Halleux, "Rupescissa," 283 as an Italian translation of the *Liber lucis*.[1] It is actually an Italian version of the Latin text found in Riccardiana MSS 942 and 984; see below.

Florence, Biblioteca Nazionale Centrale di Firenze, Magliabechi Cl. XVI, cod. 72, fols. 30r–35v (ca. 1600)

Cited in Giovanni Targioni Tozzetti, *Catalogo generale dei manoscritti Magliabechiani* (manuscript), vol. 5, fols. 228r–29r as "Joannis de Rupescissa (ut videtur) nepotis Raimundi Lulli, de compositione magni Lapidis Philosophici." This Italian-language document is titled "Ad inveniendum et Componendum 5am essentiam mineralem secundum opiniones omnium philosophorum cap. 3." It is unrelated to the *Liber lucis*, nor is it directly related to the *De famulatu*, even though it mentions the quintessence in passing. It also uses letters to stand for substances in the fashion of the pseudo-Lull. Its most notable feature is the claim that John was Lull's nephew!

Florence, Biblioteca Riccardiana, MS 923, fols. 174r–80v (16th c.)

Cited in Giovanni Lami, *Catalogus codicum manuscriptorum qui in Bibliotheca Riccardiana Florentiae adservantur* (Livorno, 1756), 248; Thorndike and Kibre, 761; and Halleux, "Rupescissa," 283. This text, described in the volume's table of contents as *Joanis de Rupe Scissa de Lapide Philosophorum seu liber lucis*, is headed on fol. 174r as *Joannis de rupe scissa de lapide philosophorum, Compendium quod liber lucis appellatum et Arcanum Arcanorum Deique optimi maximi donum*, and listed in the explicit (fol. 180v) as *liber Joannis de Rupescissa qui per eum liber lucis titulatus*. Nevertheless, this text has little overlap with the authentic *Liber lucis*. The unknown author of this text endeavored to pass it off as John's own composition in the title, explicit, and body of the text, with its repeated exhortations to the *viri evangelici*. Most of the practical and theoretical details diverge significantly from John's authentic ideas, but one section presents a modified version of part of the mercury sublimation from the *Liber lucis*. A variant version exists as Rome, Accademia dei Lincei, MS Rossi 238; see below.

1 See also *Chantilly: Le Cabinet des livres, manuscrits*, 3 vols. (Paris: Plon, 1900–11), 1:332–33, on 333.

Florence, Biblioteca Riccardiana, MS 942, fols. 21v–27v (late 15th c.)

Entitled *Raimundi Lullii De lapide philosophico minerali liber*, this text begins with a section lifted from Operation 1 of the *Liber lucis* (ll. 18–27). The remainder, however, bears no further similarity to the *Liber lucis*. The coopted section contains readings found only in **M**, **Rc**, and **W^{I}** (descendants of π^2). Like this document, **Rc** and **W^{I}** are also attributed to Lull, further suggesting that the author borrowed from a witness in that branch of the stemma (cf. next entry).[2]

Florence, Biblioteca Riccardiana, MS 984, fols. 1r–8r (ca. 1495)

This text, described simply as "ex intentione Raymundi," is part of an alchemical collection compiled by Ludovico Lazzarelli (1447–1500) around 1495.[3] It is nearly identical to the text of Riccardiana MS 942 (see above). Despite minor wording differences, the chapter headings are the same, and the same text is coopted from the *Liber lucis*, which here again contains readings found only in **M**, **Rc**, and **W^{I}**. This version, however, includes the heading *Scientia ad ostendendum que sit materia lapidis nostris*[sic] *et de modo preparationis ejus* for the coopted section, a version of which occurs otherwise only in **Rc** and **W^{I}**, thus strengthening the connection of this text with the line of descent passing through Bisticci's β.

Gotha, Forschungsbibliothek Gotha der Universität Erfurt, Chart. B 368, fols. 16r–21r (1571)

Cited in Hubert Herkommer, *Die deutsche Literatur des Mittelalters: Verfasserlexicon* (Berlin: De Gruyters, 1983), col. 726 (erroneously as part of *De famulatu*; see Benzenhöfer, 31n16) and Benzenhöfer, 24n54. This item contains a substantial portion of the *Liber lucis*, but so highly rearranged, edited, and interpolated that it must be considered a new composition based partly on John's work rather than a witness of his own writing. The text incorporated into this composition identifies the author's source as a manuscript descending through γ, probably through γ'''. The composition is titled *Die arbeith Johannis Rubicissa Honoralis Philosophus* and begins with the recipe in Operation 1 but omits most of the rest of that operation. The "demonstratio" (ll. 65–77) exists but has been moved to the end of Operation 2; its steps are rearranged and text from *De famulatu* is appended. A rewritten and substantially elaborated furnace chapter, with many illustrations, intervenes between Operations 3 and 4. Operations 5–7 follow; but again, more than half of the text of the original *Liber lucis* is omitted, much of the rest paraphrased, and much new material (some from *De famulatu*, some from practical

2 The two readings found otherwise only in **M**, **Rc**, and **W^{I}** are the insertion of the line *dicentes est omni re vilius quod nos philosophi veneramur* and the use of *cloacis veteribus* in place of *cavernis* at l. 21.

3 The text is partly edited by Mirella Brini in Eugenio Garin, et al., *Testi umanistici su l'ermetismo* (Rome: Bocca, 1955), 75–77, and the manuscript and its context are cited by Chiara Crisciani, "Hermeticism and Alchemy: The Case of Ludovico Lazzarelli," *Early Science and Medicine* 5 (2000): 145–59, on 146.

experience) inserted. While not useful as a witness, the document is worth closer investigation as an instance of the continuing practical engagement with the *Liber lucis* and its chrysopoetic methodology.

Kassel, Universitätsbibliothek Kassel-Landesbibliothek und Murhardsche Bibliothek der Stadt Kassel, 4° Ms. chem. 48, fols. 13r–15v (first q. 17th c.)[4]

Cited in Benzenhöfer, 24n54 (as fols. 11r–13v) and Broszinski, *Handschriften*, 173–74. This German-language manuscript is titled *Rupecissa Annotatio* and contains a German translation of the three sections appended to the 1561 Basel edition by Gratarolo. This is the only known manuscript version of those new sections, and was presumably translated from the printed edition, although it omits the first paragraph, the concluding poem, and an occasional line elsewhere.

Kassel, Universitätsbibliothek Kassel-Landesbibliothek und Murhardsche Bibliothek der Stadt Kassel, 4° Ms. chem. $60_{[5,3]}$, fols. 61r–v (1610s)

Cited in Broszinski, *Handschriften*, 254. This short document reprises the first two operations of the *Liber lucis*, heavily edited; this text descends through γ. This reworking, however, diverges radically from John's text in the third operation, which is where this text ends with the words "das ist di Meisterschafft."

Kassel, Universitätsbibliothek Kassel-Landesbibliothek und Murhardsche Bibliothek der Stadt Kassel, 4° Ms. chem. $60_{[7,2]}$, fols. 55r–59r (1530s)

Cited in Broszinski, *Handschriften*, 274. This German-language manuscript preserves the format of the *Liber lucis* in terms of its seven operations and an eighth titled "furnace of the secrets" (*fornax arcanorum*) and contains numerous lines translated from the authentic text. Nevertheless, the overwhelming majority of the text is by an unknown author who, although using the *Liber lucis* as a model, has written a completely different text that relies largely on ideas and processes distinct from John's. It represents yet another instance of practitioners' engagement with John's text.

Kassel, Universitätsbibliothek Kassel-Landesbibliothek und Murhardsche Bibliothek der Stadt Kassel, 4° Ms. chem. 73, fols. 115r–18v (1505)[5]

Titled *Rupicissa de lapide minerali*, this Latin text is a reworking of the *Liber lucis* by a practitioner who endeavored to follow John's work practically. It occurs in the same miscellany as **Ka**$^{5}_{d}$. The author's practical experience is clear from the many caveats he inserts regarding particular parts of the operation, for example, about the importance of drying the salts carefully before the initial sublimation. The text preserves about half of Operation 1 and contains reworked versions of

4 https://orka.bibliothek.uni-kassel.de/viewer/image/1540459690051/27/

5 Broszinski, *Handschriften*, 334–35. https://orka.bibliothek.uni-kassel.de/viewer/image/1550147333249/231/

Operations 2 through 6. While the author often paraphrases or edits John's text, he also reuses a considerable amount verbatim. These direct borrowings permit the confident identification of the author's reference text as a descendant of $\pi^{I''}$.

Incipit: "Multi multa locuntur et maxime fatui et illi qui credunt se sapienciores sapientibus..."

Explicit: "...quomodo omnes vere ignorant, et sic videbitis mirabilia a deo glorioso concreata. Laus sit Deo".

Leiden, Leiden University Library, Vossianus Chym. O. 5, fol. 184r (1470–90)

Cited in Halleux, "Rupescissa," 283. The reference is to a six-line poem beginning "Materia lapidis est una res vilis licet ubique reperta", which does no more than use one rephrased line of text borrowed from the opening of Operation 1.[6]

Leiden, Leiden University Library, Vossianus Chym. Q. 54, fols. 124r–31r (2nd half of 16th c.)[7]

This item is titled *Opus bonum Johannis Rupescissae de sulphure communi* (inc: "Recipe die minera so ich dier gezeigt hab...; expl: ...wenig ohne die gnade Gottes des allmechtigen gefunden haben") and describes the preparation of a transmutational agent. The text is written to appear a genuine composition of John, given the inclusion of the phrase "ich Johan Rupescisa hab..." (fol. 129v), but the ideas and process bear no relation to his authentic works aside from deploying the term "quintessence."

London, British Library, Harley MS 4724, fols. 106r–20r and 156v–61r (17th c.)

Cited in Halleux, "Rupescissa," 268 and 283. According to this citation, fols. 106r–20r present a Latin version of the *Liber lucis*, while 156r–61r contain an English translation of the same. In fact, *pages* 106–20 (55r–61v in the modern foliation) contain the English translation of the *Liber lucis* (L^7_e), and no folios or pages 156–61 exist in the manuscript.[8]

London, British Library, Sloane MS 353, fols. 56v–61v (15th c.)

Cited in Dorothea Waley Singer and Annie Anderson, *Catalogue of Latin and Vernacular Alchemical Manuscripts in Great Britain and Ireland*, 3 vols. (Brussels: Lamertin, 1928–31), 1:279 and Halleux, "Rupescissa," 283. The text is actually pseudo-Albert's *Semita recta* in English translation. On the misattribution, see Marguerite A. Halverson, *The Consideration of the Quintessence: An Edition*

6 Boeren, *Codices Vossiani Chymici*, 248–55, on 255.

7 Boeren, *Codices*, 223–24 on 224. The catalogue lists separately a *Multiplicatio* attributed to John, but this is only a section heading within the *Opus bonum*. See also Benzenhöfer, 36.

8 Nares et al., eds., *Catalogue of the Harleian Manuscripts in the British Museum*, 3:196.

of a Middle English Translation of John of Rupescissa's Liber de consideratione (Ph.D. diss, Michigan State University, 1998), 83–84.

London, British Library, Sloane MS 976, fols. 81v–83v (mid-15th c.)

This short and fragmentary text, in an English cursive hand on paper, is headed *Incipit tractatus qualiter fiat athanor id est furnus philosophicus*. Its first section (*De arte ignis*) is a nearly complete text of Operation 8. The treatise then continues with a second section (*De regimine ignis*), which has no relation to the *Liber lucis* or to John, but which ends with a modified version of ll. 323ff: "vides mira ingenia et subtilia revelata a domino nostro jhesu christo cui honor...." Two more sections (*De vase philosophico* and *De luto sapiencie*) then follow, the former citing "Haly" and the second Albert and "Gilbertus," before the text breaks off in mid-sentence at the bottom of fol. 83v.

Munich, Bayerische Staatsbibliothek, Clm 27000, fols. 301r–14r and 315r–25r (17th c.)

Cited in Herkommer, *Verfasserlexicon*, cols. 726 and 728 and Karl Halm, *Catalogus codicum latinorum Bibliothecae Regiae Monacensis*, vol. 2, part 4 (*Codices latinos continens 21406-27268*), 232. The first of these Latin texts, written in an Italian hand, is a supposed "Commentary" by John himself on *De famulatu* and titled *De quinta essentia commentum primum Domini Fratris Joannis de Rupe scisa*. It is written in the first-person singular and addressed to the *viri evangelici*. The second, written in the same hand and titled *Joannis à ripescisa liber lucis*, might have also been intended as a commentary supposedly by John, but this is not explicitly stated, and nothing of the authentic *Liber lucis* appears here. The incipit summarizes its purpose as "to collect together the various ways of making the beyond-precious philosophers' stone" ("Intentio mea est recolligere diversos modos praticandi supra preciosum lapidem Philosophorum").

Paris, Bibliothèque Nationale de France, MS Lat. 7147, fols. 51v–54v (ca. 1537)[9]

Cited in Halleux, "Rupescissa," 284. This anonymous text bears the simple title *Alius tractatus lapidis philosophici satis bonus*. It incorporates much of Operations 1–3 and a small portion of Operation 6 into a longer text and significantly alters the process by the addition of gold and silver. This manuscript was copied and owned by the mathematician, astronomer, and cartographer Oronce Finé (1494–1555).

9 Corbett, *Catalogue*, 1:55–64, on 60.

Paris, Bibliothèque Nationale de France, MS Lat. 14012, fol. 57v (1585)

Cited in Halleux, "Rupescissa," 284. Here only two sentences out of Operation 6 (ll. 213–19): "Alii vero miscent tria elementa...creata in rebus") are present, appended to a Latin version of George Ripley's *Compound of Alchemy.*[10]

Paris, Bibliothèque Nationale de France, MS NAL 1293, fols. 62r–67r (early 16th c.)[11]

This text is titled *Liber philosophorum et ditum per paulum Ursanum de ursinis rationeque declaravit enigmata magistri arnaldi de villa nova.* It coopts a considerable portion of the *Liber lucis*, although with significant deletions and insertions which become more substantial as the document proceeds. The text breaks off in mid-sentence in the furnace description and ends with one highly garbled illustration of the furnace. There is some resemblance in readings to the descendants of ι^2; the text incorporated here probably came from such a descendant.[12]

Paris, Bibliothèque Sainte-Geneviève, MS 2263, fols. 99r–120r (early 18th c.)

Cited in Halleux, "Rupescissa," 269n217. Although given the promising title of *Le livre de frere Jean de Rupe-scissa par lequel il declare comment se peut faire la transmutation des metaux*, this late text is not the *Liber lucis*, contains neither reference to nor quotations from it, and does not have any discernible relationship to John.

Rome, Biblioteca dell'Accademia dei Lincei, MS Rossi 238 (36.E.11), fols. 137v–46r (ca. 1600).

This text provides a somewhat variant version of the text given by Florence, Riccardiana 923 (see above). Titled *Religiosissimi Joannis Rupescisse de lapide philosophorum compendium qui liber lucis intitulatur et arcanum arcanorum Deique donum*, it contains more direct quotations and phrases taken from *De famulatu* and *Liber lucis* than does Riccardiana 923. The two are sufficiently distinct in variants that it is very unlikely that one was copied from the other, which implies that this pseudo-Rupescissa text may have been fairly well circulated in late 16th-century Italy.

Washington, DC, Catholic University Library, MS 129, fols. 97r–101v (late 16th c.)

Mentioned in Benzenhöfer, 24n54, as possibly a German version of the *Liber lucis*, but needing to be checked. Upon checking, this text, which bears the title

10 See Jennifer M. Rampling, "The Catalogue of the Ripley Corpus: Alchemical Writings Attributed to George Ripley," *Ambix* 57 (2010): 125–201, on 158; and Corbett, *Catalogue*, 1:202–7, on 203.

11 Corbett, *Catalogue*, 1:211–19, on 215.

12 For example, mercury is generated *in saxis* (**Fn**^{2} reads *in sassis*); *ad sensum* is garbled to *asensum* (descendants of ι^2 read *ascensum*); and the hair, salt water, and vinegar are excluded from the recipe for the furnace clay, as in all descendants of ι^2.

Liber lucis, is not related to John's work.[13] It begins: "Das ist ein puechl und tractat raymundi luly genant dz buech des liechts. Recipe wolgeschmachen Rotten Wein..."

Wolfenbüttel, Herzog August Bibliothek, Cod. 185 Blankenburg, fols. 119r–20r (1629)

Cited in Hans Butzmann, *Kataloge der Herzog-August-Bibliothek Wolfenbüttel 11: Die Blankenburger Handschriften*, (Frankfurt: Klostermann, 1966), 182. This is a German-language text bearing the title *Johannes Ruppescissa eine grosse Heimbligkeit des phylosophischen steins wil ich kurzlich alhier erkleren.* Inc.: "So nim die Minera die ich dir geheisset hab poch die ganz glein thue sie in ein geschirlein..."

Part 2—Other Alchemical Texts Attributed to John of Rupescissa

The second part of this appendix describes three alchemical texts that are explicitly attributed to John of Rupescissa but are not related to the *Liber lucis* and which do not have the character of intentional forgeries, nor do they contain ideas or expressions that are anachronistic or alien to John's ideas as established in his authentic works. Further research is required to determine the exact origins of these texts.

1. *Aliud opus Rubicisse de aqua vite rectificata*

Wolfenbüttel, Herzog August Bibliothek, Cod. Guelf. 66 Aug. 8°, fols. 253r–57v.[14]

Incipit: "Recipe ergo de vino albo vel rubeo claro absque fecibus et pone in concurbita cum suo allempico in cineres calidos."
Explicit: "Precor mi dilecte ne illa pandatis alicui fraudis et doli homini sed occultetis ne deveniat ad manus superborum et reproborum quia non licet margaritas jacere ante porcos."

Of the three texts considered here, the *Aliud opus* is the one most closely related to John's authentic works; it does at the very least *contain* authentic material. It is the longest of these suppositious texts (roughly 2500 words) and is related to *De famulatu* rather than to *Liber lucis*. It is presently known from only a single witness, dating from 1468, which appears in a collection of alchemical texts copied by one Stefan Hullier of Lauingen, who identifies himself as "*auriscriba*, rector of

[13] See also the description in W. J. Wilson, "Catalogue of Latin and Vernacular Alchemical Manuscripts in the United States and Canada," *Osiris* 6 (1939): 167.

[14] H. Herbst, "Ein Einband Friedrich Walthers der Bibliothek zu Wolfenbüttel," *Zeitschrift für Bücherfreunde* 21 (1929): 72–76, and Otto von Heinemann, *Die Handschriften der herzoglichen Bibliothek zu Wolfenbüttel*, 2nd series, *Die Augusteische Handschriften V* (Wolfenbüttel: Zwissler, 1908), 130–31. Mentioned by Thorndike, *HMES*, 3:369.

the Teutonic school." Nevertheless, the *Aliud opus* must have had a reasonable circulation since it is cited as one of John's compositions in the list of alchemical texts published in 1599 by the alchemical author Giovanni Battista Nazari.[15]

About half of *Aliud opus* (fols. 254v–55v and 256v–57v) is based on material from *De famulatu*. The text is sometimes epitomized, sometimes repeated nearly verbatim, and sometimes elaborated, and does not occur in the same order as it appears in *De famulatu*. The remainder of the text of *Aliud opus* has no counterpart in *De famulatu*. This additional material gives more detailed practical directions for distilling wine, records observations on the dissolution of gold in acids, and offers instructions on how to distill stronger aquaforts. The opening section, which deals with the separation of the four elements from wine (fols. 253r–54r), appears also in a text (inc: "Dixit philosophus accipe lapidem") sometimes found prefixed to manuscript copies of the Emerald Tablet commentary by Hortulanus, an author about whom almost nothing solid is known. Part of this same text also appears, with much additional material, in the first part of the pseudo-Lullian *Potestas divitiarum*.[16] The *Aliud opus*'s subsequent material on the dissolution of gold (fols. 255v–56v) does not, however, appear in these other texts but does show up (together with the material on separating wine) in a document titled "De quinta essencia et de auro potabili" found in a manuscript in Dresden dated 1463. This latter text also includes several sections taken nearly verbatim from *De famulatu* but in rearranged order and not the same ones that appear in *Aliud opus*.[17] The sections

[15] Giovanni Battista Nazari, *Della tramutatione metallica* (Brescia, 1599), 139. The *Opus aliud* from Nazari's list is repeated in Pierre Borel, *Bibliotheca chimica* (Heidelberg, 1656), 194. Nazari also recognized the *Thesaurus mundi* (section 3.4) as John's and cites an "Abbreviatio," possibly one of the abridged versions of the *Liber lucis*; the "Compendium artis" he mentions may be yet another of these shortened versions. The "De secretis secretorum" as well as the "Abbreviatio" may, however, result from a conflation of John of Rupescissa with Johannes Pauperum whose *Quaedam abbreviatio de secretis secretorum* was published at Basel in the third volume added to *Artis auriferae* in 1610, 131–39. See also Thorndike, *HMES*, 3:369.

[16] On Hortulanus and the commentary on the Emerald Tablet, see Thorndike, *HMES*, 3:176–90, 686–88, and Carmen Schmechel, "Leaven of Dough, Ferment of Gold: The Breadmaking Analogy in Medieval Metallic Transmutation," *Ambix* 71 (2024): 243–70, on 263–67. Hortulanus' commentary is presented in Julius Ruska, *Tabula smaragdina: Ein Beitrag zur Geschichte der hermetischen Literatur* (Heidelberg: Winter, 1926), 180–86, and the text prefixed to some manuscript copies (inc: "Dixit philosophus accipe lapidem") is available, for example, in Paris, BNF, Lat. MS 11202, fols. 19r–21r. The composition of this latter material by a later author as an "update" to Hortulanus is treated in Jean-Marc Mandosio, "La Tabula Smaragdina e i suoi commentari medievali," in *Hermetism from Late Antiquity to Humanism*, ed. Paolo Lucentini, Ilaria Parri, and Vittoria Perrone Compagni (Turnhout: Brepols, 2003), 681–96, on 688–89. For the *Potestas divitiarum*, it is printed (in a very corrupt form) in *BCC* 2: 866–72, see Michela Pereira, *The Alchemical Corpus Attributed to Raymond Lull* (London, Warburg, 1989), 81–82. The various relationships among the *Aliud opus*, the text prefixed to Hortulanus, and the pseudo-Lullian composition are open to question since the dates of composition of each remain uncertain; this topic will be treated further in the upcoming critical edition of *De famulatu*.

[17] Dresden, Sächsische Landesbibliothek, Staats und Universitätsbibliothek Dresden, Mscr. Dresd. C. 278, fols. 160v–63v.

from *De famulatu* in "De quinta essencia" relate to making *aqua ardens* and the quintessence of wine, while those in *Aliud opus* relate predominantly to medical remedies.

The Wolfenbüttel *Sammelband* that contains *Aliud opus* also contains a *Rosarius, id est, Ortus rosarum philosophorum* (fols. 155r–96r) that is attributed to John of Rupescissa, but this is clearly a spurious attribution since the contents are nothing like John's own ideas or style.

2. *Practica*

Venice, Biblioteca Nazionale Marciana, Lat. fondo antico 323 (=1579), fols. 186v–87v; Hamburg, Staats- und Universitätsbibliothek, Cod. Alch. 659, fols. 120v–22r; and Leiden, Leiden University Library, Vossianus Chym. Q. 2., fols. 122v–23v.

Incipit: "Recipe vitrioli romani puri et electi uncias xii salis petre cynobrii ana uncias tres terantur omnia subtiliter..."
Explicit: "Sed vere si debite facta fuerit una pars tingit ultra mille partes. Unde laus deo ejus nomen sanctum sit benedictum in eternum."

The *Practica* is a transmutational recipe that traveled alongside one of the abbreviated versions of the *Liber lucis* and is thus present in moderately divergent forms alongside $\mathbf{Z}_a$, $\mathbf{Hb}_a$, and $\mathbf{Le}^1{}_a$. It must date from before 1468 when $\mathbf{Hb}_a$ was prepared. In the Venetian witness, the recipe is somewhat edited relative to the others and introduced simply as "Here follows another practical operation of the same John of Rupescissa" (*Sequitur alia practica ejusdem Johannis rubecisse*). In the other two witnesses, the process is headed with "John of Rupescissa etc., and you ought to know that this practical operation is from that master who wrote the *Book of the Quintessence*" (*Johannes de rupecissa etc. Et debes scire quod hec practica est illius magistri qui composuit librum de quinta essencia*). I have also located the same recipe in another manuscript (dated 1464) where it is not connected with the abbreviated *Liber lucis* but is instead attributed to Arnald of Villanova as a process he gave to Boniface VIII.[18]

The process and methodology of the *Practica* are unrelated to those of the *Liber lucis*. The recipe uses precious metals, both gold and silver—a mixture prohibited by pseudo-Arnald—and employs a ferment. It involves making separate solutions of silver and gold in multiple aquaforts distilled from saltpeter, vitriol, and cinnabar. These solutions are then combined, and after several further steps, the resultant crystalline material is treated with an additional silver solution and

[18] St. Gall, Kantonbibliothek Vadiana, VadSlg MS 429, fols. 64r–65r: "Practica particularis quam scripsit Arnaldus de villa nova ad papam bonifacium."

coagulated. A small portion of the final product projected upon mercury provides an argyropoetic "medicine"; substituting a gold solution gives a chrysopoetic one. The final aquafort is said "to have within itself the quintessence and the fiery sphere" (*habere in se quintam essenciam et speram igneam inclusas*), which recalls John's famous invocations of the quintessence in both *De famulatu* and the *Liber lucis*, and thus may be a cause for the attribution of this text to the friar.

3. *Operatio fratris Johannis de Rubescissa*
Vatican City, Biblioteca Apostolica Vaticana, MS Palatinus latinus 1216, fols. 62r–v.[19]

Incipit: "Recipe limature lune et ferri ana [ounce]i salis armoniaci"
Explicit: "...et iterum sine sale communi funde et post fac quod volueris quia est [luna] bona."

Known only in a single witness, this short *Operatio* promises a multiplication of silver. It is part of a collection of alchemical and medical texts and recipes produced in Schaffhausen that bears the date 1389 (on fol. 51v), which would make this recipe earlier than any surviving witness of the *Liber lucis*. If the date is correct, the *Operatio* would document a late fourteenth-century linkage of John's name to transmutational processes, whether or not the text is authentically his. The process involves treating a silver-containing amalgam with molten salts to increase the quantity of the precious metal. There are no features of the process that either reprise or conflict with the ideas of the *Liber lucis*.

[19] Ludwig Schuba, *Die medizinischen Handschriften der Codices Palatini Latini in der Vatikanischen Bibliothek* (Wiesbaden: Reichert, 1981), 210–18.

SOURCES CITED

Abate, Giuseppe, "Le *Constitutiones Generales Antique* dei Frati Minori nella redazione Assisana del 1279," *Miscellanea Francescana* 35 (1935): 58–100 on 80.

Alessandrini, Ada, *Cimeli Lincei a Montpellier* (Rome: Accademia Nazionale dei Lincei, 1978).

Antolín, Guillermo, *Catálogo de los códices latinos de la Real Biblioteca del Escorial*, 5 vols. (Madrid: Imprenta Helénica, 1910–23).

Aromatico, Andrea and Marcella Peruzzi, *Giovanni da Rupescissa: Il libro della luce* (Venice: Marsilio, 1997).

Aromatico, Andrea and Marcella Peruzzi, *Medicamenti, pozioni e incantesimi del ricettario magico urbinate* (Milan: Roche, 1994).

Bandini, Angelo Maria, *Bibliotheca Leopoldina Laurentiana*, 2 vols (Florence, 1792).

Beaujouan, Guy, *Manuscrits scientifiques médiévaux de l'Université de Salamanque et de ses colegios mayores* (Bordeaux: Féret, 1962).

Benzenhöfer, Udo, *Johannes de Rupescissa* Liber de consideratione quintae essentiae omnium rerum *deutsch* (Stuttgart: Franz Steiner Verlag, 1989).

Biblioteca chemica curiosa, ed. Jean Jacques Manget, 2 vols. (Geneva, 1702); reprint ed., Sala Bolognese: Arnoldo Forni, 1976.

Bignami-Odier, Jeanne, *Études sur Jean de Roquetaillade* (Paris: Vrin, 1952).

Bihl, Michael, "Statua generalia ordinis edita in capitulis generalibus celebratis Narbonae an. 1260, Assisii an. 1279, atque Parisiis an. 1292," *AFH* 34 (1941): 13–94 and 284–358.

Black, William Henry, *Catalogue of the Manuscripts Bequeathed...by Elias Ashmole* (Oxford: Oxford University Press, 1845).

Boeren, P. C., *Codices Vossiani Chymici* (Leiden: Leiden University Press, 1975).

Boilloux, Marc, *Étude d'un commentaire prophetique du XIV siecle: Jean de Roquetaillade et l'Oracle de Cyrille* (Thesis presented at l'École des Chartes, Paris, 1993).

Bonus, Petrus, *Margarita preciosa novella*, in *BCC* 2: 64.

Borel, Pierre, *Biblioteca chemica* (Paris, 1656).

Broszinski, Hartmut, *Die Handschriften der Universitätsbibliothek Kassel Landesbibliothek und Murhardsche Bibliothek der Stadt Kassel: Manuscripta chemica in quarto* (Wiesbaden: Harrassowitz, 2011).

Broszinski, Hartmut, *Katalog der Alchemica in Folio* (typescript at Kassel library).

Broszinski, Hartmut, "Spuren klösterliche Alchemie in Kasseler Handschriften des 15. und 16. Jahrhunderts," 823–47 in *Nova de veteribus: Mittel- und neulateinische Studien für Paul Gerhard Schmidt*, eds. Andreas Bihrer, Elisabeth Stein, and Paul Gerhard Schmitt (Munich: Saur, 2004).

Brunel, Clovis, *Bibliographie des manuscrits littéraires en ancien provençal* (Liège, 1935)

Bullarium Franciscanum, vol. 6, ed. Conrad Eubel (Rome: Vatican Press, 1902).

Burr, David, "Bonaventure, Olivi, and Franciscan Eschatology," *Collectanea Franciscana* 53 (1983): 23–40

Burr, David, "Olivi, Apocalyptic Expectation, and Visionary Experience," *Traditio* 41 (1985): 273–88.

Burr, David, *The Spiritual Franciscans* (University Park, PA: Pennsylvania State University Press, 2001).

Butzmann, Hans, *Kataloge der Herzog-August-Bibliothek Wolfenbüttel 11: Die Blankenburger Handschriften* (Frankfurt: Klostermann, 1966).

Butzmann, Hans, *Kataloge der Herzog-August-Bibliothek Wolfenbüttel 15: Die mittelalterlichen Handschriften der Gruppen Extravagantes, Novi und Novissimi* (Frankfurt: Klostermann, 1972).

Calvet, Antoine, "Alchimie et joachimisme dans les *alchemica* pseudo-Arnaldiens," 93–107 in *Alchemie et philosophie à la Renaissance*, eds. Jean-Claude Margolin et Sylvain Matton (Paris: Vrin, 1993).

Calvet, Antoine, "Le *De arte alchemica* (inc: Dixit Arturus explicator huius operis) est-il une oeuvre authentique de Richard de Fournival?" in Joelle Ducos and Christopher Lucken, eds. *Richard de Fournival et les sciences au XIIIe siècle*, Micrologus Library 88 (Florence: SISMEL, 2018).

Calvet, Antoine, "Étude d'un texte alchimique latin du XIVe siècle: Le *Rosarius philosophorum* attribué au médecin Arnaud de Villeneuve (ob. 1311)," *Early Science and Medicine* 11 (2006): 162–206.

Calvet, Antoine, *L'alchimie au Moyen Âge: XII*e*-XV*e *siècles* (Paris: Vrin, 2018).

Calvet, Antoine, *Les oeuvres alchimiques attribuées à Arnaud de Villeneuve: Grand oeuvre, médecine et prophétie au Moyen-Âge, Textes et Travaux de Chrysopoeia* 11 (Paris: SÉHA, 2011).

Calvet, Antoine, *Le rosier alchimique de Montpellier: Lo rosari (XIVe siècle)* (Paris: Presses de l'Université de Paris-Sorbonne, 1997)

Calvet, Antoine, "Le *Tractatus parabolicus* du pseudo-Arnaud de Villeneuve, présentation, édition et traduction," *Chrysopoeia* 5 (1997): 145–71.

Camps, Jean-Baptiste, *Les manuscrits occitans à la Bibliothèque nationale de France* (Paris: Enssib, 2010).

Carlini, Armando G., "Constitutiones generales...anno 1316 Assisii," *AFH* 4 (1911): 269–302.

Casamassima, Emanuele, *I manoscritti della biblioteca communale di Poppi* (Milan: Giunta Regionale Toscana, 1993).

Casteen, Elizabeth. "John of Rupescissa's Letter *Reverendissime pater* in the Aftermath of the Black Death," *Franciscana* 6 (2004): 139–84.

Castel-Branco, Nuno, ed. *Religious Experience and the Foundations of Modern Science*, (Cambridge: Cambridge University Press, 2026).

Catálogo de Manuscritos de la Biblioteca Universitaria de Salamanca, 6 vols. (Salamanca: Ediciones universidades, 2002).

Catalogue générale des manuscrits des bibliothèques publiques de France, vol 12: *Orléans* (Paris: Plon, 1889).

Catalogue of Harleian Manuscripts in the British Museum, 4 vols. (London, 1808–12).

Catalogue of the Manuscripts Preserved in the Library of the University of Cambridge, 5 vols. (Cambridge: Cambridge University Press, 1858).

Catalogue of the Stowe Manuscripts in the British Museum, 2 vols. (London: British Museum, 1895–96).

Chantilly: Le Cabinet des livres, manuscrits, 3 vols. (Paris: Plon, 1900–11).

Ciola, Graziana, "Giovanni di Rupescissa: autobiografia, profezia e leggenda," 517–78 in *The Medieval Legends of Philosophers and Scholars, Micrologus* 21 (Florence: SISMEL, 2013).

Constitutiones generales Ordinis Fratrum Minorum, eds. Cesare Cenci and Georges Mailleux, 2 vols. (Grottaferrata/Rome, 2007–10).

Corbett, James, *Catalogue des manuscrits alchimiques latins 1: Manuscrits des bibliothèques publiques de Paris* (Brussels: Union Académique Internationale, 1939).

Corbett, James, *Catalogue des manuscrits alchimiques latins 2: Manuscrits des bibliothèques publiques des départements français* (Brussels: Union Académique Internationale, 1951).

Cortesi, Paolo, ed., *Giovanni da Rupescissa: Sulla preparazione della vera pietra dei filosofi*, (Rome: Atanor, s.d.).

Crisciani, Chiara, "Alchimia e potere: presenze francescane (secoli XIII-XIV)" in *I Francescani e la politica, Franciscana 13* (Palermo: Officina di Studi Medievali, 2007), 223–35.

Crisciani, Chiara, "I Francescani, alchimia e profezia," 239–79 in *I Francescani e le scienze* (Spoleto: Fondazione Centro Italiano di Studi sull'alto Medioevo, 2012).
Crisciani, Chiara, "Giovanni di Rupescissa: sapere, alchimia e profezia," 239–79 in *I Francescani e le scienze: Atti del XXXIX Convegno internazionale* (Spoleto: Centro Italiano di Studi Sull'Alto Medioevo, 2012).
Crisciani, Chiara, "Hermeticism and Alchemy: The Case of Ludovico Lazzarelli," *Early Science and Medicine* 5 (2000): 145–59.
Crisciani, Chiara, "History, Novelty, and Progress in Scholastic Medicine," *Osiris* 6 (1990): 118–39.
Crisciani, Chiara, "Note sull'alchimia 'francescana' nel sec. XIII" in *Atti del XXV Congresso Nazionale di Filosofia* (Rome: Società filosofica italians, 1980), 214–20.
Crombie, Alistair, "Some Attitudes to Scientific Progress: Ancient, Medieval, and Early Modern," *History of Science* 13 (1975): 213–30.
Cross, Richard, *The Physics of Duns Scotus: The Scientific Context of a Theological Vision* (Oxford: Clarendon Press, 1998).
Dapsens, Marion, *Les* Masā'il Khālid li-Maryānus al-rāhib *dans leurs versions arabe et latine: Éditions critiques et traductions* (Florence: SISMEL, 2024).
de Dmitrewski, Michel, "Fr. Bernard Délicieux OFM: sa lutte contra l'inquisition," *AFH* 18 (1925): 1–32.
de Ricci, Seymour, *Census of Medieval and Renaissance Manuscripts in the United States and Canada*, 3 vols. (New York: H. W. Wilson, 1935–40).
Declaratio lapidis physici Avicennae filio suo Aboali, in *BCC*, 1: 633–36.
Dedieu, Hugues, "Les ministres provinciaux d'Aquitaine, des origines à la division de l'ordre," *AFH* 76 (1983): 129–214 and 646–70.
Delmeulle, Guillaume, *Aux prémices de l'alchimie latine: Édition critique, traduction et étude du* Liber perfecti magisterii (Ph.D. diss., Université catholique de Louvain, 2025).
DeLorme, F. M., "L''Ars faciendi sermones' de Géraud du Pescher," *Antonianum* 19 (1944): 169–98.
DeVun, Leah, *Prophecy, Alchemy, and the End of Time: John of Rupescissa in the Late Middle Ages* (New York: Columbia University Press, 2009).
Dutschke, C. W., *Guide to Medieval and Renaissance Manuscripts in the Huntington Library*, 2 vols. (San Marino, CA: Huntington Library, 1989).
Emden, A. B., *A Biographical Register of the University of Oxford to A.D. 1500*, 3 vols. (Oxford: Clarendon, 1957).
Engle, Paul, *Conciatore: The Life and Times of 17th Century Glassmaker Antonio Neri* (Hubbardston, MA: Heiden & Engle, 2014).
Espelt, Jose Perarnau, "Los manuscritos lulianos en las Bibliotecas Casanatense y Angélica," *Anthologica annua* 21 (1974): 185–248.
Faucon, Maurice, *La librairie des papes d'Avignon*, 2 vols. (Paris: Thoren, 1887).
Ferrario, Gabriele, *The Book of Alums and Salts of Pseudo-Rāzī*, *Ambix* 70 (Supplement 1, 2023).
Ferrarius, *De lapide philosophorum*, in *Verae alchemiae*, 2 vols. (Basel, 1561), 2: 232–37.
Fill, Hauke, *Katalog der Handschriften des Benediktinerstiftes Kremsmünster*, 2 vols. (Vienna: Österreichischen Akademie der Wissenschaften, 2000).
Fors, Hjalmar, Lawrence M. Principe, and H. Otto Sibum, "From the Library to the Laboratory and Back Again: Experiment as a Tool for Historians of Science," *Ambix* 63 (2016): 85–97.
Franciscains d'Oc: Les Spirituels ca. 1280-1324, *Les cahiers de Fanjeux* 10 (Toulouse: Privat, 1975).
Frühmorgen-Vos, Hella and Norbert H. Ott, with Ulrike Bodemann and Gisela Fischer-Heetfeld, *Katalog der deutschsprachigen illustrierten Handschriften des Mittelalters* (Munich: Beck, 1991).

Garrod, Raphaële and Alexander Marr, eds., *Descartes and the* Ingenium: *The Embodied Soul in Cartesianism* (Leiden: Brill, 2021).
Gentile, Luigi, *I codici Palatini*, 3 vols. (Rome, 1890).
"Géraud du Pescher, frère mineur," *Histoire littéraire de la France* 36 (1927): 614–17 and 656.
Hagen, Hermann, *Catalogus codicum Bernensium* (Bern, 1875).
Hagen, Paul, *Naturwissenschaftliche Handschriften*, Cat. manuscr. 106 (Lübeck 1932; in manuscript, ed. Anne Harnisch, 2018).
Halleux, Robert, "Ouvrages alchimiques de Jean de Rupescissa," in *Histoire litteraire de la France*, 41: 241–77 (Paris: Academie des Inscriptions et Belles-Lettres, 1981).
Hauke, Hermann, *Die mittelalterlichen Handschriften in der Abtei Ottobeuren* (Wiesbaden: Harrassowitz, 1974).
Hauke, Hermann, *Katalog der lateinischen Handschriften der Bayerischen Staatsbibliothek München*, Volume 4, Part 5: *Clm* 27270–27499 (Wiesbaden: Harrassowitz, 1975).
Heptas alchymica, das ist, Auszerlesene philosophische Tractätlein von dem magisterio lapidis, ed. Theophilus Neander (Halle, 1621).
Herbst, H., "Ein Einband Friedrich Walthers der Bibliothek zu Wolfenbüttel," *Zeitschrift für Bücherfreunde* 21 (1929): 72–76.
Heuer, Wolfgang, *Wilhelm Vernuken: Ein Bildhauer und Baumeister des Manierismus in Deutschland* (Ph.D diss., Universität Düsseldorf, 2000).
Heyne, Sirka, *Die mittelalterlichen Handschriften der Universitätsbibliothek Marburg* (Wiesbaden: Harrassowitz, 2002).
Hofinger, Benedict, "Verzeichniss der Handschriften der Bibliothek des Stiftes Schlierbach," in *Xenia Bernardina 2, Handschriften-Verzeichnisse der Cistercienser-Stifte der Österreichisch-ungarischen Ordensprovinz*, 2 vols. (Vienna, 1891).
Holmes, Frederic L., *Antoine Lavoisier: The Next Crucial Year, or The Sources of His Quantitative Method in Chemistry* (Princeton, NJ: Princeton University Press, 1998).
Jager, Isabelle, *Catalogue des manuscrits latins* 1-376 (Geneva: Bibliothèque de Genève, 2016).
James, Montague Rhodes, *A Descriptive Catalogue of the Latin Manuscripts in the John Rylands Library at Manchester* (Manchester: Manchester University Press, 1921); expanded reprint ed., Frank Taylor, ed. (Munich: Kraus Reprint, 1980).
James, Montague Rhodes, *A Descriptive Catalogue of the Manuscripts in the Library of Corpus Christi College Cambridge*, 2 vols. (Cambridge: Cambridge University Press, 1912).
James, Montague Rhodes, *The Western Manuscripts in the Library of Trinity College, Cambridge*, 4 vols. (Cambridge: Cambridge University Press, 1900–04).
Jammer, Max, *Concepts of Mass* (Cambridge, MA: Harvard University Press, 1961).
John of Rupescissa, *De confectione veri lapidis philosophorum*, in *Verae alchemiae artisque metallicae*, 2 vols. (Basel, 1561), 2: 226–31.
John of Rupescissa, *Liber lucis*, in *Secreta alchimiae magnalia D. Thomae Aquinatis... accessit et Johannis de Rupescissa liber lucis...* (Cologne, 1579).
John of Rupescissa, *Liber ostensor quod adesse festinant tempora, Édition critique*, sous la direction d'André Vauchez par Clémence Thévenaz Modestin et Christine Morerod-Fattebert (Rome: École Française de Rome, 2005).
John of Rupescissa, *Vade mecum in tribulatione: Edizione critica a cura di Elena Tealdi; introduzione storica a cura di Robert E. Lerner e Gian Luca Potestà* (Milan: Vita e Pensiero, 2015).
Jørgensen, Ellen, *Catalogus codicum latinorum medii aevi bibliothecae regiae Hafniensis* (Copenhagen: Gyldendals, 1926).
Josten, C. H., "The Text of John Dastin's Letter to Pope John XXII," *Ambix* 4 (1949): 34–51.

Karpenko, Vladimir and Ivo Purš, "Sebald Schwertzer: Between Metallurgy and Alchemy," 617–90 in *Alchemy and Rudolf II.*

Karpenko, Vladimir and Ivo Purš, "Some Notes on the Early History of Nitric Acid: 1300–1700," *Bulletin of the History of Chemistry* 34 (2009): 105–16.

Kaup, Matthias, *John of Rupescissa's* Vade mecum in tribulacione *(1356): A Late Medieval Eschatological Manual for the Forthcoming Thirteen Years of Horror and Hardship* (New York: Routledge, 2017).

Kelmers, A. D., L. Maya, D. N. Browning, and W. Davis, "The Chloride-Catalyzed Decomposition of Ammonium Nitrate in Nitric Acid Media at 100°C," *Journal of Inorganic Nuclear Chemistry* 41 (1979): 1583–88.

Kircher, Athanasius, *Mundus subterraneus* (Amsterdam, 1678).

Kraus, Paul, *Jābir ibn Hayyān: Contribution à l'histoire des idées scientifiques dans l'Islam, Volume I: Le Corpus des écrits jābiriens, Mémoires de l'Institut d'Égypte* 44 (1943): xxvii–xxxiii.

le Bel, Jean, *Chronique*, eds. Jules Viard and Eugène Déprez, 2 vols. (Paris: Renouard, 1905).

Le manuscrit trouvé dans l'armoire, *Oliviana* 3 (2009); https://journals.openedition.org/oliviana/322

Leitschuh, Friedrich and Hans Fischer, *Katalog der Handschriften der Königlichen Bibliothek zu Bamberg*, 3 vols. (Bamberg: Buchner, 1887–1912).

Lerner, Robert E., "John the Astonishing," *Oliviana* 3, (2009); https://oliviana.revues.org/335

Lerner, Robert E., "Life, Works, and Fortune of the Prophet Immured," 25–39 in Giovanni di Rupescissa, *Vade mecum in tribulatione*, ed. Elena Tealdi (Milan: Vita e Pensiero, 2015).

Lerner, Robert E., "Prophet and Pope: John of Rupescissa confronts Clement VI," *Franciscana* 23 (2021): 97–122.

Lerner, Robert E., "The Pope and the Doctor," *The Yale Review* 78 (1988): 62–79.

Lerner, Robert E., *The Powers of Prophecy: The Cedar of Lebanon Vision from the Mongol Onslaught to the Dawn of the Enlightenment* (Berkeley: University of California Press, 1983; reprint ed., Ithaca, NY: Cornell University Press, 2009).

Lerner, Robert E., "Yet another work by John of Rupescissa," *Oliviana* 5 (2016); https://journals.openedition.org/oliviana/825?lang=en

Lerner, Robert E. and Christine Morerod-Fattebert, *Johannes de Rupescissa*, Liber secretorum eventuum*: Édition critique, traduction et introduction historique* (Fribourg: Éditions Universitaires, 1994).

Lerner, Robert E. and Pavlina Rychterová, *John of Rupescissa, Vade mecum in tribulatione Translated into Medieval Vernaculars* (Milan: Vita e Pensiero, 2019).

Maclean, Ian, "Heterodoxy in Natural Philosophy and Medicine: Pietro Pomponazzi, Guglielmo Gratarolo, Girolamo Cardano," in *Heterodoxy in Early Modern Science and Religion*, eds. John Brooke and Ian Maclean (Oxford: Oxford University Press, 2005).

Magnus, Philip, Martin J. Slater, and Lawrence M. Principe, "Stereospecific Dicobalt Octacarbonyl Mediated Enyne Cyclization for the Synthesis of Methyl Deoxynorpentalenolactone H," *Journal of Organic Chemistry* 54 (1989): 5148–53.

Mandosio, Jean-Marc, "La Tabula Smaragdina e i suoi commentari medievali," 681–96 in Paolo Lucentini, Ilaria Parri, and Vittoria Perrone Compagni, eds., *Hermetism from Late Antiquity to Humanism* (Turnhout: Brepols, 2003).

Mandosio, Jean-Marc and Carla DiMartino, "La *Météorologie* d'Avicenne (*Kitāb al-Shifā'* V) et sa diffusion dans le monde latin," *Miscellanea mediaevalia* 33: *Wissen über Grenzen: Arabisches Wissen und lateinisches Mittelalter* (2006): 406–24.

Matus, Zachary, *Franciscans and the Elixir of Life* (Philadelphia: University of Pennsylvania Press, 2017).

Mazzatinti, Francesca and Maria Luisa Tanganelli, *Catalogo dei manoscritti della Biblioteca Riccardiana di Firenze: Ricc. 421-520* (Rome: Istituto Poligraphico e Zecca dello Stato, 2022).
McGinn, Bernard, *Apocalyptic Spirituality* (New York: Paulist Press, 1979).
McGinn, Bernard, *The Calabrian Abbot: Joachim of Fiore in the History of Western Thought* (New York: Macmillan, 1985).
McWilliams, J. A., "St. Albert the Great and the Plurality of Forms," *The Modern Schoolman* 9 (1932): 43–44 and 59–61.
Melnikov, P., et al., "Thermal decomposition mechanism of iron(III)nitrate," *Journal of Thermal Analysis and Calorimetry* 115 (2014): 145–51.
Mencherini, S., "Constitutiones generales...a capitulo Perpiniani anno 1331," *AFH* 2 (1908): 269–92, 412–30, and 575–599.
Menhardt, Hermann, *Handschriftenverzeichnisse Österreichischer Bibliotheken, Kärnten*: vol. 1, Klagenfurt, Maria Saal, Friesach (Vienna: Österreichischen Staatsdruckerei, 1927).
Meuthen, Erich, *Die letzten Jahre des Nikolas von Kues* (Cologne: Westdeutscher Verlag, 1958).
Molland, George, "Medieval Ideas of Scientific Progress," *Journal of the History of Ideas* 39 (1978): 561–77.
Moorat, S. A. J., *Catalogue of Western Manuscripts on Medicine and Science in the Wellcome Historical Medical Library*, 2 vols. (London: Wellcome Institute for the History of Medicine, 1962–73).
Morienus, *A Testament of Alchemy*, ed. and trans. by Lee Stavenhagen, (Hanover, NH: Brandeis University Press, 1974).
Morton, Jonathan, "Making Sense of Ingenium: Translating Thought in Twelfth-Century Latin Text on Cognition," 90–110 in Katja Krause, Maria Auxent, Dror Weil, eds., *Premodern Experience of the Natural World in Translation* (New York: Routledge, 2022).
Moureau, Sébastien, *Le* De anima *alchimique du pseudo-Avicenna*, 2 vols. (Florence: SISMEL, 2016).
Moureau, Sébastien, "*Elixir atque fermentum*: New Investigations about the Link between pseudo-Avicenna's Alchemical *De anima* and Roger Bacon: Alchemical and Medical Doctrines," *Traditio* 68 (2013): 277–325.
Moureau, Sébastien, "*Min al-Kīmīyā' ad Alchimiam*: The Transmission of Alchemy from the Arab-Muslim World to the Latin West in the Middle Ages," 87–141 in *The Diffusion of the Islamic Sciences in the Western World* (Florence: SISMEL, 2020).
Multhauf, Robert P., "John of Rupescissa and the Origin of Medical Chemistry," *Isis* 45 (1954): 359–67.
Nares, Robert, Stebbing Shaw, Joseph Planta, and Francis Douce, eds., *A Catalogue of the Harleian Manuscripts in the British Museum*, 4 vols (London: Eyre and Strahan, 1808–12).
Nazari, Giovanni Battista, *Della tramutatione metallica* (Brescia, 1599).
Newman, William R., *Atoms and Alchemy: Chymistry and the Experimental Origins of the Scientific Revolution* (Chicago: University of Chicago Press, 2006).
Newman, William R., "Brian Vickers on Alchemy and the Occult: A Response," *Perspectives on Science* 17 (2009): 482–506.
Newman, William R., *Gehennical Fire: The Lives of George Starkey, and American Alchemist in the Scientific Revolution* (Cambridge, MA: Harvard University Press, 1994).
Newman, William R., "The Genesis of the *Summa perfectionis*," *Archives Internationales d'Histoire des Sciences* 35 (1985): 240–302.
Newman, William R., "Mercury and Sulphur among the High Medieval Alchemists; From Rāzī and Avicenna to Albertus Magnus and Pseudo-Roger Bacon," *Ambix* 61 (2014): 327–44.

Newman, William R., "An Overview of Roger Bacon's Alchemy," 317–36 in *Roger Bacon and the Sciences: Commemorative Essays*, ed. Jeremiah Hackett (Leiden: Brill, 1996).
Newman, William R., *Promethean Ambitions: Alchemy and the Quest to Perfect Nature* (Chicago: University of Chicago Press, 2004).
Newman, William R., *The Summa Perfectionis of Pseudo-Geber: A Critical Edition, Translation and Study* (Leiden: Brill, 1991).
Newman, William R., "Technology and Alchemical Debate in the Late Middle Ages," *Isis* 80 (1989): 423–45.
Newman, William R. and Lawrence M. Principe, *Alchemy Tried in the Fire: Starkey, Boyle, and the Fate of Helmontian Chymistry* (Chicago: University of Chicago Press, 2002).
Pabst, Bernhard, *Atomtheorien des lateinischen Mittelalters* (Darmstadt: Wissenschaftliche Buchgesellschaft, 1994).
Patschovsky, Alexander, *Ein kurialer Ketzerprozeß: Die Verurteilung der Franziskanerspiritualen Giovanni di Castiglione und Francesco d'Arquata* (Wiesbaden: Harrassowitz Verlag, 2018).
Paulus, Julian, *Katalog der alchemischen Handschriften der Staats- und Universitätsbibliothek Hamburg* (unpublished typescript, 1995).
Pereira, Michela, "Alchemy and the Use of Vernacular Languages in the Late Middle Ages," *Speculum* 74 (1999): 336–56.
Pereira, Michela, "Arnoldo da Villanova e l'alchimia: Un'indagine preliminare," *Actes de la Trobada international d'estudis sobre Arnau de Vilanova*, ed. Josep Perarnau, 2 vols. (Institut d'Estudis Catalans: Barcelona, 1995), 2: 95–174.
Pereira, Michela, "I Francescani e l'alchimia," *Convivium Assisiense* 10 (2008): 117–57.
Pereira, Michela, "Mater Medicinarum: English Physicians and the Alchemical Elixir in the Fifteenth Century," 26–52 in Roger French, Jon Arrizabalaga, Andrew Cunningham, and Luis Garcia-Ballester, eds. *Medicine from the Black Death to the French Disease* (Burlington, VT; Ashgate, 1998).
Pereira, Michela, "*Medicina* in the Alchemical Writings Attributed to Raimond Lull (14th-17th Centuries)," 1–15 in Piyo Rattansi and Antonio Clericuzio, eds. *Alchemy and Chemistry in the 16th and 17th Centuries* (Dordrecht: Springer, 1994).
Pereira, Michela, "*I Septem Tractatus Hermetis*," 651–79 in *Hermetism from Late Antiquity to Humanism*, eds. P. Lucentini, I. Parri, and V. Perrone Compagni (Turnout: Brepols, 2003).
Pereira, Michela and Barbara Spaggiari, *Il Testamentum alchemico attribuito a Raimondo Lullo* (Florence: SISMEL, 1999).
Peruzzi, E., "Un trattato di alchimia attribuito ad al-Kindī," 21–38 in *Scritti in onore di Eugenio Garin* (Pisa: Scuola normale superiore, 1987).
Pfeil, Brigitte, *Katalog der deutschen und niederländischen Handschriften des Mittelalters in der Universitäts- und Landesbibliothek Sachsen-Anhalt in Halle (Saale)*, 2 vols. (Halle: ULSAB, 2007).
Piron, Sylvain, "Les *studia* franciscains de Provence et d'Aquitaine," 303–58 in *Philosophy and Theology in the Studia of the Religious Orders and at Papal and Royal Courts* (Turnhout: Brepols, 2012).
Piron, Sylvain, "Un traité inconnu de Jean de Roquetaillade," *Revue d'histoire des textes* 4 (2009): 299–307.
Podlaha, Antonín, *Soupis rukopisů knihovny metropolitní kapitoly Pražské, F-P* (Prague: Czech Academy, 1922).
Potestà, Gian Luca, "L'anno del antichristo: Il calcolo di Arnaldo di Villanova nella letteratura teologica e profetica del XIV secolo," *Rivista di storia del Cristianismo* 4 (2007): 431–64.

Potestà, Gian Luca, "Il profeta degli Antichristi, del reparator e del millennio," 41–60 in *VM* (Tealdi).

Potestà, Gian Luca, "'Gli spiriti dei profeti sono soggetti ai profeti': Da Giovanni di Rupescissa a Pietro Galatino," 47–59 in ed. Vittorio Scotti Douglas, *"Per una severa maestra". Dono a Daniela Romagnoli* (Fidenza: Mattioli, 2014); https://publicatt.unicatt.it/retrieve/e309db6c-26c9-0599-e053-3705fe0a55db/Galatino%20lettore%20di%20Rupescissa.pdf

Principe, Lawrence M., *The Aspiring Adept: Robert Boyle and His Alchemical Quest* (Princeton, NJ: Princeton University Press, 1998).

Principe, Lawrence M., "Bibliotheca Philosophica Hermetica MS 338: The 'Missing Link' in the Transmission of John of Rupescissa's *Liber lucis*," *Quaerendo* 53 (2023): 187–97.

Principe, Lawrence M., "Chemical Translation and the Role of Impurities in Alchemy: Examples from Basil Valentine's *Triumph-Wagen*," *Ambix* 34 (1987): 21–30.

Principe, Lawrence M., "Chymical Exotica in the Seventeenth Century, or, How to Make the Bologna Stone," *Ambix* 63 (2016): 118–44.

Principe, Lawrence M., "The Development of the Basil Valentine Corpus and Biography: Pseudepigraphic Corpora and Paracelsian Ideas," *Early Science and Medicine* 24 (2019): 549–572; reprinted in *Pseudo-Paracelsus: Forgery and Early Modern Alchemy, Medicine, and Natural Philosophy*, eds. Didier Kahn and Hiro Hirai (Leiden: Brill, 2021), 137–59.

Principe, Lawrence M., "Friars Minor and *Opus Magnum*: Why Did So Many Medieval Franciscans Practice Alchemy?" in *Religious Experience and the Foundations of Modern Science*, ed. Nuno Castel-Branco (Cambridge: Cambridge University Press, 2026).

Principe, Lawrence M., *The Secrets of Alchemy* (Chicago: University of Chicago Press, 2013).

Principe, Lawrence M., *The Transmutations of Chymistry: Wilhelm Homberg and the Académie Royale des Sciences* (Chicago: University of Chicago Press, 2020).

Principe, Lawrence M., and William R. Newman, "Some Problems with the Historiography of Alchemy," 385–431 in *Secrets of Nature: Astrology and Alchemy in Early Modern Europe*, ed. Newman and Anthony Grafton, (Cambridge, MA: MIT Press, 2001).

pseudo-Aristotle, *De perfecto magisterio*, *BCC* 1: 638–59.

Purš, Ivo and Vladimir Karpenko "Alchemy at the Aristocratic Courts of the Lands of the Bohemian Crown, 47–92 in *Alchemy and Rudolf II: Exploring the Secrets of Nature in Central Europe in the 16th and 17th Centuries*, eds. Ivo Purš and Vladimir Karpenko (Prague: Artefactum, 2016).

Quantin, Maximilien, "Notice sur P. Bureteau religieux celestin, chroniqueur et poète," *Bulletin de la société des sciences historiques et naturelles de l'Yonne* 29:(1875): 407–41.

Rampling, Jennifer M., "The Catalogue of the Ripley Corpus: Alchemical Writings Attributed to George Ripley," *Ambix* 57 (2010): 125–201.

Rampling, Jennifer M., *The Experimental Fire: Inventing English Alchemy, 1300-1700* (Chicago: University of Chicago Press, 2020).

Rampling, Jennifer M., "Spirited Matter and Ingenious Nature: Accounting for Alchemical Change," 13–24 in *Ingenuity in the Making*, eds. Richard J. Oosterhoff, José Ramón Marcaida, Alexander Marr (Pittsburgh: University of Pittsburgh Press, 2021).

Rampling, Jennifer M., "Citrination and Its Discontents: Yellow as a Sign of Alchemical Change," *Ambix* 71 (2024): 73–97.

Reeves, Marjorie, *The Influence of Prophecy in the Later Middle Ages: A Study of Joachimism* (Oxford: Clarendon Press, 1969).

Rinotas, Athanasios, "The *Sciant artifices* in the Work of Albert the Great: Towards Two Kinds of Transmutations," *Early Science and Medicine* 27 (2022): 57–82.

Rischpler, Susanne, *Die Handschriften und Drucke des 15. und frühen 16. Jahrhunderts der Staatsbibliothek Bamberg* (*Katalog der illuminierten Handschriften der Staatsbibliothek Bamberg*, 4), forthcoming.

Ritter von Aschbach, Joseph, *Die Wiener Universität und ihre Gelehrten, 1520 bis 1565* (Vienna, 1888), 251–53.

Rubino, Elisa and Samuela Pagani, "Il *De mineralibus* di Avicenna tradotto da Alfredo di Shareshill," *Bulletin de philosophie médiévale* 58 (2016): 23–87.

Ruska, Julius, "Die Alchemie des Avicenna," *Isis* 21 (1934): 14–51 on 45–48.

Ruska, Julius, *Al-Rāzī's Buch Geheimnis der Geheimnisse* (Berlin: Springer, 1937).

Ruska, Julius, *Das Buch der Alaune und Salze* (Berlin: Verlag Chemie, 1935).

Ruska, Julius, *Tabula smaragdina: Ein Beitrag zur Geschichte der hermetischen Literatur* (Heidelberg: Winter, 1926).

Santi, Francesco, "La vision de la fin du temps chez Arnaud de Villeneuve: Contenu théologique et expérience mystique," *Cahiers de Fanjeaux* 27 (1992): 107–27.

Schäfer, Karl Heinrich, *Die Ausgaben der apostolischen Kammer unter Johann XXII* (Paderborn: Schöningh Verlag, 1911).

Schäfer, Karl Heinrich, *Die Ausgaben der apostolischen Kammer unter Benedikt XII, Klemens VI, und Innocenz VI* (Paderborn: Schöningh Verlag, 1914).

Schäfer, Karl Heinrich, *Die Ausgaben der apostolischen Kammer unter den Päpsten Urban V und Gregor XI* (Paderborn: Schöningh Verlag, 1937).

Scherer, Carl, "Der niederländische Bildhauer Wilhelm Vernuken in hessischen Diensten," *Repertorium für Kunstwissenschaft* 31 (1908): 218–26.

Schmechel, Carmen, "Leaven of Dough, Ferment of Gold: The Breadmaking Analogy in Medieval Metallic Transmutation," *Ambix* 71 (2024): 243–70.

Schmitt, Wolfram, "Eine handschriftliche Sammulung alchemistischer Traktate aus Böhmen, *Stifter-Jahresbericht* 7 (1962): 177–195.

Schuba, Ludwig, *Die medizinischen Handschriften der Codices Palatini Latini in der Vatikanischen Bibliothek* (Wiesbaden: Reichert, 1981).

Secret, François, "Littérature et alchimie," *Bibliothèque d'Humanisme et Renaissance* 35 (1973): 499–531.

Singer, Dorithea W., *Catalogue of Latin and Vernacular Manuscripts*, 3 vols. (Brussels: Lamertin, 1928–31), 1: 228.

Sorbelli, Albano, *Inventari dei manoscritti delle biblioteche d'Italia*, vol. 28: Torino (Florence: Olschki, 1923).

Sorbelli, Albano, *Inventari dei manoscritti delle biblioteche d'Italia*, vol. 21: Bologna (Florence: Olschki, 1914).

Steele, Robert, *Opera hactenus inedita Rogeri Baconi*, Volume 5: *Secretum Secretorum* (Oxford: Clarendon, 1920).

Štemberka, Otto and Pavel Zahradník, *Cesty Vlachů: Mysterium Antonia z Florencie* (Vrchlabí: Green Mango, 2022).

Sudhoff, Karl, *Versuch einer Kritik der Echtheit der Paracelsischen Schriften, Theil 2: Paracelsus-Handschriften* (Berlin, 1899).

Summary Catalogue of Western Manuscripts in the Bodleian Library, 7 vols (Oxford: Clarendon Press, 1895–1953).

Tabulae codicum manu scriptorum praeter graecos et orientales in Bibliotheca Palatina Vindobonensi asservatorum, vol. 4: *Cod.* 5001–*Cod.*6500 (Vienna, 1870).

Talbot, C. H. and E. A. Hammond, *The Medical Practitioners of Medieval England: A Biographical Register* (London: Wellcome Historical Medical Library, 1965).

Targoni Tozzetti, Giovanni, *Catalogo generale dei manoscritti Magliabechiana*, 14 vols. (Manuscript: Florence, Biblioteca Nazionale Centrale, Sala Manoscritti e Rari, Cat. 45).

Tealdi, Elena, "Political Propheticism: John of Rupescissa's Figure of the End Times Emperor and its Evolution," 441–63 in *Cultures of Eschatology*, Vol 1: *Empires and*

Scriptural Authorities in Medieval Christian, Islamic and Buddhist Communities, eds. Veronika Wieser, Vincent Eltschinger and Johann Heiss (Berlin: DeGruyter, 2020).
Telle, Joachim, "Johannes von Teschen," in *Die deutsche Literatur des Mittelalters: Verfasserlexicon* (1983) 4: 774–76.
Thesaurus mundi, 294–320 in *Heptas alchymica, das ist, Auszerlesene philosophische Tractätlein von dem magisterio lapidis*, ed. Theophilus Neander (Halle, 1621).
Thomson, Rodney M., *A Descriptive Catalogue of the Medieval Manuscripts of Corpus Christi College Oxford* (Oxford: Brewer, 2011).
Thorndike, Lynn, "An Alchemical Manuscript: Klagenfurt, Bischöfl. Bibl. XXIX.d.24," *Ambix* 11 (1963): 95–96.
Thorndike, Lynn, "Bisticius (or Bistichius) of Florence," *Romanic Review* 19 (1929): 244–49.
Thorndike, Lynn, *A History of Magic and Experimental Science*, 8 vols. (New York: Columbia University Press, 1934–58).
Thorndike, Lynn, *Science and Thought in the Fifteenth Century* (New York: Columbia University Press, 1929).
Timmerman, Anke, "Alchemy in Cambridge: An Annotated Catalogue of Alchemical Texts and Illustrations in Cambridge Repositories," *Nuncius* 30 (2015): 345–511.
Tönnies, Bernhard, *Die Handschriften der Thüringer Universitäts- und Landesbibliothek Jena*, Band 1: *Die mittelalterlichen lateinischen Handschriften der Electoralis-Gruppe* (Wiesbaden: Harrassowitz, 2002).
Torrents, J. Massó, *Manuscritos catalanes de la bibliotheca de S. M.* (Barcelona, 1888).
Tractat Bapsts Innocentii, Vom Stein der Philosophorum, 227–32 in Joachim Tancke, *Promptuarium alchemiae* (Leipzig, 1610).
Valentinelli, Giuseppe, *Bibliotheca manuscripta ad S. Marci Venetiarum: Codices mss. latini* (Venice, 1872).
van Weert, Gus and Yuxing Shang, "Iron control in nitrate hydrometallurgy by (auto) decomposition of iron (II) nitrate," *Hydrometallurgy* 33 (1993): 255–71.
Verae alchemiae artisque metallicae, ed. Guglielmo Gratarolo, 2 vols. (Basel, 1561).
Vogt, Berard, "The *forma corporeitatis* of Duns Scotus and Modern Science," *Franciscan Studies* 3 (1943): 47–62.
von Heinemann, Otto, *Die Handschriften der herzoglichen Bibliothek zu Wolfenbüttel*, 2nd series, *Die Augusteische Handschriften V* (Wolfenbüttel: Zwissler, 1908).
von Heinemann, Otto, *Die Handschriften der herzoglichen Bibliothek zu Wolfenbüttel: Die Helmstedter Handschriften*, 3 vols. (Wolfenbüttel, 1884–88).
Walther, Hans, *Alphabetisches Verzeichnis der Versanfänge mittellateinischer Dichtungen* (Göttingen: Vandenhoeck & Ruprecht, 1959).
Webb, J. R., "'Knowledge Will be Manifold': Daniel 12: 4 and the Idea of Intellectual Progress in the Middle Ages," *Speculum* 89 (2014): 307–57.
Weszprémi, Stephan, *Succincta medicorum Hungariae et Transilvaniae biographia* (Vienna, 1787).
Wieczorek-Ciurowa, K. and A. J. Kozak, "The Thermal Decomposition of $Fe(NO_3)_3{\cdot}9H_2O$," *Journal of Thermal Analysis and Calorimetry* 58 (1999): 647–51.
Wilson, William Jerome, "An Alchemical Manuscript by Arnaldus de Bruxella," *Osiris* 2 (1936): 220–405.
Wilson, William Jerome, *Catalogue of Latin and Vernacular Alchemical Manuscripts in the United States and Canada*, *Osiris* 6 (1939): 1–836.
Witten, Laurence C. and Richard Pachella, *Alchemy and the Occult: A Catalogue of Books and Manuscripts from the Collection of Paul and Mary Mellon Given to the Yale University Library*, 4 vols (New Haven, CT: Yale University Library, 1968–77).

Wunderle, Elisabeth, *Katalog der mittelalterlichen lateinischen Papierhandschriften aus den Sammlungen der Herzog-von-Sachsen-Coburg-und-Gotha'schen Stiftung für Kunst und Naturwissenschaft* (Wiesbaden: Otto Harrassowitz Verlag, 2002).

Yates, Donald, *Descriptive Inventories of Manuscripts Microfilmed for the Hill Monastic Manuscript Library: Austrian Libraries*, 3 vols. (Collegeville, MN: Hill Monastic Library, 1981).

Zacour, Norman P. and Rudolf Hirsch, *Catalogue of Manuscripts in the Libraries of the University of Pennsylvania to 1800* (Philadelphia: University of Pennsylvania Press, 1965).

Zavalloni, Roberto, *Richard de Mediavilla et la controverse sur la pluralité des formes* (Louvain: Institut supérieur de philosophie, 1951), 213–503.

Ziegler, Joseph, *Medicine and Religion c. 1300: The Case of Arnau de Vilanova* (Oxford: Clarendon, 1998).

Zuber, Mike A., "Alchemical Promise, the Fraud Narrative, and the History of Science from Below: A German Adept's Encounter with Robert Boyle and Ambrose Godfrey," *Ambix* 68 (2021): 28–48.

Zuber, Mike A. and Leigh T. I. Penman, "Robert Boyle's Anonymous 'Crosey-Crucian' Identified: The German Alchemist and Religious Dissenter Peter Moritz," *Notes and Records of the Royal Society of London* 74 (2020): 95–103.

INDEX OF MANUSCRIPTS

GENERAL INDEX

Note: Italicized page numbers refer to material in the text of the *Liber lucis*

For Product Safety Concerns and Information please contact our EU representative GPSR@taylorandfrancis.com Taylor & Francis Verlag GmbH, Kaufingerstraße 24, 80331 München, Germany

Batch number: 10726459

Printed by Printforce, the Netherlands